BEGONIAS
of Peninsular Malaysia

BEGONIAS
of Peninsular Malaysia

Ruth Kiew

Botanical illustrations
Zainal Mustafa

Watercolour paintings
Wendy Gibbs

Principal photographer
Yap Kok-Sun
with
Ali Ibrahim and Tan Jiew-Hoe

Natural History Publications (Borneo)
Kota Kinabalu

in association with

Singapore Botanic Gardens
National Parks Board
Singapore

2005

Dedicated to
Tan Jiew-Hoe
who made this book a reality

Published by

Natural History Publications (Borneo) Sdn. Bhd. (216807-X)
A913, 9th Floor, Phase 1, Wisma Merdeka
P.O. Box 15566
88864 Kota Kinabalu, Sabah, Malaysia
Tel: 088-233098 Fax: 088-240768
e-mail: info@nhpborneo.com
website: www.nhpborneo.com

in association with

Singapore Botanic Gardens
National Parks Board
1 Cluny Road
Singapore 259569

Copyright ©2005 Natural History Publications (Borneo) Sdn. Bhd. and Singapore Botanic Gardens
Photographs copyright ©2005 as credited.
First published 2005

Date of publication: 1 February 2005

All rights reserved. No part of this publication may be reproduced,
stored in a retrieval system, or transmitted in any form or
by any means, electronic, mechanical, photocopying, recording,
or otherwise, without the prior permission of the copyright owners.

Begonias of Peninsular Malaysia by Ruth Kiew

ISBN 983-812-086-3

Half-title page: The Tortoise Shell Begonia, *Begonia kingiana* (Photo: Yap Kok-Sun).
Frontispiece: Jiew-Hoe's Begonia, *Begonia jiewhoei* (Photo: Yap Kok-Sun).
Facing Foreword: The Cave Begonia, *Begonia jayaensis* (Photo: Yap Kok-Sun).
Facing Preface: The Tortoise Shell Begonia, *Begonia kingiana* (Photo: Yap Kok-Sun).

Printed in Malaysia.

CONTENTS

Foreword ix
Preface xi

Chapter 1: Introduction 1

Chapter 2: The Plant 5
    Habit 5
    Leaves 5
    Flowers 10
    Fruits 13
    Seeds 14
    Vegetative propagation 17
    Seasonality 17
    Pests 18
    Habitats 18
    Distribution 19
    Conservation 20

Chapter 3: Peninsular Malaysian Begonias 25
    History 25
    The genus *Begonia* 26
    Identification 28
    Key to Peninsular Malaysian species of *Begonia* 30
    The species 35
        1. *Begonia eiromischa* 35
        2. *Begonia kingiana* 37
        3. *Begonia ignorata* 43
        4. *Begonia tigrina* 50
        5. *Begonia sibthorpioides* 54
        6. *Begonia sinuata* var. *sinuata* 58
            Begonia sinuata var. *pantiensis* 64
        7. *Begonia martabanica* var. *pseudoclivalis* 68
        8. *Begonia carnosula* 70

9. *Begonia integrifolia* 74
10. *Begonia phoeniogramma* 85
11. *Begonia variabilis* 92
12. *Begonia elisabethae* 98
13. *Begonia thaipingensis* 102
14. *Begonia longifolia* 107
15. *Begonia holttumii* 112
16. *Begonia isopteroidea* 116
17. *Begonia wrayi* 118
18. *Begonia jiewhoei* 123
19. *Begonia barbellata* 129
20. *Begonia decora* 133
21. *Begonia wyepingiana* 140
22. *Begonia vallicola* 145
23. *Begonia alpina* 149
24. *Begonia koksunii* 153
25. *Begonia pavonina* 158
26. *Begonia klossii* 164
27. *Begonia paupercula* 170
28. *Begonia maxwelliana* 174
29. *Begonia venusta* 179
30. *Begonia fraseri* 184
31. *Begonia longicaulis* 188
32. *Begonia lowiana* 192
33. *Begonia jayaensis* 197
34. *Begonia corneri* 202
35. *Begonia forbesii* 206
36. *Begonia lengguanii* 211
37. *Begonia rajah* 216
38. *Begonia reginula* 218
39. *Begonia yappii* 223
40. *Begonia foxworthyi* 227
41. *Begonia nurii* 233
42. *Begonia scortechinii* 238
43. *Begonia perakensis* 240
44. *Begonia rhoephila* 245
45. *Begonia abdullahpieei* 252
46. *Begonia rhyacophila* 256
47. *Begonia aequilateralis* 261
48. *Begonia tampinica* 265
49. *Begonia praetermissa* 268

50. *Begonia herveyana* 272
51. *Begonia nubicola* 275
52. *Begonia rheifolia* 280
53. *Begonia cucullata* 286
54. *Begonia hirtella* 289

Acknowledgements 294
Appendix: Begonia Sections in Peninsular Malaysia 295
Glossary 300
References 303
Index 305

Foreword

The beauty of tropical forests is depicted in hues and tones of green. Flashes of exuberant colour, pattern and form are subtly woven into a verdant tapestry. Exemplary of this unique manifestation of tropical beauty is the genus *Begonia* with more than 1500 pantropical species. Horticulture has been gifted with the perfect bloom of the tuberous begonias from the Andes, the flamboyant palette of Rex begonias from the Himalayas, and the drama of angel-wing, star-leaf and iron-cross begonias from both Old World and New World countries. The potential of Peninsular Malaysia, an Asian centre of distribution of *Begonia* species, to contribute to the horticulturist's treasure trove has scarcely been tapped. This publication seeks to redress the under-representation.

The 52 indigenous species described and depicted in this book capture the subtle beauty and variety in shape, texture, colour and pattern of foliage found in the Malaysian begonias. The fragility suggested by the delicate and exquisitely textured foliage of Malaysian begonias often reflects the perilous purchase on survival of their habitats. Yet Malaysian begonias can be remarkably durable, surviving in small populations in the same niche for centuries if undisturbed. Unfortunately, many begonia habitats have been pushed to the brink of existence by human activity. Some hope for the preservation of such beauty lies in ex-situ conservation of these species through collaborative effort of the plant collector, the horticulturist, and the gardener. This book will hopefully facilitate the process.

I liken this work, 'Begonias of Peninsular Malaysia', to a beautiful chalice in use, a well-crafted vessel enhancing the value of its precious contents. The craftsman in this case is the genius behind Natural History Publications (Borneo). The publisher has combined the eye for beauty and balance of an artist, a draftsman, an artisan and a craftsman in producing a quality tome that captures the attention of its intended audience. It delivers the information contained in the text in an accessible and pleasing manner.

The other partners in the production of this work who rendered the paintings, drawings and photographs have done justice to their subjects. The images of the habitats also allow armchair travelers to enjoy the beauty of the Malaysian forest, while deriving some gardening tips for the growing of the beautiful species shown. Anchoring the book is the text. This is the painstaking work of a botanist with the knowledge, experience and wisdom gained from careful fieldwork, observation with a keen eye, and the dedication that is rewarded with the discovery of facts and finds otherwise hidden from the world. It bespeaks a familiarity with the forests of Malaysia, and an affinity for nature.

The final bit of alchemy that is required to bring such work to light lies in that traditional relationship of an enlightened patron with the professional scientist that has been behind so many valuable publications in the world of Botany. In this particular venture, the patron is also an avid plant explorer and collector in his own right. In pursuing his interests, he has made possible the publication of several invaluable books by botanists whose contributions to the world of science would not have otherwise been possible.

It now leaves me the pleasure to recommend 'Begonias of Peninsular Malaysia' to you, for I know it will please your eye and mind, as it did mine.

Dr Tan Wee-Kiat
CEO
National Parks Board

Preface

The purpose of this book is to interest local naturalists in wild begonias and provide them with a means to identify them easily. It is therefore written in a simple style and is profusely illustrated with botanical drawings, photographs and some watercolour paintings. However, the book retains scientific standards with complete nomenclature including full scientific names, synonyms (or 'old' names), full descriptions, citation of herbarium specimens examined, analytic botanical drawings and a taxonomic key for identification.

The research for this book involved many weeks of exploration in forests, limestone hills and mountains in all corners of the Peninsula, often with the photographer in tow, over a period of two years. Twelve new species and one new variety are described. Of the 52 native species, an amazing 47 species were photographed in their natural habitat during field work. Some of these are species that had not been found for many years and were feared extinct.

Asian begonias are renowned for their beautiful foliage. The most famous ones from Malaysia that entered the trade were *Begonia rajah* and *Begonia decora*. They have now contributed their virtues by being parents of a number of horticultural varieties. Many others have potential although they are not generally easy to grow.

All begonias are delicate plants that in nature require undisturbed habitats in forests or on limestone hills to survive. Already one is probably extinct and several others are critically endangered as the forest where they live is currently being destroyed. This book fulfills the need to document Malaysia's natural heritage, especially in view of forest clearance and quarrying of hills.

Dr Ruth Kiew has been working on begonias and other herbaceous forest plants for over 30 years but it was Tan Jiew-Hoe, plant enthusiast, explorer, collector, gardener and benefactor who provided the impetus and means to get the book written and published to such high standards.

The team must be congratulated on producing a most beautiful scientific book. It will surely be a benchmark for how such books can be written and published.

Dr Chin See-Chung
Director
Singapore Botanic Gardens

Chapter 1

INTRODUCTION

Begonias display an amazing variety of shapes, colours, patterns and textures in their leaves rarely seen in other groups of plants. Their hallmark is the very asymmetric leaf where one side is much larger than the other. The beautiful leaves account for their popularity as ornamental plants and their great commercial importance in the horticultural trade. Today over 10,000 kinds of hybrids and cultivars are available. Indeed, while the hybrids are well known, for many people it comes as a surprise to learn that there are wild begonias in Malaysian forests.

Begonias are found wild throughout tropical and subtropical Asia, Africa and America. To date more than 1,500 species have been named with many more new species waiting to be discovered. In Peninsular Malaysia, 52 native species are known. These are all confined to shaded habitats either in primary forests or on limestone hills. Although several native species are extremely beautiful, all the commonly cultivated species in Malaysia are exotics (Kiew & Kee, 2002). The Rex Begonias (hybrids and cultivars based on *Begonia rex* from Assam,

(Opposite). Map of Peninsular Malaysia. (Right). *Begonia kingiana*.

1

INTRODUCTION

India), the Wax-flower Begonia, *B. cucullata*, from South America, the Brazilian Angel-wing Begonia (*B. coccinea*), the Star-leaf Begonia (*B. heracleifolia* from Mexico), as well as the Iron-cross Begonia (*B. masoniana* from southern China) are popular cultivated species. The Bearded Begonia, *B. hirtella*, from the West Indies and South America is well established along the roadsides in Malaysian hill stations having become a successful garden escape since it was introduced as a garden plant in 1896. *B. cucullata* has more recently escaped from gardens and has become established along drains.

Asia and Malaysia have played their part in the horticultural trade. The first Asian species, the painted-leaf begonia, *Begonia rex,* was introduced by chance in a consignment of orchids sent from Assam, India, to Belgium in 1857. This begonia proved easy to hybridize and plants with an extraordinary range of leaf colours and textures have been produced from it. In 1895, it was hybridized with the decorative begonia, *B. decora*, from Cameron Highlands to produce a range of hybrids with rich shades of red, purple and bronze. *Begonia decora* was also hybridized with the Sumatran *B. goegoensis* to produce the Big Decora hybrid. The most famous Malaysian species in cultivation is *B. rajah*. Prized for its chocolate-brown blotched leaves, it won a Royal Horticultural Society First Class Certificate when it was first introduced into England in 1894.

In Malaysia, begonias are forest plants well adapted to the shady, humid conditions of the forest understorey. When judged by the number of species, they are one of the most successful herbaceous groups in this environment. The current tally is 52 begonias in Peninsular Malaysia and this is only exceeded by four other genera of herbs, namely the orchids *Bulbophyllum, Dendrobium* and *Eria* with about 130, 85 and 50 species, respectively, and *Henckelia* (Gesneriaceae, the African Violet family) with about 90 species.

As the illustrations in this book show, many Malaysian species are extremely decorative but their potential as ornamental plants has yet to be realized.

Locally, their most frequent use is in cooking. The acids in the leaves (oxalic acids, the acids that make *belimbing* (*Averrhoa belimbi,* Oxalidaceae) and star-fruit (*A. carambola*) sour) impart an appetizing sour taste. The leaves are used to wrap fish before baking or roasting or are chopped fine and mixed with *belachan* (prawn paste) and chili to make a *sambal* for cooking prawns or fish. I have met a surprising number of people in Malaysia who have told me that as children they ate the refreshingly sour begonia flowers.

In common with many forest herbs, there is no commonly used Malay name for begonias, although in some areas the large-leaved ones are generically referred to *pokok riang*.

(Opposite). *Begonia rex* Putzeys (Reproduced from the Botanical Magazine, 1859).

Chapter 2

THE PLANT

Habit

In Peninsular Malaysia, begonias are mostly herbs with a few shrubby species with woody stems. All have fine, fibrous roots. The tallest species are the shrubby, cane-like begonias with erect stems more than half a metre tall; the smallest are the rhizomatous begonias. The majority produce a thick, fleshy rhizome that clings to rocks with leaves that are close together and form a rosette or tuft of leaves. Only *Begonia corneri* and *B. thaipingensis* produce a long, thin rhizome that creeps over the soil surface and have widely spaced leaves. The rhizomatous begonias range in size from the smallest, the Shilling Begonia *B. sibthorpioides*, with leaves just 1 cm long, to the largest, the Rhubarb-leaved Begonia *B. rheifolia*, with leaves up to 34 cm long.

There are three types of erect begonias, which all have leaves widely spaced on the stem. The ones with soft, succulent stems like the Sparkling Begonia, *B. sinuata*, often have a small tuber about 5–10 mm wide at the base. These stems are weak and tend to fall and root. However, the Spotted Begonia, *B. integrifolia*, can grow up to 27 cm tall when growing against earth banks.

In contrast, stems of the cane-like begonias are bamboo-like, jointed and woody. There are just six cane-like begonias in Peninsular Malaysia, *B. holttumii*, *B. isopteroidea*, *B. jiewhoei*, *B. longifolia*, *B. lowiana* and *B. wrayi*. The erect stems of *B. jiewhoei* that grows on limestone hills eventually become pendent and hang down the cliff face. Another unusual erect species is the Bristly Begonia, *B. barbellata*, that has a short, erect, wiry stem up to 30 cm tall.

The third type of stem is seen in the semi-erect begonias, such as the Cabbage-leaved Begonia, *B. venusta*. It is really a variant of the cane-like ones as their stems are bamboo-like but they are not self-supporting and readily fall and root at the nodes, only the top part remaining erect.

Leaves

The begonia's asymmetric leaf shape is so characteristic that several other unrelated species with similar lop-sided leaves are named '*begoniifolia*' (begonia-leaved). In Peninsular Malaysia, *Pentaphragma begoniifolia* (Pentaphragmataceae), *Sonerila begoniifolia* (Melastomataceae) and *Zippelia begoniifolia* (Piperaceae) are three such examples.

While most Malaysian begonias do have asymmetric leaves, a few have equal-sided ones. These can be broad and heart-shaped like that of the Rhubarb-leaved Begonia, *B. rheifolia*, which has leaves 34 cm wide, or narrow and just as few centimetres wide, such as those of *B. alpina*, *B. perakensis* and *B. rhoephila* or they can be almost round as are the leaves of the Shilling Begonia, *B. sibthorpioides* or the Diminutive Limestone Begonia, *B. nurii*. Only *B. kingiana*, *B. ignorata*, *B. tigrina* and the extinct *B. eiromischa* have peltate leaves where the stalk is attached above the base of the leaf blade.

(Opposite). *Begonia rajah*.

THE PLANT

Just two species, the Rhubarb-leaved Begonia and the Cloud Begonia, *B. nubicola*, produce leaves of different shapes during their lifetime. They start life with slender, narrow leaves and later produce broadly ovate ones in mature flowering plants.

The leaf margin in most species is not toothed or is very finely toothed but in a few begonias, like *B. reginula*, it is shallowly scalloped. In Peninsular Malaysia, only *B. sinuata* var. *pantiensis* has deeply jagged leaves.

Begonia leaves are unrivalled in the range of colours, patterns, textures and hairs they display and Malaysian begonias are no exception. Colours range from light to dark green to bronzy or purple brown to almost black or iridescent blue or green. Apart from the blue leaves, this range of colours is caused by the presence of purple or red pigments in the lower layers of the leaf. Without the purple coloration, the upper leaf surface appears light green; but in combination with the purple or red pigment it appears dark-green, bronze, liver-coloured or even black while the undersurface appears purple or reddish.

For over a hundred years, scientists have been speculating about the reason why red or purple coloration is so common on the underside of leaves of many tropical herbs that grow in deep shade. One suggestion is that it is an adaptation to increase the absorption of what little light (only about 1–3 per cent of sunlight) that penetrates to the forest floor. In spite of modern instrumentation, all

(Above). The deep red underside of a *Begonia wyepingiana* leaf blade. (Opposite). Blue iridescent leaves of the Peacock Begonia, *Begonia pavonina*.

efforts to prove this have failed. A novel idea is that far from being an adaptation to increase photosynthesis in low light conditions, it is quite the opposite and serves to prevent photo-inhibition (Lee, 2001). Photo-inhibition occurs when deep shade plants are suddenly exposed to full sunlight, for example when a sunbeam penetrates directly through holes in the canopy to the forest floor. This burst of light can 'knock out' the photosynthetic system of deep shade plants. Yet another idea is that the chemical nature of the purple and red pigments protects the leaf against insect attack but again this has yet to be proven.

Some Malaysian begonias have blue leaves. The most dramatic of these is the Peacock Begonia, *B. pavonina,* whose leaves change colour from a brilliant iridescent blue to bright green depending on the angle of light on the leaf. The blue colour is not, however, due to pigments but to iridoplasts, special chloroplasts that are made up of minute layers of microfibrils regularly arranged that alter the reflectance properties of the leaf (Lee, 2001). Like pigmentation, the possession of iridoplasts has not been proven to enhance photosynthesis. Other begonias with leaves that sometimes have a bluish tinge include *B. alpina*, *B. carnosula*, *B. integrifolia* and *B. thaipingensis*.

The striking patterns of begonia leaves are one of their greatest attractions in horticulture and are amply illustrated in this book. The two common forms of patterns are the uneven distribution of the pigment, so that there are paler or light green bands, patches or blotches contrasting with dark green, bronzy or reddish areas, and the presence of silvery spots. The silvery spots are not caused by pigments but by raised air-filled cells. The adaptation of these spots, if any, is not known although it has been suggested that the white spots mimic insect eggs and butterflies will avoid laying their eggs on a leaf that already has eggs that would hatch before theirs. Indeed, in some other groups of plants, it has been observed that butterflies do not lay eggs on spotted leaves.

However, this cannot be the whole story as variation in leaf pattern is observed in a single species or population. It is quite common, not only in begonias, for the leaves of young plants to be spotted, while those of mature plants are quite plain. Variation within a single population can be so strikingly different that plants appear to belong to a different species. Examples of these polymorphic species (i.e., species exhibiting more than one state for a particular character) are the spotted begonia, *B. integrifolia*, and the red striped begonia, *B. phoeniogramma*. A population often includes plants with light green leaves (without spots or purple coloration), dark green leaves (with red or purple pigmentation but no spots), light green leaves with white spots, and dark green leaves with white spots. In horticulture it is the dark-coloured, spotted plants that are selected over the plain ones.

Most begonias have extremely thin leaves (Sheue *et al.*, 2003). For example, the Decorative Begonia, *B. decora*, has leaves about 170 µm thick; the Peacock Begonia, *B. pavonina*, about 185 µm thick and the Sparkling Begonia, *B. sinuata*, 195 µm; while the more succulent begonias, such as the Cabbage-leaved

Variation (polymorphism) of leaf colour and pattern in a single population of *Begonia integrifolia*.

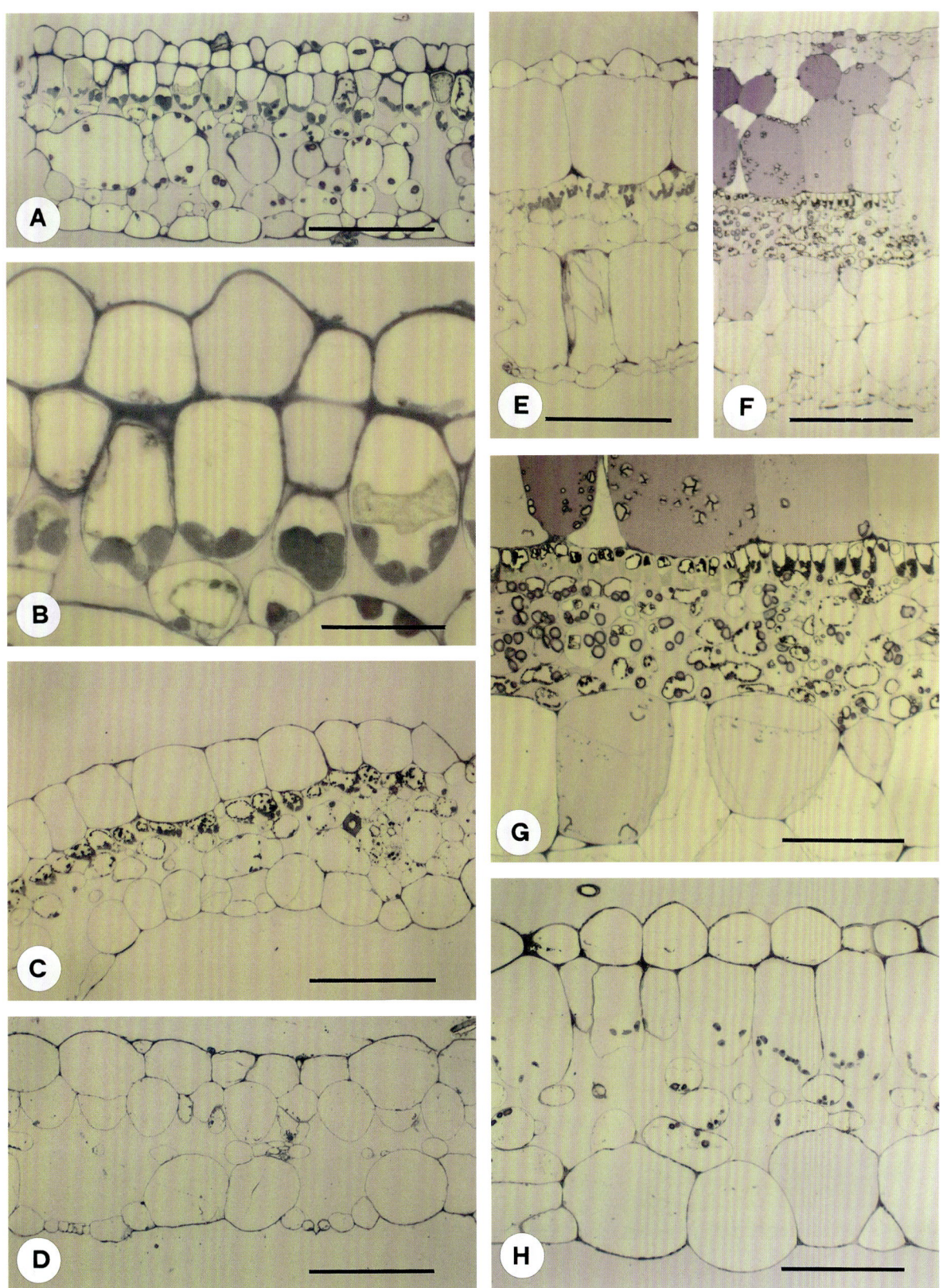

Cross-sections of begonia leaves.
A. *Begonia pavonina*. **B.** Upper layers of *B. pavonina* leaf. **C.** *B. decora*. **D.** *B. sinuata*. **E.** The black-leaved hybrid *Begonia* 'Kinbrook'. **F.** *B. kingiana*. **G.** Centre photosynthetic tissue of *B. kingiana*. **H.** *B. thaipingensis*. (bar=20 μm in A, C, D, E, G & H; bar=5 μm in B; bar=50 μm in F. (Reproduced from Sheue *et al.*, 2003)

THE PLANT

begonia, *B. venusta*, and the Taiping Begonia, *B. thaipingensis*, have leaves about 310 µm and 360 µm thick, respectively. The fleshy leaves of the Tortoise Shell Begonia, *B. kingiana*, are exceptional in being about 880 µm thick, which is still only a fraction of a millimetre thick (1000 µm =1 mm).

These thin leaves rapidly wilt outside the rain forest, where humidity is high and rarely drops below 80 per cent in the day and at night rises to 100 per cent. Rapid wilting occurs because begonia leaves are not supported by mechanically strong fibres or cell walls. Instead they are supported by large, thin-walled cells in the outer layers (the epidermis) which, when fully turgid, are rigid and support the leaf (Kiew, 1988). When water is lost from these cells, they become flaccid and the leaf floppy, resulting in wilting. In *B. decora*, *B. sinuata* and *B. venusta*, it is the upper epidermal layer that makes up between 50 and 65 per cent of the leaf thickness and these cells function to support the leaf, not to store water.

In contrast, the thick, succulent leaf of *B. kingiana* is remarkable for having a multiple epidermis three cells thick on both surfaces (Sheue *et al.*, 2003), making up 85 per cent of its leaf thickness. Water storage is a prime function for this species, which grows on dry vertical limestone cliff faces. Even in spells of hot, dry weather, its leaves do not show signs of wilting.

The surface of begonia leaves ranges from the glossy, almost rubbery leaves of the Cabbage-leaved Begonia, *B. venusta,* to the rough surface of *B. decora*, where the hairs sit on raised cones, which appear as pits on the lower surface. The glossy surface is extremely effective in shedding rainwater. In heavy rain, I watched raindrops fall on leaves of *B. maxwelliana*, run down to the leaf base where they formed a pool of water; the leaf blade then swivelled so that the tip pointed down and the water was shot off the elongated leaf tip, after which the leaf returned to its normal horizontal position. Shedding rainwater is important because a film of water not only cuts down the light that the leaf can absorb for photosynthesis but it also encourages the growth of micro-organisms (bacteria, fungi and algae). These at best shade the leaf surface and reduce photosynthesis; at worst they can cause disease. The elongated leaf tip is termed a 'drip tip' and is characteristic of some rainforest plants.

Some of the most attractive begonia leaves are the velvety ones of *B. thaipingensis* or the literally sparkling ones of *B. sinuata*. The velvety or sparkling effects are caused by light-scattering from the upper epidermal cells that are conical in shape and project above the leaf surface. In the case of the Sparkling Begonia, Lee (2001) reported experiments that show that these conical epidermis cells are adapted to low light conditions because they act like lenses and increase the capture of light by ten to twenty times compared with those of the usual flat leaf surface.

Most species are not conspicuously hairy. The notable exceptions are the Decorative Begonia, *B. decora*, the Cave Begonia, *B. jayaensis*, Low's Begonia, *B. lowiana*, and the Red-haired Begonia, *B. wyepingiana*, which are all densely hairy. The hairs of *B. decora* and *B. wyepingiana* are long and particularly attractive in being magenta coloured. The soft, transparent hairs of *B. jayaensis* and *B. lowiana* are distinctly sticky to touch, which may serve as a deterrent to herbivores. Among Malaysian begonias, *B. sinuata* is unique in being covered in tiny, star-like hairs.

Flowers

The arrangement of flowers in begonias is basically cymose, which means that the flowers are produced in threes with the centre one opening first. In most begonias, the three-flower pattern is repeated several times to produce a many-flowered inflorescence with two main branches. In a few species, such as Holttum's Begonia, *B. holttumii*, and the Common Malayan Begonia, *B. wrayi*, the main axis is long with short side branches, which is termed racemose. When identifying begonias,

THE PLANT

Arrangement of the flowers.
A. Simple Cyme. **B.** Many-flowered Cyme. **C.** Many-flowered Raceme.
(O=Male flowers. ▽=Female flowers)

it is important to note whether the inflorescence is terminal (arises from the top of the stem), for example in the Spotted Begonia, *B. integrifolia*, or is axillary (arises from the leaf axils) as in the rhizomatous species.

In rhizomatous begonias, which have fruits with one wing larger than the other two, such as the Decorative Begonia, *B. decora*, the developing flowers are enveloped by two large, pale green leafy structures called bracts. These drop off as the flowers expand and open. Flowering begins when the inflorescence is about as long as the leaf stalk so the flowers are held just above the leaves. After flowering, the inflorescence stalk continues to grow, sometimes almost doubling its length, so that the fruits are held well above the leaves. In rhizomatous species with fruits with wings of equal size, such as *B. ignorata*, the bracts are small and do not protect the developing flowers. They do not fall off. In these begonias, the inflorescence reaches its maximum height before flowering.

The flowers of most Malaysian begonias are small, measuring about 1–2 cm across; the largest, at 4–5 cm across, belongs to the Cabbage-leaved Begonia, *B. venusta*. Begonia flowers do not have green sepals. Instead the sepals are coloured and look like petals. Together these sepals and petals are called tepals. The tepals are usually pale pink or white, though in some populations of the Tortoise Shell Begonia, *B. kingiana*, they are a striking bright scarlet. The Red Striped Begonia, *B. phoeniogramma*, and the Variable Begonia, *B. variabilis*, are unusual in their red-striped tepals. Many of the white-flowered begonias are nevertheless attractive because the tepals have a scintillating surface. The tepals are usually rounded—only the tepals of the female flowers of *B. wrayi* have a jagged margin, and the tepals of Low's begonia, *B. lowiana*, have pointed tips. Most begonias have hairless tepals, but a few, like the Cave Begonia, *B. jayaensis*, *B. lowiana* and *B. wyepingiana*, have a few sparse long hairs on the outer surface.

Begonias are monoecious, that means that they produce separate male and female flowers on the same plant, frequently on the same inflorescence, but in a few cases, as in the Berry Begonia,

THE PLANT

B. longifolia, the female flowers are in a separate, few-flowered inflorescence. Per plant, many more male flowers are produced than female. Depending on the species, the male flowers may open first (a condition called protandry) followed by the female. In protogynous species, the female flowers open first. Since they open at different times, there are effectively male and female phases of flowering, which promotes cross-pollination between plants.

The male flowers have two larger rounded outer tepals, orientated top and bottom. Some species have an additional two tepals that are much narrower and are orientated left and right. The many stamens are joined at the base, which sometimes forms a short stalk, and the pale to golden yellow or less usually greenish yellow anthers are arranged in a spherical or hemispherical cluster. Pollen is liberated through slits along the side of the anther or rarely, as in the Shilling Begonia, through slits on the inner face of the anther.

Begonia pollen grains range in size from 16–35 by 8–14 μm. They are tricolporate, which means they have three grooves each with a pore. The outer layer, the exine, is thin and is patterned with extremely fine parallel ridges (Berg, 1985).

The female flower usually has five tepals of more or less the same shape but with the outer ones slightly larger than the inner. The ovary develops below the tepals, i.e., it is inferior. Inside, it has two or three chambers called locules and, except for the Berry Begonia, *B. longifolia*, has three wings outside. Within the locules, the many tiny ovules cover the surface of the placentas (plate-like structures) that protrude into the locule. In Malaysian begonias, there are either one or two placentas per locule. The number of locules and placentas are important for grouping the species into sections (see Appendix). The two or three styles are positioned conspicuously above the tepals and each forks to produce either four or six branches. When receptive, the stigma is moist and glistens. (It is made up of a band of papillose, i.e., glandular, cells). Depending on the shape of the style branch, the stigma is spiral or U-shaped.

In male flowers, pollen is the reward for the pollinator. The female flowers, however, offer no reward. (There is no nectar in either the male or female flower). The pollinator is thought to visit the female flower by mistake because the colour of the tepals of the male and female flower and the anthers and stigmas is almost the same. This type of pollination is known as 'pollination by deceit'. It is certainly effective as fruits with a multitude of seeds are almost always found.

Single-sexed flowers of begonia.
A. Male flowers with many stamens. **B.** Female flowers with inferior ovary, styles and stigmas.

THE PLANT

At Cameron Highlands, I have watched pollen being avidly collected from the male flowers of *B. decora* and *B. venusta* by stingless bees (*Trigona* sp.), which also visit the female flowers, and so effect pollination. I have also observed the Malayan honeybee, *Apis cerana*, collecting pollen from flowers of *B. foxworthyi*.

Artificial crosses between species of begonia are very easy to make and have given rise to thousands of hybrids. However, under natural conditions this happens only very rarely mainly because species do not grow together either because they are isolated geographically or because they have difference ecological requirements. The only example of a natural hybrid in Peninsular Malaysia (p. 138) is found where the Decorative Begonia, *B. decora*, and the Cabbage-leaved Begonia, *B. venusta*, grow together at Cameron Highlands (Teo & Kiew, 1999).

Trigona bee collecting pollen from the male flowers of *Begonia decora*.

Fruits

Fruit type in begonia is clearly related to the mode of seed dispersal. Seeds in the fleshy green fruits of the Berry Begonia *B. longifolia* are presumably eaten and dispersed by herbivores as the fruit does not split to release the seeds. All other Malaysian begonias have dry winged capsules that split between the locules and wings to release the seeds.

Three types of capsules are found in Malaysian begonias. One is rather rare and is found in the cane-like begonias, *B. holttumii* and *B. wrayi*, which have fruits with three equal-sized wings and a short, stiff, reflexed stalk that holds the fruit below the leaves. The seeds will only be released if

13

THE PLANT

the plant is shaken by heavy rain or animals brushing past. A second type, seen in *B. foxworthyi*, also has three thin wings of equal size but it dangles down on a long, fine, hair-like stalk. The slightest movement, whether by water drops hitting the plant or air movement, will shake the seeds out of the capsule like a shaking a pepper pot.

The third type is the splash cup. In these capsules, there is one large, thick, fibrous wing and two short curved ones. When ripe, the large, heavy wing hangs down like a keel of a boat and the two short wings are curved and upright and form a cup. The seeds are dispersed by large water drops dripping from high up in the canopy after rain that hit the cup with great ballistic force bouncing the tiny seeds out (Savile & Hayhoe, 1978). Dispersal of seeds from the capsule appears to be local as the great majority of species occupy very narrow geographic ranges. In fact, geographical isolation has probably played an important role in the speciation of begonias. The exception is the Berry Begonia. Tebbitt (2003) attributes its wide distribution from India to Java to its being animal dispersed.

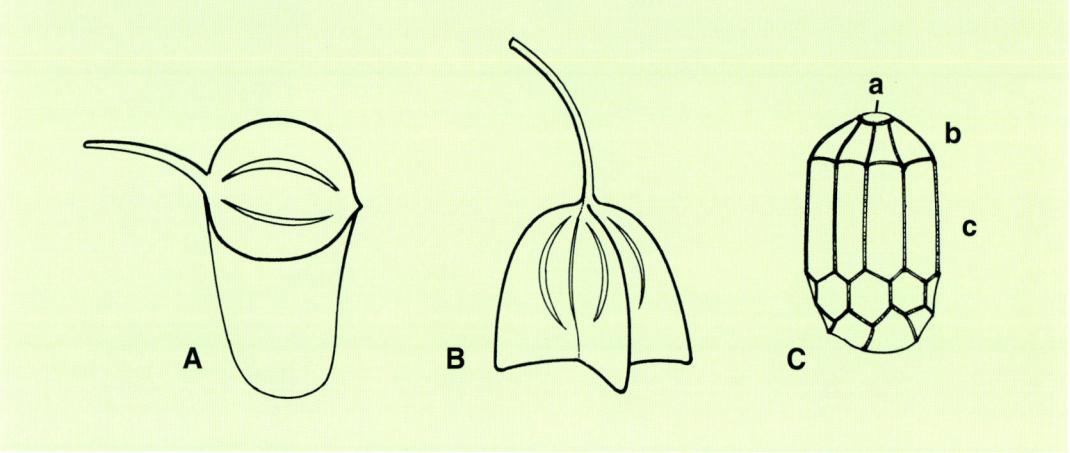

Fruits and seed of begonias
A. A splash cup. **B.** A dangling capsule. **C.** A seed.
(**a.** Seed stalk scar. **b.** Lid. **c.** Collar cells)

Seeds

Begonia seeds are so tiny and light they are called dust seeds. It has been estimated that between 30,000 and 70,000 begonia seeds are needed to weigh one gram. A single capsule produces hundreds of seeds. Malaysian begonia seeds are all brown, barrel-shaped and about 0.3–0.6 mm long. The base consists of the lid that narrows to the stalk that attaches the seed to the fruit. Above the lid is a ring of elongated cells called the collar cells, which are a unique feature of begonia seeds. The rest of the seed is covered in polygonal cells. As the seed matures, the thin outer wall collapses and the centre of the cells becomes concave while the cell walls stand out as ridges. This gives the seed its characteristic sculptured surface. When shaken from the fruit, this uneven surface creates micro-turbulence, which slows the rate of fall and potentially increases the distance the seeds can be dispersed from the mother plant. This is an advantage in the rainforest understorey where there is almost no wind movement except along torrential streams and near waterfalls or before thunderstorms. The rough surface also helps the seed to catch onto vertical rock surfaces.

THE PLANT

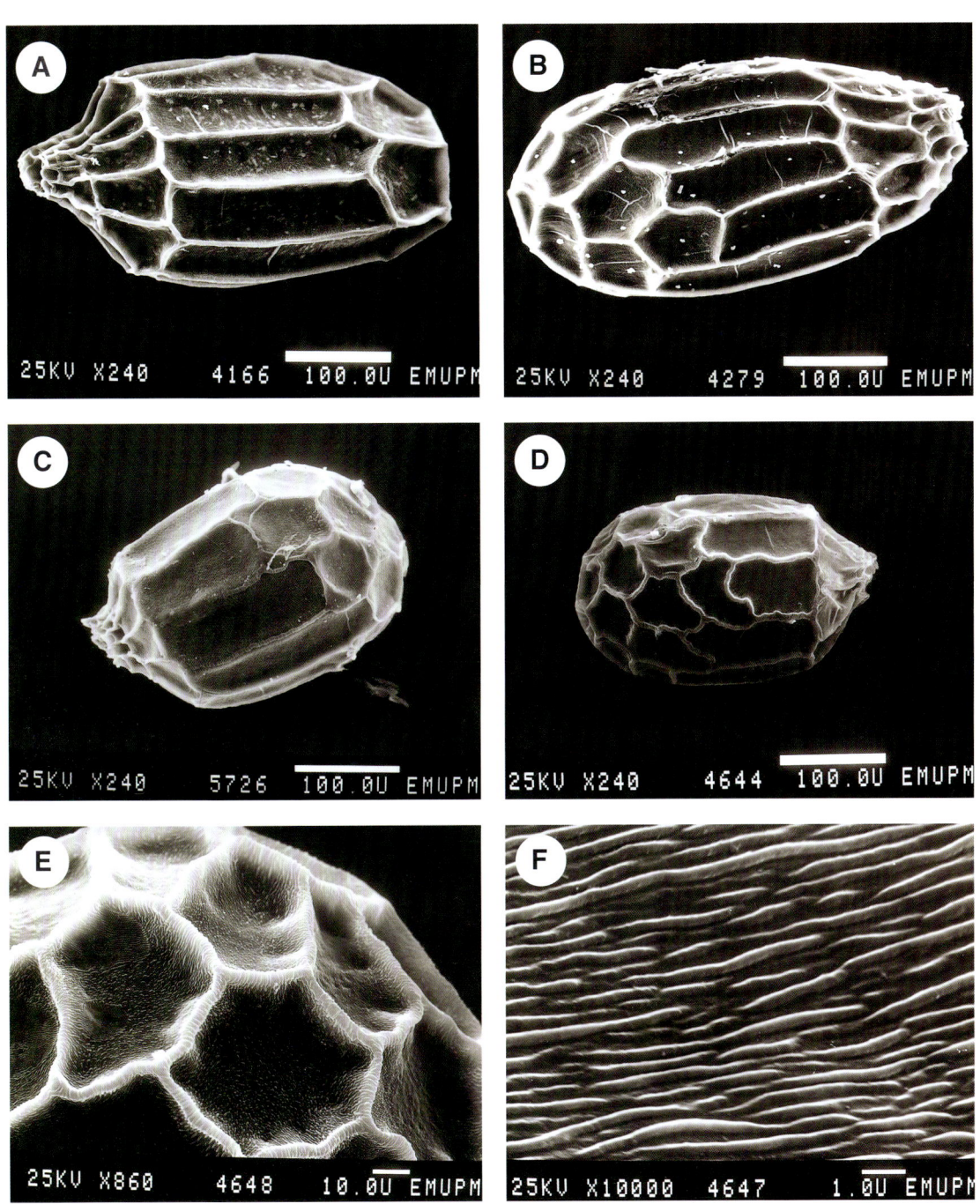

Begonia seeds.
A. *Begonia holttumii*. **B.** *B. nurii*. **C.** *B. longifolia*. **D–F**. *B. sinuata*. **E.** Polygonal cells showing sunken outer cell wall. **F.** Micro-patterning on the sunken cell wall. *(R. Kiew & Noraini Munasib)*

15

THE PLANT

This ability is demonstrated when you try to shake begonia seeds off a piece of paper; some just cling on!

The seeds of the Berry Begonia *B. longifolia* are not released from the fruit but they are similar in having a sculptured surface like those produced in dry capsules. This is unlike the African begonias with berries where the seeds have a smooth seed coat. This suggests that *B. longifolia* with its fleshy berry has evolved recently from species with dry capsules as the sculptured seed coat is adapted to being carried by water drops or air movements.

By the time the seeds are shed, the embryo, surrounded by a single layer of endosperm, completely fills the seed. Food is stored in the embryo in the form of oil and aleurone grains, a type of protein (Boesewinkel & de Lange, 1983). When shed, the seeds germinate rapidly. Seeds of the Decorative Begonia, *B. decora*, and the Cabbage-leaved Begonia, *B. venusta*, for example, germinate within 4–12 days (Teo & Kiew, 1999). If seeds are not shed, they will germinate in the fruit while still attached to the plant. When the seed germinates, the lid falls off and the walls between the collar cells split as the root of the seedling emerges. The unique collar cells therefore function to release the developing seedling from the tough seed coat.

When the seedling germinates from the tiny seeds, it is very small, just a few millimetres tall. This probably explains why so many begonias grow on rocks or on steep banks where leaf litter does not collect. On the forest floor, tiny seedlings would be smothered by fallen leaves. However, the hundreds of seeds produced compensate for their high mortality.

Seedlings of *Begonia paupercula* germinating in the old fruit.

THE PLANT

Plantlets of *Begonia decora* developing from an old fallen leaf.

Vegetative propagation

A few begonias propagate themselves by vegetative means. Some populations of the Sparkling Begonia, *B. sinuata,* produce a bulbil at the base of the leaf blade that looks like a small tuber. This bulbil grows to form a new plant. The most remarkable begonia is *B. elisabethae* where the elongated tip of the leaf arches, touches the soil surface, roots and produces a new plantlet at the leaf tip. It also produces new plants from the veins ending on the leaf margin.

In several species, such as *B. decora, B. ignorata, B. pavonina* and *B. wyepingiana*, when an old leaf begins to decompose on the damp forest floor, new plantlets develop from the veins where the leaf blade separates from the leaf stalk as the leaf blade rots. This unusual ability of a severed leaf surface to remain active and produce new tissue is exploited in propagating begonias artificially by leaf cuttings.

Seasonality

The only seasonal begonia in Peninsular Malaysia is the Sparkling Begonia, *B. sinuata*, which behaves as an annual plant by dying down towards the end of the year in the rainy season. The new population then regenerates from seed, tubers or bulbils. It is common to see rock faces covered by hundreds of even-aged seedlings or immature single-leaved plants. By the middle of the year most populations are in flower.

The extreme north of the Peninsula experiences a more monsoonal climate and here begonias like the Spotted Begonia, *B. integrifolia*, are reported to die down in the dry season. Further south, the Spotted Begonia and the closely related *B. phoeniogramma* do not show distinct seasonal patterns in their growth. However, these soft-stemmed begonias, like the Spotted Begonia, are probably short-lived. Seedlings are common in their populations.

The rhizomatous and cane-like begonias are probably perennials, although there is no information on how rapidly they grow. It is an enigma how *B. holttumii* and *B. wrayi* reproduce and are dispersed because they are usually encountered as scattered individuals and seedlings are rarely found. In spite of this, they are some of the most widespread species in the Peninsula.

Except for the annual Sparkling Begonia, begonias flower throughout the year, although rather fewer flowers are found in the rainy season at the end of the year because the tepals are prone to fall in heavy rain.

THE PLANT

Pests

Wild begonias are remarkably free from insect attack or disease. Usually, their leaves are perfect with no sign of holes or ragged edges. In thirty years of looking at begonias, I have only seen begonias defoliated by caterpillars on four occasions (*B. holttumii, B. klossii, B. longifolia* and *B. paupercula*).

The most remarkable case of herbivory was finding on the Bristly Begonia, *B. barbellata,* a female stick insect of *Abrosoma festinatum* that K.S. Yap observed and photographed emitting a strange blue light.

Begonia holttumii defoliated by caterpillars.

Habitats

Malaysian begonias are either plants of primary rain forest or the shaded base of limestone hills. Only *B. rhyacophila* grows fully exposed to the sun on waterfalls. None is weedy or grows in secondary forests.

Substrates

Six species, *Begonia ignorata, B. jayaensis, B. jiewhoei, B. kingiana, B. nurii* and *B. tigrina*, are restricted to limestone hills, where they grow around the deeply shaded, damp base of the cliff either directly on the cliff face or on scattered boulders around the hills. Only the succulent-leaved *B. kingiana* can persist during drought periods; the soft-leaved species, like the Cave Begonia, *B. jayaensis*, which uniquely grows on soil inside caves as well as on the cliffs, die down, then regenerate from seed when rain comes.

THE PLANT

A further four species, *B. foxworthyi, B. integrifolia, B. phoeniogramma* and *B. variabilis*, grow on other rock types as well as on limestone hills. Apart from *B. foxworthyi*, the other three also grow on earth banks in forest.

In the forest most begonias grow on rocks associated with small streams and waterfalls. The most common rock type is granite but they are also found on quartzite and shale. These begonias have rhizomes that cling tenaciously to the rock surface.

Some grow on rock faces beside small, shaded waterfalls, such as *B. forbesii, B. lengguani, B. reginula* and *B. sinuata*, but most, such as *B. alpina, B. klossii, B. perakensis* and *B. rheifolia*, grow on large or small boulders in or beside shallow streams. These begonias grow well above the flood zone and are usually found on the downstream side of rocks away from the water current. They have broad leaves. *Begonia rhoephila*, however, is remarkable in that it grows on rocks close to the water level where it is susceptible to torrential floods. It is a true rheophyte with narrow, streamlined leaves that are adapted to resist being torn by strong water currents in times of flood.

The Cloud Begonia, *B. nubicola*, which grows on coastal hills, is unique in growing on gigantic boulders at the headwaters of streams that only have running water after rain. It grows below the summit at an altitude exactly where cloud gathers mid-morning. Presumably this cloud provides sufficient moisture for its growth.

Rather few begonias grow in soil on the flat forest floor. Examples include the cane-like begonias, *B. holttumii, B. longifolia* and *B. wrayi* and the creeping *B. thaipingensis*. The Bristly Begonia, *B. barbellata,* is unusual in that it grows in swampy places on the muddy or sandy edges of slow-flowing streams. *Begonia elisabethae* is also found in this habitat. (There are no truly aquatic begonias).

Most begonias that grow on soil are found on steep earth banks or slopes where leaf litter does not collect, which would otherwise smother the tiny seedlings. A few, like *B. corneri* and *B. yappii*, are found on river banks, but most are found on steep slopes particularly in hill (*B. carnosula* and *B. fraseri*) or montane forest (*B. decora, B. lowiana* and *B. venusta*).

Altitude

Begonias are found from sea level to the summit of mountains, but individual species live within rather narrow altitudinal ranges. *Begonia thaipingensis* is an exception in growing between 100 and 1,000 m altitude. The Sparkling Begonia, *B. sinuata,* and the Spotted Begonia, *B. integrifolia*, are the only begonias that actually grow at sea level—in Langkawi on a steep forested bank above a beach or on limestone sea cliffs, respectively.

Altitudinal zonation in begonias can be seen at Cameron Highlands and is one of the reasons for the wealth of begonias found there. Below 1,000 m grow *B. carnosula* and *B. wrayi,* above 1,000 m in hill forest *B. longifolia* and *B. pavonina* are found, while in montane forest above 1,500 m *B. decora* and *B. venusta* are common with *B. lowiana* growing below the summits at almost 2000 m.

Distribution

The majority of the begonias (86%) are endemic to the Peninsula. Only four represent the Asian element in the flora, being widely distributed in Asia. The distribution of the Berry Begonia, *Begonia longifolia*, extends from the Himalayas (India) to south China and Vietnam and south through Thailand, Peninsular Malaysia, Sumatra to Java. The Spotted Begonia, *B. integrifolia*, is found in India and both it and *B. martabanica* occur in Myanmar and Thailand and reach their

THE PLANT

southern limits in Perak. The Sparkling Begonia *Begonia sinuata*, is the most widespread begonia in the Peninsula, being found in every state. It also grows in Thailand and Vietnam.

A few begonias cross the border into Peninsular Thailand. The Bristly Begonia, *B. barbellata* and the Common Malayan Begonia, *B. wrayi* just enter Thailand in the east. *Begonia elisabethae*, on the other hand, is more common in Peninsular Thailand and just reaches into Malaysia in northern Kedah.

The Bornean element is very poorly represented in the Peninsula by just two begonias, both endemic. *B. barbellata* belongs to the group of begonias with short leaf stalks and narrow, almost symmetric, non-oblique leaves that is very rich in species in Borneo. Its distribution east of the Main Range from Kelantan to Johore is characteristic of the Bornean element of the Peninsula's flora (Corner, 1960). The other species, *B. jiewhoei*, is the only cane-like begonia that grows on limestone in Peninsular Malaysia where it has been found on two hills north of Gua Musang, Kelantan. This group is well represented on Bornean limestone (Kiew, 2001; Kiew & Geri, 2003).

Of the six endemic species restricted to limestone, only *B. kingiana* is widespread and found on almost all limestone hills. The others, *B. ignorata*, *B. jayaensis*, *B. jiewhoei*, *B. nurii* and *B. tigrina*, all have their distributions centred on Kelantan limestone.

The Main Range is particularly rich in begonias with 20 species recorded. This is in part due to the wide range of suitable habitats, in particular an abundance of rocky streams, and in part due to the wide range of altitude. Many of these begonias have very local distributions, being confined to a single valley or peak. Of these species, only *B. maxwelliana* has been found on the east side of the Main Range. With the present state of knowledge, it is not possible to judge whether this is because the Main Range acts as a barrier for dispersal from west to east or whether it is an artifact, because almost no botanical exploration has been carried out on the east side of the Main Range.

Several begonias that are found east of the Main Range in Kelantan, Trengganu and Pahang do not reach the foothills of the Main Range. These include the group of small begonias with round leaves, *B. corneri*, *B. lengguanii*, *B. reginula* and *B. yappii*, the large leaved *B. nubicola* and *B. rheifolia*, as well as *B. longicaulis* (the only endemic begonia on Gunung Tahan) and *B. barbellata*.

The begonias found on east coast islands have unusual disjunct distributions. *Begonia herveyana* found on Pulau Tinggi is also known from Malacca, *B. rheifolia* on Pulau Tioman is known from the Gunung Tahan area of Kelantan, Pahang and Trengganu, and *B. longifolia*, also from Pulau Tioman, is otherwise known in Peninsular Malaysia only from mountains in the Main Range.

Only four begonias are widespread. *Begonia sinuata* is found in all states; *B. wrayi*, which is common in the north, extends as far south as Negri Sembilan; *B. holttumii* ranges from Penang, Pahang and Trengganu south to Johore; and *B. barbellata* is found east of the Main Range from Kelantan to Johore.

It is conspicuous that the number of begonias in the south is very small and comprises, apart from *B. rajah*, only widespread species, *B. barbellata*, *B. holttumii* and *B. sinuata*. (There are no native begonias in Singapore). This is in part due to the Main Range not extending as far south as Johore and the absence of limestone in Johore. However, it is strange that no begonias have been reported from Gunung Ledang, which otherwise has a rich flora including many endemics.

Conservation

The greatest threat to the continued survival of native begonias is habitat disturbance or destruction. Begonias are particularly vulnerable as they do not adapt to exposure to light, heat and

THE PLANT

Habitat destruction is the greatest danger to the survival of begonias.

low humidity that follows, for example, when the tree canopy is destroyed. All the begonias in the Peninsula grow in primary forest or on limestone hills.

Threats to their habitats include the opening of the forest canopy and the silting of streams due to logging activities, clear-felling for agriculture or other development activities, disturbance along streams associated with the development of recreation forests; destruction of vegetation on mountain tops for telecommunication installations; and on limestone hills quarrying, burning the vegetation and temple building all cause irreparable damage (Kiew, 1997).

Their vulnerability to extinction is compounded by the fact that many are extremely local—26 species are known from a single locality and some of their populations are in addition small. Destruction of a small area will therefore result in the extinction of these species.

On the other hand, provided the habitat remains undisturbed even small populations are able to survive. For example, *B. forbesii* collected once in 1888 still grows in its original locality!

THE PLANT

Begonia eiromischa, long extinct from its only known habitat in Penang. (Reproduced with permission of the Singapore Botanic Gardens)

Only eight begonias grow within legally protected areas: four, *B. longicaulis, B. ignorata, B. rheifolia* and *B. sinuata*, in Taman Negara; two, *B. integrifolia* and *B. kingiana* in the Perlis State Park; two, *B. holttumii* and *B. sinuata* in the Endau-Rompin State Park; and the extremely rare *B. lengguanii* in the Bukit Rengit Wildlife Reserve.

One begonia, *B. eiromischa* from Penang, is certainly extinct. It was always an extremely rare species known from a single rocky stream, which as early as 1900 had been eliminated by farming. We can be sure it is extinct because we know where it used to grow and searching the area showed that the habitat no longer exists (Kiew, 1989).

THE PLANT

Critically endangered species are those that face a very high risk of extinction in the wild in the immediate future. These include the 25 species known from a single locality. (*B. lengguanii* and *B. longicaulis* that occur in protected areas are not considered threatened). Three of these are of particular concern because their habitat is currently in danger of disturbance from agricultural activities (*B. aequilateralis* and *B. forbesii*) and disturbance associated with recreational forests (*B. tampinica*). Two, *B. aequilateralis* and *B. tampinica*, have extremely small populations of less than fifty individuals—we could only find eight plants of *B. aequilateralis*! *Begonia forbesii* is also particularly vulnerable as it only grows on a rock face measuring about 5 by 3 m.

Apart from the widespread species (*B. barbellata*, *B. holttumii*, *B. sinuata* and *B. wrayi*), species with populations outside Malaysia (*B. elisabethae*, *B. integrifolia* and *B. longifolia*) and those in protected areas, the remaining 15 species are considered endangered because they consist of a few fragmented populations occupying less that 500 km^2 and their populations are comparatively small (less than 2,500 mature plants).

It is not only the extinction of species that is cause for concern. A particular concern for a group like begonia with potential in the horticultural trade is the extinction of especially beautiful forms. Over the years I have noticed the disappearance of the populations of the tortoise-shell patterned *B. kingiana* around Ipoh, the golden-leaved form of *B. thaipingensis* at Genting Highlands, and the black-leaved form of *B. ignorata* at Gunung Senyum. The entire waterfall where the population of *B. reginula* with particularly finely variegated leaves grew has been bulldozed out of existence during the clear-felling of the area.

In situ (i.e., on site) conservation is always best because it conserves the total community including the insect pollinators, soil micro-organisms, as well as the exact conditions of micro-climate. This is the reason that a healthy population of *B. forbesii* with many seedlings, young plants and plants in flower, fruit and seed continues to exist on a single rock face measuring just 15 m^2.

Where species are threatened by immediate habitat change or destruction and are likely to become extinct in the wild, the only way they can continue to exist is in cultivation (*ex situ* conservation). *Ex situ* conservation has the added advantage that the begonias can be on display in botanic gardens where they can be appreciated by the public for their intrinsic beauty or for their rarity and they can be multiplied and made available for interested people to grow. Many Malaysian species need exacting conditions for their growth, in particular constant high humidity or cool night temperatures for mountain species that are not found in city environments. Enthusiasts can play an important role in conservation as witnessed by the fate of *B. rajah* that was introduced into cultivation through the Singapore Botanic Gardens. For more than a hundred years, it only survived in cultivation nurtured by begonia enthusiasts (Kiew, 1989). A start should be made to cultivate as many begonias as possible as pristine forest areas are diminishing rapidly.

The conservation-friendly way to collect begonias from the wild is to collect a small number of ripe fruits or a few leaves as this does not deplete the wild population. Each fruit contains hundreds of tiny seeds, which can be germinated on wet sand in a plant propagator (Kiew & Kee, 2002). The leaves should be cut along the base and the veins pressed on to damp sand. As happens in the wild, new plants will grow from the cut veins. The plants should be grown in a terrarium to maintain high humidity conditions. As begonias hybridize readily, it is important to keep the species separated in cultivation to ensure that pollination does not occur between species and they remain genetically pure. Many of the *B. rajah* plants available in the market are obviously hybrids!

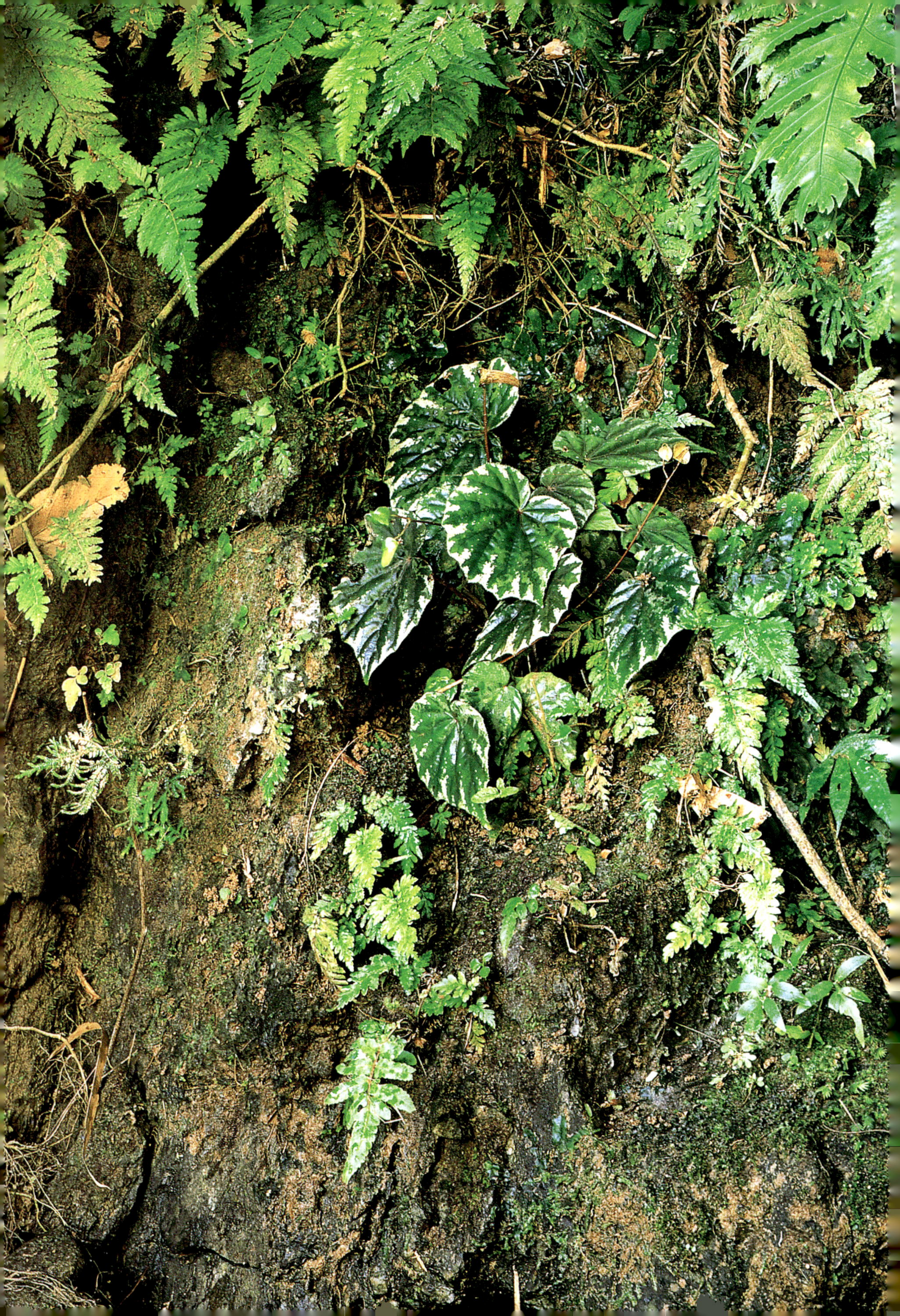

Chapter 3

PENINSULAR MALAYSIAN BEGONIAS

History

It was Charles Plumier, a Franciscan monk and botanist, who first named six begonias he had discovered in the Antilles Islands in honour of Michel Bégon, his travelling companion and at that time the French Governor of Santo Domingo in the West Indies. Since then, species have been collected from all continents in the tropical and subtropical regions of the world and the number of species named exceeds 1,500 (Doorenbos *et al.*, 1998).

The first two species described from Peninsular Malaysia, *Begonia guttata* (now recognized as *B. integrifolia*) and *B. sinuata*, were collected from Penang by Nathaniel Wallich in 1822. Since that time, three botanists stand out in the Malayan begonia scene. Sir George King, Superintendent of the Royal Botanic Garden Calcutta, in 1902 wrote an account of 18 begonias sent to him mostly from the west coast of the Peninsula of which 14 species he described as new species. *Begonia kingiana* is named in his honour.

The next account was written in 1922 by Henry N. Ridley, the first Director of the Botanic Gardens Singapore, and included 34 species of which 17 were new. Ridley was an indefatigable collector who made expeditions to many then very remote areas and collected a great many begonias himself.

Edgar Irmscher, Professor of Botany at Hamburg, studied begonias worldwide for the ambitious *Die natürlichen Pflanzenfamilien* (The Natural Plant Families, a German project of many volumes that set out to describe all the plant families of the world). In 1929, he wrote in German a comprehensive account of the 44 begonias then known from Peninsular Malaysia, based on his study of dried specimens from the Singapore Botanic Gardens Herbarium. He described 9 new species and 11 new varieties.

The first illustrated account appeared in 1949, reprinted in 1959 as M.R. Henderson's *Malayan Wild Flowers*. It included line drawings of 12 of the more common species. Henderson was the fourth Director of the Botanic Gardens Singapore. Our exploration has led to the discovery of a further 13 new species and 1 new variety bringing the current total to 54 begonia species for Peninsular Malaysia, of which two are garden escapes of South American origin.

(Opposite) Kok-Sun's Begonia, *Begonia koksunii*.

The Genus *Begonia*

Begonia belongs to the family Begoniaceae, which includes one other small genus: *Hillebrandia* with a single species from Hawai'i. *Hillebrandia* is distinct in having female flowers with ten tepals and a half-inferior ovary with parietal (as opposed to axillary) placentation.

Molecular studies place the Begoniaceae with the Datiscaceae, a family that includes the majestic trees *Octomeles sumatrana* from Borneo and Sumatra and *Tetrameles nudiflora*, which in the Peninsula grows in Langkawi. Both families have male and female flowers, inferior ovaries with many ovules, capsular fruits and minute seeds with an oily embryo and little or no endosperm. Together with the cucumber family, Cucurbitaceae, and the Malaysian Anisophyllaceae, they are included in the order Cucurbitales.

Begonia L.

Linnaeus, Species Plantarum 2 (1753) 1056; King, JASB 7 (1902) 56; Ridley, FMP 1 (1922) 853; Irmscher, MIABH 8 (1929) 86; Henderson MWF Dicot. (1959) 157.

(Characters not seen in Peninsular Malaysian species are in parentheses).

Succulent herbs, sometimes shrubs (root climbers or epiphytes); perennial, rarely annual (*B. sinuata*). **Stems** rhizomatous, rarely creeping (*B. corneri*, *B. thaipingensis*), if erect either cane-like and woody or soft and succulent; sometimes with a small tuber at the base. Stipules large, sometimes persisting. **Leaves** usually alternate or spiral, rarely opposite (*B. sibthorpioides*), in rhizomatous species tufted or in a rosette, in erect begonias distant, with or without hairs, only *B. sinuata* has star-like hairs; leaf stalk present, succulent; blade simple, rarely lobed (*B. sinuata* var. *pantiensis*), (palmately divided), usually oblique and asymmetric with the broader side with an enlarged basal lobe, less frequently not oblique, i.e., the midrib is in line with the stalk, and almost symmetric, rarely peltate (*B. eiromischa*, *B. ignorata*, *B. kingiana* and *B. tigrina*), often strikingly variegated, venation usually palmate-pinnate or palmate, rarely pinnate, margin toothed, scalloped or not toothed; rarely with a bulbil at the base of the blade (*B. sinuata*). **Inflorescences** erect, usually axillary, sometimes terminal, with many male and a few female flowers, in most species with two main branches (cymose), sometimes with a long axis (racemose), usually much branched, in most species male flowers open before the female, in a few the female flowers open first. Bracts either large, leaf-like and enclosing the developing inflorescence before falling as the flowers open or small and persistent; rarely with small bracteoles (*B. jayaensis*). **Flowers** unisexual, (rarely scented), tepals free, white, pink or rarely red (*B. kingiana*) or with red stripes (*B. phoeniogramma*, *B. variabilis*), (rarely yellow or orange), tip rounded, rarely pointed (*B. lowiana*), without nectaries. **Male flowers** with 2 or 4 tepals in opposite pairs, not toothed, inner two tepals if present much narrower than the outer two; stamens (4 to) many, often joined at the base and forming a cluster, anthers yellow with 2 anther sacs, which usually open by lateral slits or in *B. sibthorpioides* by slits on the inner face of the anther (or by pores), (sometimes with a connective at the top), without a vestige of the ovary. **Female flowers** with an inferior ovary with 3 wings or without wings, locules 2 or 3, each locule with 1 or 2 axile placentas (or placentation parietal), ovules many, tepals (2 to) 5, overlapping, inner tepals smaller than the outer, not toothed or rarely toothed (*B. wrayi*), styles 2 or 3 (4–6), joined at base, generally forked, not persistent in the fruit, styles and stigmas yellow, stigma papillose forming a spiral or U-shaped band. **Fruits** 3-winged capsules or rarely a fleshy, unwinged berry (*B. longifolia*); capsules either a splash cup with one wing much longer and very fibrous and 2 wings shorter and thinner or wings equal and thin, splitting between the locules and wings to release the seeds. **Seeds** barrel-shaped, many, minute, 0.25–0.75 mm long, with conspicuous collar cells, embryo oily with little endosperm.

Distribution: More than 1,500 species have been named from tropical and subtropical regions worldwide. The primary centre of diversity is South America as far north as Mexico. (The tuberous begonias grow in the Andes). The secondary centre of diversity is Asia, from the Himalayas and southern China and throughout Malesia. Africa is relatively poorer in species than America or Asia.

Within Malesia, Borneo (probably with about 300 species) and New Guinea are centres of diversity. Currently, 52 native species and 2 garden escapes are recorded for Peninsular Malaysia.

Habitat: In Peninsular Malaysia, begonias are terrestrial, frequently on rocks associated with small streams, rarely as rheophytes (*B. rhyacophila*, *B. rhoephila*), either in rain forest from the lowlands to mountains or on limestone hills, always in the shade. (None are weedy species found in secondary forests or open areas).

Notes: With such a large number of species in the genus, it is only natural to group similar species together. These infrageneric groups are called sections and currently 63 sections are recognised worldwide (Doorenbos *et al.*, 1998) of which 8 occur in Peninsular Malaysia (see Appendix). Some of the characters used to define sections are conspicuous, such as whether the stem is erect or rhizomatous, whether the leaves are asymmetric or not, or the number and relative size of the fruit wings; others are more difficult to see such as the placenta type of the ovary. For the purpose of this book, the grouping into sections is rather academic and reference to them can be found in the Appendix.

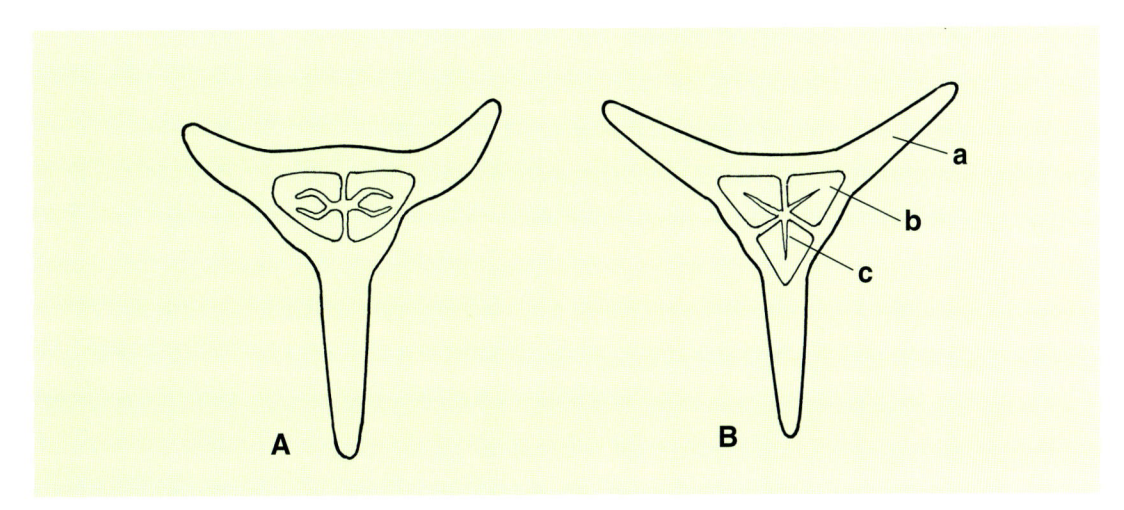

Cross-sections of begonia fruits.
A. A splash cup with one wing much longer than the other two and with two locules each with two placentas. **B.** A dangling fruit with three wings of equal size and three locules each with one placenta. (**a.** Wing; **b.** Locule; **c.** Placenta)

Identification

Important and conspicuous characters that are used to identify species include the habit (whether the stem is rhizomatous or erect), leaf shape (peltate, symmetric or asymmetric) and venation (palmate, pinnate or palmate-pinnate) and whether the leaf is oblique or not, the number of tepals in the male flower (two or four), fruit characters (whether it is a berry or a capsule) and for capsules whether the wings are the same size and shape or one is much larger than the others.

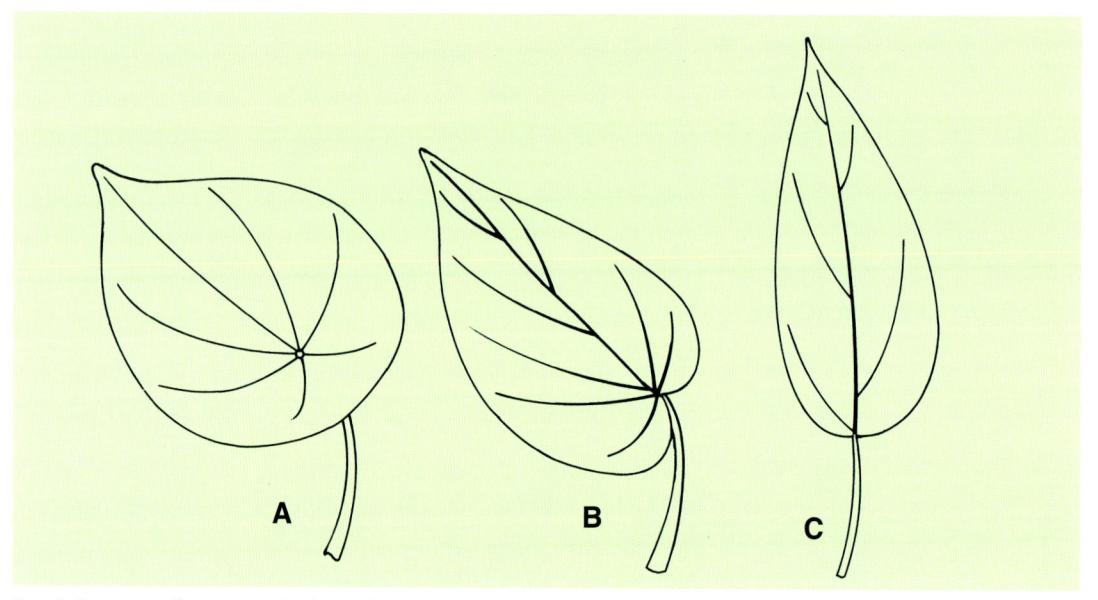

Leaf shapes and venation in begonia
A. A peltate leaf blade with palmate venation. **B.** An oblique, asymmetric leaf with palmate-pinnate venation. **C.** A symmetric leaf, not oblique, with pinnate venation.

A word is required about how the asymmetric leaf is measured. Length refers to the length of the midrib (a), width to the widest part of the blade (b). width of the broadest side (c), and length of the larger basal lobe (d) take care of the asymmetry of the leaf.

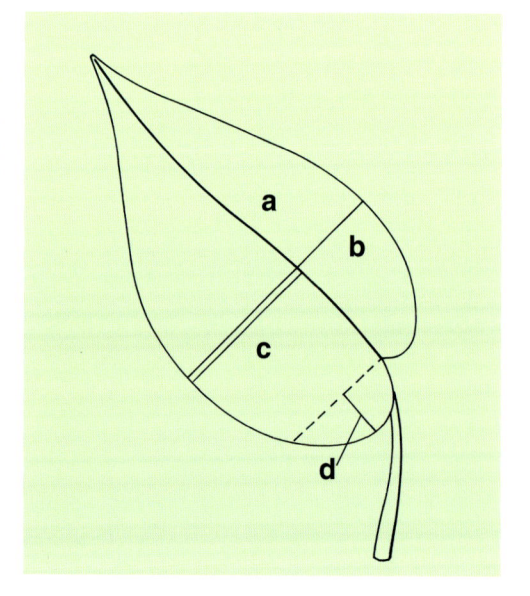

The Species

Fifty two begonia species are native to Peninsular Malaysia and two species are garden escapes. Thirteen of these species and one variety are here described as new, most having come to light as a result of recent field work. Field work has also been important in assessing variation within species and this has led to the reduction of some species to synonymy, most notably for the very widespread and variable *Begonia integrifolia*, which absorbs six species (*B. clivalis*, *B. curtisii*, *B. debilis*, *B. guttata*, *B. haniffii* and *B. leucantha*). Yet another three species described from a single specimen have proved to fall within the variation of an already described species (*B. collina* is synonymous with *B. rhoephila*; *B. robinsonii* with *B. pavonina*; and *B. tiomanensis* with *B. rheifolia*).

Re-visiting the area from where the species was first collected has led to the re-discovery of species that have not been collected for more than a hundred years (*B. forbesii* and *B. paupercula*), eighty years (*B. klossii* and *B. tampinica*) and sixty years (*B. corneri* and *B. yappii*). All of these are extremely local species. However, in spite of extensive field work, one species (*B. eiromischa*) was not re-found and is extinct as its habitat no longer exists and a further two (*B. scortechinii* and *B. isopteroidea*) and one variety (*B. martabanica* var. *pseudoclivalis*) were not re-located but their habitat remains intact.

The species are arranged below based on similarity to enable easy comparison.

References that apply to many species have been abbreviated as follows:

FBI – Flora of British India
FMP – Flora of the Malay Peninsula
JASB – Journal of the Asiatic Society of Bengal
JSBRAS – Journal of the Straits Branch of the Royal Asiatic Society
JFMSM – Journal of the Federated Malay States Museums
MIABH – Mitteilungen aus dem Institut fuer allgemeine Botanik in Hamburg
MWF Dicot – Malayan Wild Flowers Dicotyledons

The acronyms for herbaria where the specimens are deposited include:

BM – Natural History Museum of London; CGE — University of Cambridge, K – Royal Botanic Gardens Kew, KEP – Forest Research Institute Malaysia, KLU – University of Malaya in Kuala Lumpur, L – The National Herbarium of the Netherlands (Leiden University Branch), SING – Singapore Botanic Gardens, SINU – University of Singapore, UKMB – Universiti Kebangsaan Malaysia in Bangi.

PENINSULAR MALAYSIAN BEGONIAS

Key to Peninsular Malaysian species of *Begonia*

1a. Stems creeping on the soil surface. Leaves not tufted, up to 5 cm apart
2a. Leaves kidney-shaped, wider than long, 3–6 × 4–6.5 cm, base heart-shaped. Inflorescences 12.5–27 cm long. Fruits with one wing longer than the other two ...... **13. B. thaipingensis**
2b. Leaves broadly oval to ovate, longer than wide, 7.5–11 × 6.5–9 cm, base rounded. Inflorescences short 2–3 cm long. Fruits with 3 equal-sized wings .... **34. B. corneri**

1b. Stems rhizomatous with tufted leaves or stems erect with well-spaced leaves

3a. Leaves peltate (leaf stalk from above the base of the leaf)
4a. Leaf blades thickly succulent, leaf stalks without hairs .... **2. B. kingiana**
4b. Leaf blades thin and soft, leaf stalks hairy
5a. Leaf blades wider than long, stalks sparsely hairy. Male flowers with two tepals .... **4. B. tigrina**
5b. Leaf blades longer than wide, stalks densely hairy. Male flowers with four tepals
6a. Leaf margins densely fringed by hairs, not toothed. Fruits with 3 equal wings. Always on limestone .... **3. B. ignorata**
6b. Leaf margins minutely toothed, not fringed by hairs. Fruits with 1 long and 2 short wings. On granite rocks .... **1. B. eiromischa**

3b. Leaves not peltate (leaf stalk from the base of leaf)

7a. Leaf blades densely hairy above
8a. Leaf blades extremely narrow, at least 4 times longer than wide .... **42. B. scortechinii**
8b. Leaf blades up to 3 times longer than wide
9a. Stems erect
10a. Large plant 30–130 cm tall. Leaf blades large 16 × 14–19.5 × 18.5 cm. Tepals rosy red and pointed. Fruits larger 12 × 23–22 × 35 mm .... **32. B. lowiana**
10b. Smaller plant up to 22 cm tall. Leaf blades small 6 × 6.5–10 × 11 cm. Tepals whitish or whitish green, tip rounded. Fruits smaller 7 × 4–11 × 19 mm .... **33. B. jayaensis**
9b. Stems rhizomatous
11a. Leaf blades not oblique (midrib is in line with leaf stalk), base not heart-shaped, basal lobes 0.3–0.7 mm long, blades not pitted beneath .... **22. B. vallicola**
11b. Leaf blades oblique (midrib is at an angle with leaf stalk), base heart-shaped, basal lobes 2–4 cm long, blades pitted beneath
12a. Leaf blades conspicuously variegated, upper side dark green with yellow veins, 6.5–12 × 5.5–10.5 cm. Male flowers without hairs on the outer surface .... **20. B. decora**
12b. Leaf blades not variegated, upper side plain dark green, 10–15 × 9–14.5 cm. Male flowers hairy outside .... **21. B. wyepingiana**

7b. Leaf blades not densely hairy above or with sparse microscopic hairs or bristles

13a. Leaf blades extremely narrow, 3–4 times longer than wide
14a. Leaf bases heart-shaped .... **43. B. perakensis** var. **perakensis**
14b. Leaf blades narrowed to base .... **44. B. rhoephila**

13b. Leaf blades up to 3 times longer than wide

15a. Stems erect and free-standing or if on limestone pendent

16a. Stems woody, not succulent, without a tuber at the base
17a. Leaf stalks short, up to 2 cm long, leaf blades narrow about 2.5 times longer than wide, basal lobe scarcely developed, blade six times longer than the basal lobes
18a. Stems densely bristly with reddish brown hairs, up to 30 cm tall. Flowers in short clusters. Fruits oblong, 12–15 × 4.5–6.75 mm .... **19. B. barbellata**
18b. Stems without hairs, 50–130 cm tall. Flowers in inflorescences 4.5–9 cm long. Fruits 20–23 × 14–20 mm .... **17. B. wrayi**
17b. Leaf stalks longer than 4 cm, leaf blades broad, less than twice as long as wide, basal lobe well developed
19a. Stems 25–80 cm becoming pendent. Leaf blades small up to 8 cm long (5–8 × 3–7 cm), variegated. Only on limestone .... **18. B. jiewhoei**
19b. Stems up to 100–175 cm, always erect. Leaf blades more than 8 cm long (8–29.5 × 3.5–13 cm), plain green. Never on limestone
20a. Leaf blades less than twice as long as wide, broadly ovate, (9 × 6–21 × 13 cm). Fruit a capsule with 3 equal wings .... **15. B. holttumii**
20b. Leaf blades more than twice as long as wide. Fruit a berry or a capsule with 1 wing longer than the other two
21a. Leaf blades shorter 8 × 3.5–11 × 5 cm, basal lobe 2.5–3 cm long. Fruit a dry capsule with 3 unequal wings .... **16. B. isopteroidea**
21b. Leaf blades longer 10 × 4–29.5 × 13 cm, basal lobe 3–6 cm long. Fruit a berry without wings .... **14. B. longifolia**

16b. Stems soft and succulent often with a small tuber at the base
22a. Leaves not oblique
23a. Leaf blades asymmetric with one basal lobe larger than the other
24a. Leaf stalks without hairs .... **53. B. cucullata**
24b. Leaf stalks woolly with long white hairs .... **54. B. hirtella**
23b. Leaf blades symmetric, basal lobes equal
25a. Leaves with the stalk shorter than the blade, hairs star-like .... **6. B. sinuata**
25b. Leaves with the stalk longer than the blade, hairs straight
26a. Leaf blades round, tiny, up to 12 mm long .... **5. B. sibthorpioides**
26b. Leaf blades longer than wide and more than 25 mm long .... **7. B. martabanica**

22b. Leaves oblique and unequal-sided
27a. Leaf stalks longer than the blades, venation palmate .... **8. B. carnosula**
27b. Leaf stalks shorter or same length as the blades, venation palmate-pinnate
28a. Blades less than twice as long as wide, tip not elongated
29a. Plants hairy. Tepals almost rotund, plain white or pale pink, veins rarely pale red and then not ribbed .... **9. B. integrifolia**
29b. Plants without hairs. Tepals elliptic, white or pink, veins always red and ribbed .... **10. B. phoeniogramma**
28b. Blades more than twice as long as wide, tip elongated

30a. Leaf blades with one basal lobe much larger often equalling a third of the blade length, variegated with pale green or silver spots, leaf tip not rooting .... **11. B. variabilis**
30b. Leaf blades with basal lobes almost equal in length, plain green, leaf tip rooting and producing a plantlet .... **12. B. elisabethae**

15b. Plant rhizomatous, if erect shoots are produced then they are not self-supporting

31a. Leaves not oblique (midrib straight with the leaf stalk), blades almost equal-sided

32a. Leaf blades up to 14 cm long
33a. Leaf blades narrowed to the base .... **45. B. abdullahpieei**
33b. Leaf bases heart-shaped or rounded at the base
34a. Leaf bases unequal with one lobe larger .... **53. B. cucullata**
34b. Leaf bases equal and heart-shaped
35a. Leaf blades more than twice as long as wide .... **43. B. perakensis** var. **conjugens**
35b. Leaf blades less than twice as long as wide
36a. Leaf blades less than 1.5 times longer than wide, basal lobes up to 10 mm long .... **46. B. rhyacophila**
36b. Leaf blades narrower (more than 1.5 times longer than wide), basal lobes 10–20 mm long .... **23. B. alpina**

32b. Leaf blades more than 14 cm long
37a. Leaf blades broadest at the base, 24–38.5 × 18–29.5 cm .... **52. B. rheifolia**
37b. Leaf blades broadest in the middle of the leaf
38a. Leaf blades narrowly elliptic, more than twice as long as wide
39a. Leaf blades with long scattered hairs above, male tepals hairy .... **49. B. praetermissa**
39b. Leaf blades hairless above, male tepals without hairs
40a. Leaf stalks 9–15 cm long, blades up to 18 cm long, narrowed to the base, margins finely toothed .... **47. B. aequilateralis**
40b. Leaf stalks 13–22 cm long, blades more than 20 cm long, rounded at the base, margins not toothed .... **48. B. tampinica**

38b. Leaf blades broadly elliptic to slightly ovate, less than twice as long as wide
41a. Leaf blades up to 25 cm long, leaf stalks to 16 cm .... **50. B. herveyana**
41b. Leaf blades 16–33 cm long, leaf stalks 18 cm long .... **51. B. nubicola**

31b. Leaves clearly oblique (midrib at an angle to the leaf stalk), asymmetric (leaf blades unequal-sided)

42a. Leaf blades with palmate venation. Fruits with 3 equal wings
43a. Leaf blades rotund, wider than long
44a. Leaf tips pointed, basal lobes not overlapping. Inflorescences with many flowers. Male flowers usually with 2 tepals, sometimes with 4 .... **40. B. foxworthyi**
44b. Leaf tips rounded, basal lobes usually overlapping. Inflorescences few–flowered. Male flowers always with 4 tepals .... **41. B. nurii**

43b. Leaf blades ovate or oval, longer than wide or as long as wide
45a. Leaf blades 7.5–10 × 6–11.5 cm or larger. West of the Main Range .... **34. B. forbesii**
45b. Leaf blades up to 7.5 × 6 cm. East of the Main Range
46a. Leaf blades very asymmetric (broad side more than half the leaf width), some plants in the population variegated
47a. Leaf blade rounded, 7–15 × 6–15. Inflorescence 10–25 cm long. Male flowers with 4 tepals .... **37. B. rajah**
47b. Leaf blade tapered to the tip, 5.5–7.5 × 5.5–7.5 cm. Inflorescence up to 10 cm long. Male flowers with 2 tepals .... **38. B. reginula**
46b. Leaf blades slightly asymmetric (broad side about same width as the narrow side), not variegated
48a. Leaf blades oval (widest in the middle of the leaf), longer than wide, margins not toothed .... **39. B. yappii**
48b. Leaf blades ovate or rotund (widest towards the base), as wide as long, margins finely toothed .... **36. B. lengguanii**

42b. Leaf blades with palmate-pinnate venation. Fruits with one wing longer than the other two

49a. Stems rhizomatous clinging to rocks. Leaves tufted
50a. Leaf blades up to 14 cm long
51a. Leaf blades up to 6 × 5.5 cm long, variegated silver along the border and between the veins .... **24. B. koksunii**
51b. Leaf blades more than 8 × 6 cm, not variegated
52a. Leaf blades more or less equal-sided, basal lobe small 1–2 cm long, margins unequally toothed with longer teeth at the veins endings, green sometimes with a bluish tinge .... **23. B. alpina**
52b. Leaf blades very unequal-sided, basal lobe large and rounded, 2–5 cm long, margins minutely and equally toothed, in life bright iridescent blue or green .... **25. B. pavonina**
50b. Leaf blades more than 14 cm long
53a. Leaf blades longer than wide, stalks woolly with long hairs 2–3 mm long .... **26. B. klossii**
53b. Leaf blades wider than long, stalks without hairs or with short velvety hairs
54a. Veins on lower leaf surface with dense, rusty-brown, velvety hairs .... **28. B. maxwelliana**
54b. Veins on leaf lower surface without hairs .... **27. B. paupercula**

49b. Stems at first rhizomatous, then producing semi-erect stems with widely spaced leaves
55a. Leaf stalks more than twice the length of the leaf blade. Male flowers large with the outer tepals 18–36 × 17–38 mm .... **29. B. venusta**
55b. Leaf stalks less than twice the length of the leaf blade. Male flowers smaller with outer tepals 8–20 × 6–20 mm
56a. Leaf blades as long as wide, without hairs .... **31. B. longicaulis**
56b. Leaf blades longer than wide with dense hairs on lower surface of the veins
57a. Hairs on the lower surface of veins dense and reddish-brown .... **28. B. maxwelliana**
57b. Hairs on the lower surface of veins minutely velvety and pale fawn .... **30. B. fraseri**

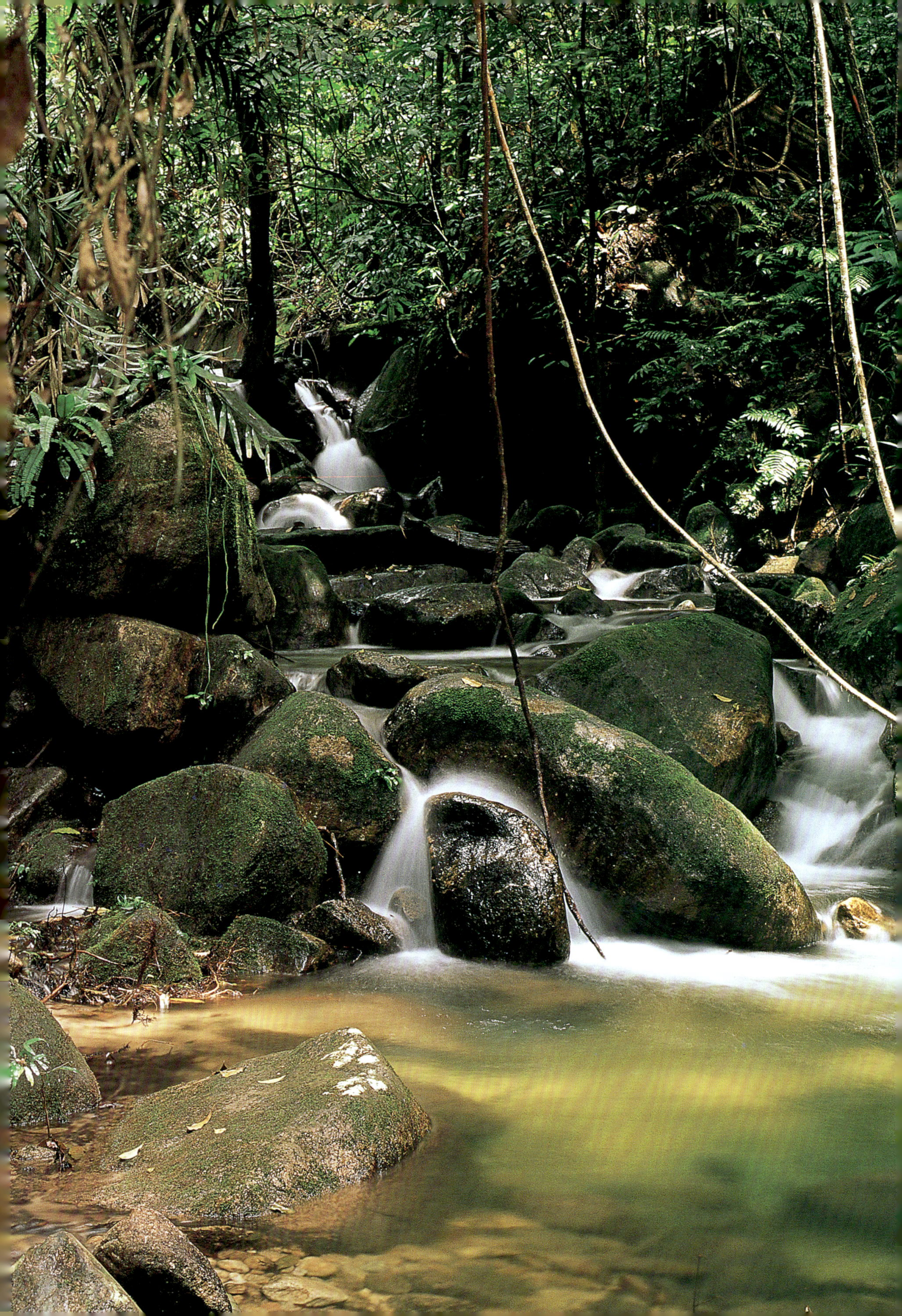

THE SPECIES

1. THE WOOLLY-STALKED BEGONIA
Begonia eiromischa Ridl.
(Greek, *eiromischa*=woolly stalk)

Ridley, JSBRAS 75 (1917) 36, FMP 1 (1922) 860 & Fig. 70; Irmscher, MIABH 8 (1929) 108; Kiew, Nature Malaysiana 14 (1989) 64. **Type:** *Curtis 1028*, November 1898, Pulau Betong [Boetong], Penang (lecto SING, here designated; isolecto SING).

Stem rhizomatous, succulent, unbranched, rooting at nodes, stout, 6–10 mm thick; without a tuber. Stipules without hairs, narrowly triangular, 12–20 × 4–8 mm, margin not toothed, tip narrowing, ending in a hair, persistent. **Leaves** tufted; stalk with dense woolly red hairs, 7–17 cm long, 2–3.5 mm thick, grooved above; blade very oblique, plain deep green, succulent, drying thinly leathery, asymmetric, peltate or sometimes almost deeply heart-shaped, rounded, as broad as long or broader than long, 7–9 × 7–12 cm, broad side 6.5–7 cm wide, base shallowly heart-shaped, *c.* 1 cm from stalk to base, margin minutely toothed, tip abruptly pointed up to 1 cm long; venation palmate, (3–)4–5 pairs, branching *c.* halfway to margin, sparsely hairy above, beneath prominent and densely hairy. **Inflorescences** axillary, red, without hairs, erect, longer than the leaves, branched, 12–20 cm long, branches (2–)3–4 cm long, elongating to 30 cm in fruit with a stalk up to 24 cm long, male flowers *c.* 12, female flowers 2, male flowers open first. Bracts soon falling. **Male flowers** with a stalk 5–11 mm long; tepals 4, rose pink, margin not toothed, tip rounded, without hairs, outer two ovate, 6.5–8.5 × 6–7 mm, inner two narrowly linear, *c.* 3.5 × 1 mm; stamens many, cluster globose, *c.* 2 × 3 mm, not stalked; filaments *c.* 1 mm long; anthers narrowly obovate, 0.75–1.25 mm long, tip slightly notched, opening by slits. **Female flowers** with a stalk *c.* 7 mm long; ovary 6–7 × 2–2.5 mm, wings 3, unequal, locules 2, placenta one per locule; tepals 3, rosy pink, without hairs, margin not toothed, tip rounded, outer rotund, *c.* 5 × 6 mm, innermost similar but smaller, *c.* 3 × 1.25 mm; styles 3, styles and stigma *c.* 3 mm long. **Fruits** pendent on a fine stalk 14–16 mm long; capsule 12–13 × 14–17 mm, hairless, locules 2, wings 3 unequal, thinly fibrous, larger broad, rounded 8.5–12 mm wide, smaller two very short, narrowed to tip, 1–2.5 mm wide, splitting between the locules and wings. **Seeds** barrel-shaped, *c.* 0.3 mm long, collar cells a quarter of the seed length.

DISTRIBUTION. Endemic in Peninsular Malaysia—Pulau Betong Reserve, Penang.

HABITAT. "Abundant on [granite] rocks in this one place only" (*Curtis 1028*) at *c.* 170 m altitude.

OTHER SPECIMENS. PENANG—Pulau Betong Reserve, *Curtis 1028* (SING), *Curtis s.n.*, Nov 1898 (SING).

NOTES. The Woolly-stalked Begonia was always very rare and known from just one locality. Its habitat no longer exists, having been overtaken by farms (Kiew, 1989), so *Begonia eiromischa* is now extinct. Ridley described it from plants growing in the Waterfall Gardens, Penang, in 1886 and from a watercolour painting.

It is a remarkable begonia in its peltate leaves with very hairy stalks. It is most closely related to *Begonia kingiana*, a limestone begonia, which also has peltate leaves and fruits with one long wing and two short wings, but plants of *B. kingiana* are quite hairless.

(Opposite). There is hardly a forest stream in Peninsular Malaysia without one or two wild begonias growing on the mossy rocks or steep banks.

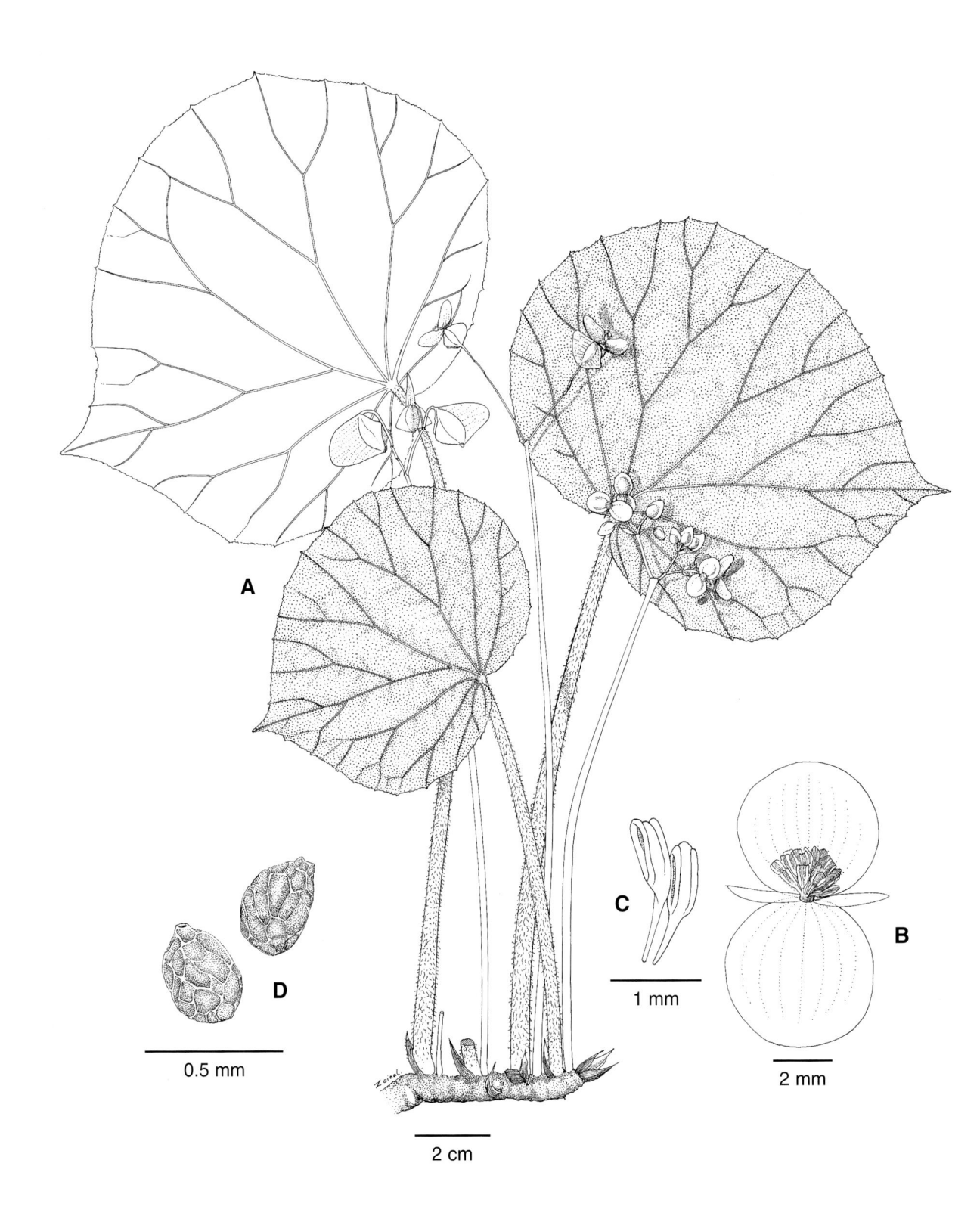

Begonia eiromischa Ridl. **A.** The plant. **B.** Male flower. **C.** Stamens. **D.** Seed. (*Curtis 1028*)

2. THE TORTOISE SHELL BEGONIA
Begonia kingiana Irmsch.
(Sir George King, Superintendent of Royal Botanic Garden, Calcutta, 1871–1898;
co-author of the 5-volume *Materials for a Flora of the Malay Peninsula*)

Irmscher, MIABH 8 (1929) 106 & Fig. 3; Henderson, MWF Dicot (1959) 160 & Fig 3; Chin, Gard. Bull. Singapore 30 (1977) 98. **Type:** *Ridley 9689*, September 1890, Kuala Dipang, Perak (lecto SING, here designated). **Synonym:** *B. hasskarlii sensu* King, JASB 71 (1902) 62 *non* Zoll. et Mor.; Ridley, FMP 1 (1922) 860 *pro pte*.

Stem rhizomatous, brown, succulent, without hairs, unbranched, creeping vertically and rooting on cliff faces, slender, up to 11 cm long, 5–7 mm thick; without a tuber. Stipules reddish, without hairs, narrowly triangular, 7–12 × 4–6 mm, margin not toothed, tip acute ending in a long hair, persistent. **Leaves** tufted, almost touching, held against rock surface with the blade pendent; stalk pale brown, without hairs, (3–)10(–21) cm long, round in cross-section, 1–2 mm thick; blade oblique, plain dull dark green, grey-green or almost black and beneath white, pale red or red or above variegated brownish between paler veins and purple beneath, thickly succulent and brittle in life, drying thinly leathery, peltate, round or sometimes broadly oval, asymmetric, 4–10 × 2.5–10 cm, broad side 1.5–6 cm wide, base rounded, 1–4 cm from stalk to the base, margin not toothed, wavy, tip shortly acute up to 1 cm long; venation palmate, 3–5 pairs of veins, branching halfway

Typical tower karst form of the limestone hills where *Begonia kingiana* grows.

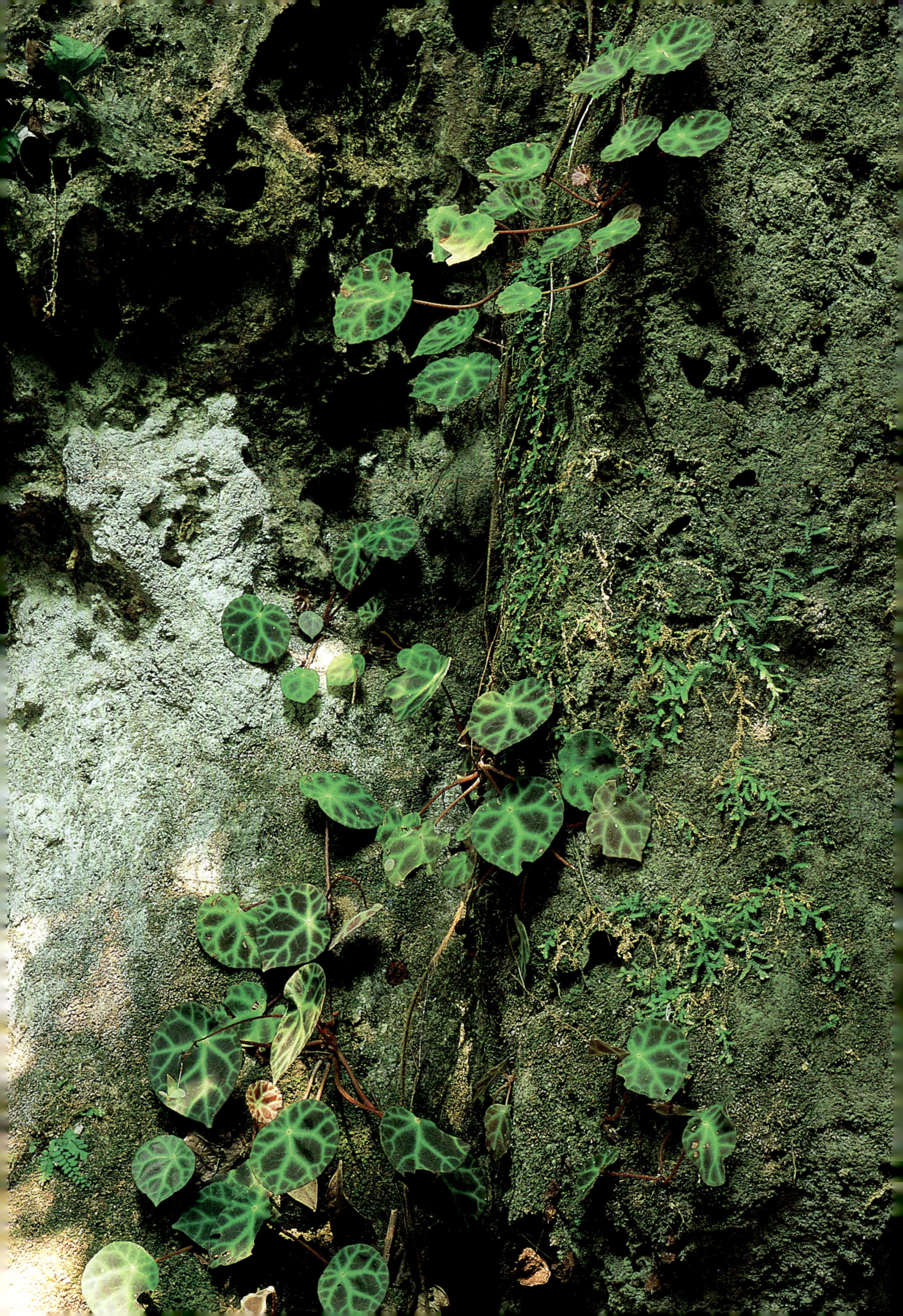

The range of colour forms seen in different populations of *Begonia kingiana*.
(Opposite) *Begonia kingiana* grows on vertical limestone cliffs.

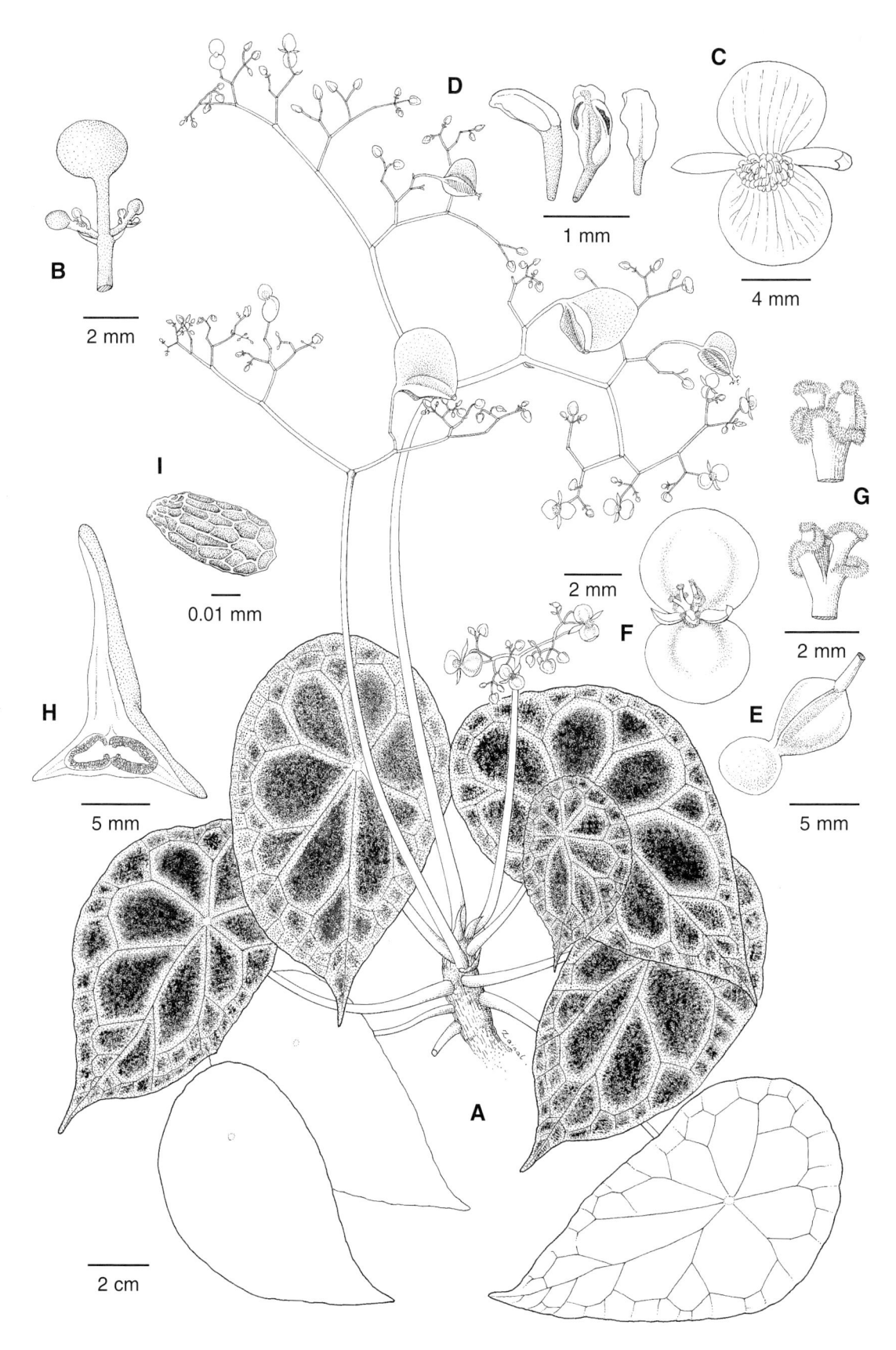

Begonia kingiana Irmsch. **A.** The plant. **B.** Inflorescence branch. **C.** Male flower. **D.** Stamens. **E.** Female bud. **F.** Female flower. **G.** Styles and stigmas. **H.** T.S. ovary. **I.** Seed. (*RK 5196*)

BEGONIA KINGIANA

to margin, slightly prominent above, beneath impressed, the same colour as the blade, hairless. **Inflorescences** axillary, scarlet or purplish, without hairs, longer than the leaves, 7.5–47 cm long with a stalk 5–33 cm long, branching up to 7 times, branches 1–8 cm long, male flowers many, female flowers few, male flowers open first. Bract pair red, deflexed, *c.* 4 × 1 mm, margin not toothed, soon falling. **Male flowers** with a reddish stalk 5–12 mm long; tepals 4, margin not toothed, tip rounded, hairless, outer two white or pale pink or deep scarlet outside and paler inside, broadly oval, 4–6 × 3–6 mm, inner two white, narrowly oval, 2–4 × 1–1.5 mm; stamens many, cluster globose, *c.* 2 mm across, without a stalk; filaments *c.* 0.5 mm long; anthers golden yellow, narrowly obovate, 1–1.5 mm, tip deeply notched, opening by slits. **Female flowers** with a pale red stalk 8–13 mm long; ovary green or red, wings 3, unequal, largest wing reddish, 6–8 × 7–9 mm, locules 2, placenta one per locule; tepals 4, margin not toothed, tip rounded, hairless, outer deep pink, darker outside, ovate, 4–7 × 4.5–7 mm, innermost tepals white, narrowly obovate, 3.5–5 × 1–2 mm; styles 3, styles and stigmas yellow, 2–3 mm long, stigmas spiral. **Fruits** pendent on a stiff stalk 12–22 mm long; capsule (7–)11–15 × (9–)21 mm, hairless, locules 2, wings 3, thick, larger 8–12 mm wide, narrowed to the tip, smaller two 2–4 mm wide, splitting between the locules and wings. **Seeds** barrel-shaped, *c.* 0.4 mm long, collar cells a quarter to a fifth of the seed length.

DISTRIBUTION. Endemic in Peninsular Malaysia—Langkawi, Perak, Selangor, Pahang, Kelantan and Trengganu.

HABITAT. Restricted to limestone, it grows in deep shade below the tree canopy on vertical cliffs or in fissures or on boulders at the base of the cliff. It is quite common and found on most limestone hills.

OTHER SPECIMENS. KEDAH—Langkawi, *Kiew RK 4927* (SING); KELANTAN—Aring FR, *Anthonysamy SA 1085* (SING), Bukit Tapah, *Boey 352* (KLU), Gua Musang, *Chin et al. 4545* (SING), *Kiew RK 5196* (SING), *UNESCO Limestone Expedition 395* (SING), Gua Teja, *Henderson SFN 29705* (SING), Gunung Renayang, *Kiew & Anthonysamy RK 2865* (SING), Kualu Betis, *Kiew RK 5252* (SING), Sungai Bring, *Kiew & Anthonysamy RK*

Begonia kingiana. (Top). Male flower. (Middle). Female flower and fruit. (Bottom). Female bud.

BEGONIA KINGIANA

2905 (KEP), Sungai Jenera, *Kiew & Anthonysamy RK 2895* (KEP, SING), Sungai Kerteh, *Md Nur & Foxworthy SFN 11984* (SING); PAHANG—Gua Hijau, *Kiew RK 3134* (SING), Gunung Jebak Puyuh, *Kiew RK 2152* (SING); PERAK- Ampang, *Burkill & Haniff SFN 13922* (SING), Gopeng, *Burkill s.n.* (SING), Gunung Kandu, *Kiew RK 4494* (SING), Gunung Rapat, *Chin 808* (SING), Kinta, *Curtis s.n.* (SING), Kuala Dipang, *Curtis s.n.* (SING), Tambun, *Burkill SFN 6300* (SING), *Chin 845* (SING); SELANGOR—Batu Caves, *Burkill SFN 2263* (SING), *Kiew RK 1341* (KEP), *RK 4714* (SING), *Ridley 314* (SING); TRENGGANU—Batu Biwa, *Kiew RK 2286* (KEP), *RK 2322* (SING).

NOTES. Of all the begonias in the Peninsula, *Begonia kingiana* has the most thick and succulent leaves. It shows a great deal of variation in colour and leaf shape between populations. Two of the most attractive are the ones with variegated leaves that look like tortoise shells, hence the common name, and the ones with almost black leaves and bright scarlet inflorescences and flowers. Some populations on the east coast have leaves that are more elongate and the tip more attenuate than the usual rounded leaf shape.

It is most closely related to *Begonia eiromischa* but is quite distinct from it in its hairless petioles. There are another two peltate begonias that grow on limestone, *B. ignorata* and *B. tigrina*, but their leaves are thin and their fruits have three thin, equal wings.

The Tortoise Shell Begonia, *Begonia kingiana*.

3. THE FRINGED LIMESTONE BEGONIA
Begonia ignorata Irmsch.
(Latin, *ignorare*=unrecognized)

Irmscher, MIABH 8 (1929) 97; Kiew, Malayan Naturalist 38(3) 32 & Fig. **Type:** *Ridley 2442*, 1891, Kota Gelanggi, Pahang (lecto SING, designated by Irmscher; iso K). **Synonym:** *Begonia hasskarlii* Zoll. et Mor. var. *hirsuta* Ridl., FMP 1 (1922) 860. **Type:** *Ridley 2442*, 1891, Kota Gelanggi, Pahang (lecto SING, designated by Irmscher; iso K).

Stems rhizomatous, reddish, succulent becoming wiry, up to 30 cm long with leaves in the top 6–10 cm, slender 7–8 mm thick, hairs sparse, long and white, unbranched; without a tuber. Stipules reddish, hairy especially on the margin, narrowly lanceolate, 4–10 × 2–6 mm, margin not toothed, tip pointed, ending in a hair, persisting as a papery remnant. **Leaves** tufted, up to 5–7 mm apart;

(Above). *Begonia ignorata*. The black-leaved form; (Bottom left to right). Male flower, female flower and fruits. (Following page). The Fringed Limestone Begonia, *Begonia ignorata*.

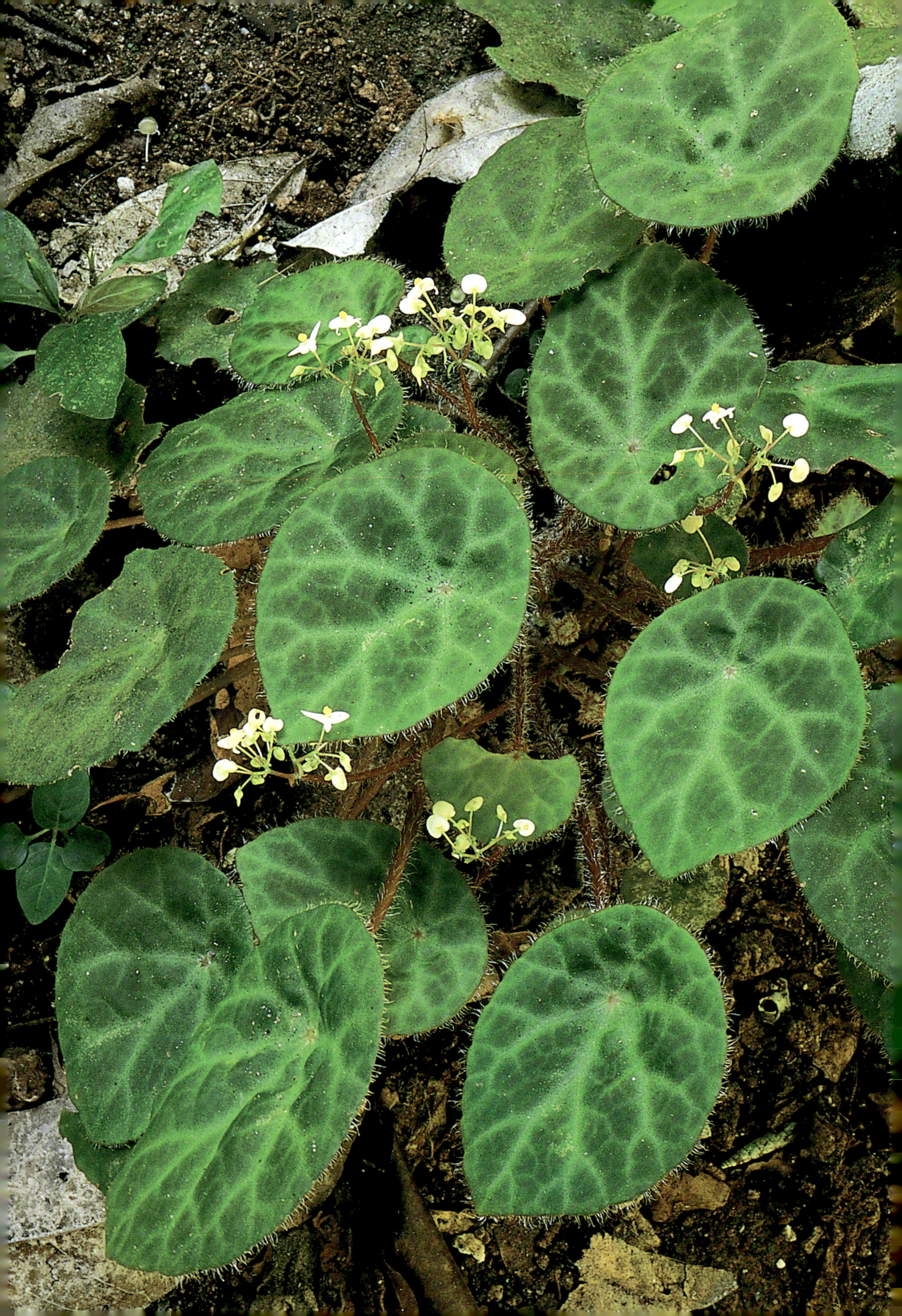

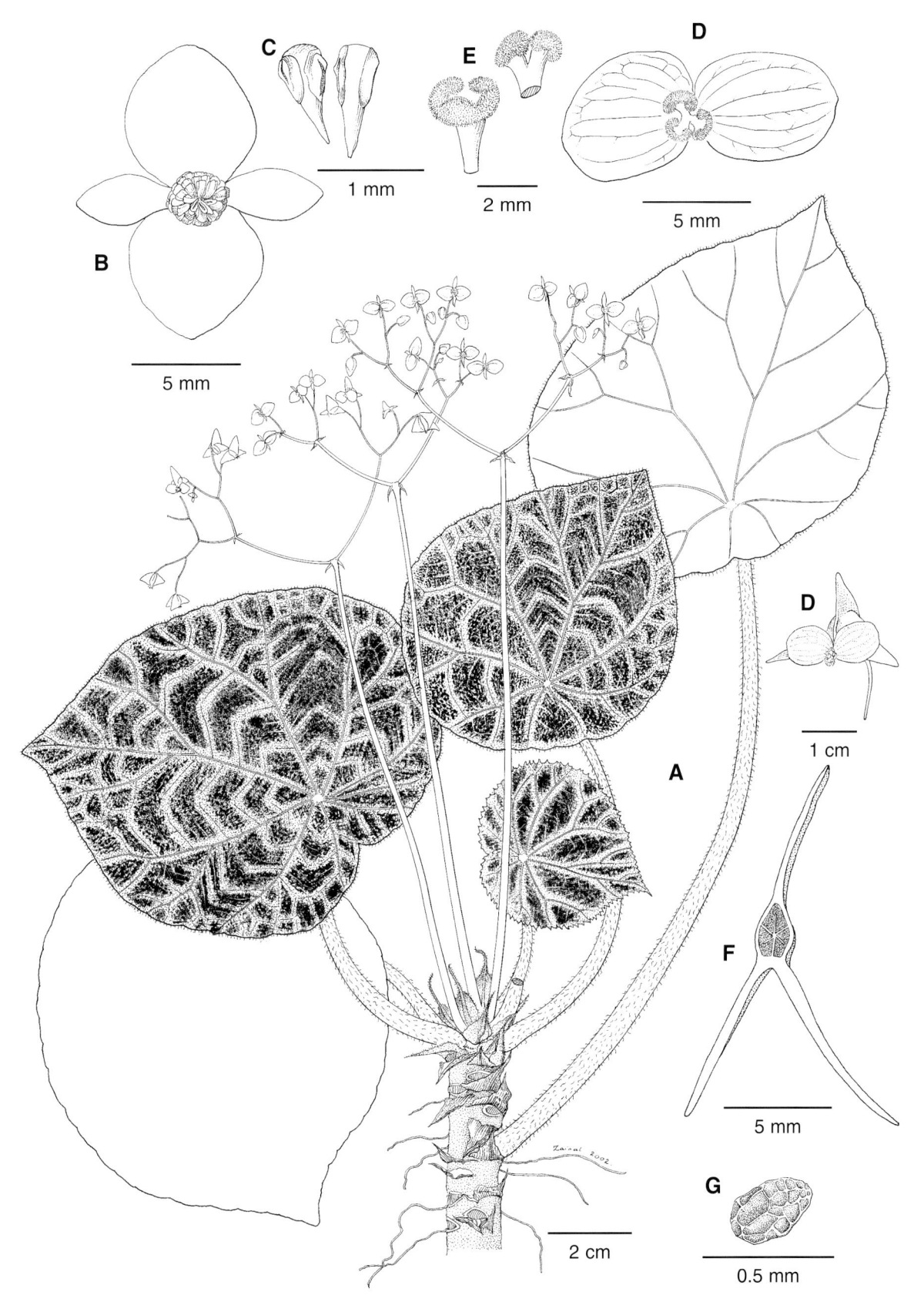

Begonia ignorata Irmsch. **A.** The black-leaved form. **B.** Male flower. **C.** Stamens. **D.** Female flower; **E.** Styles and stigmas. **F.** T.S. ovary. **G.** Seed. (*RK 5079*)

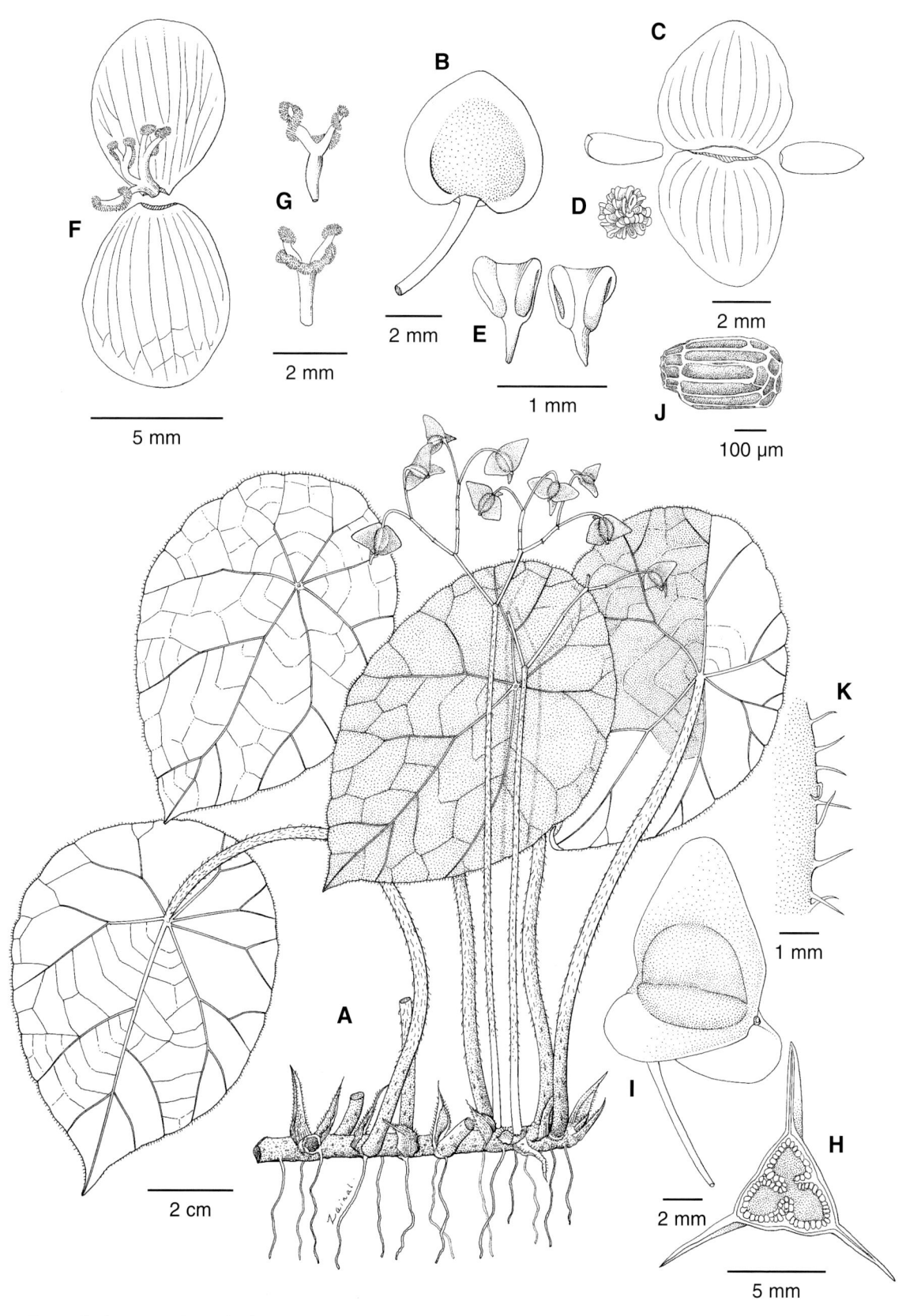

Begonia ignorata Irmsch. **A.** The plant. **B.** Male bud. **C.** Male flower. **D.** Stamen cluster. **E.** Stamens. **F.** Female flower. **G.** Styles and stigmas. **H.** T.S. ovary. **I.** Fruit. **J.** Seed. **K.** The leaf margin. (*RK 4919*)

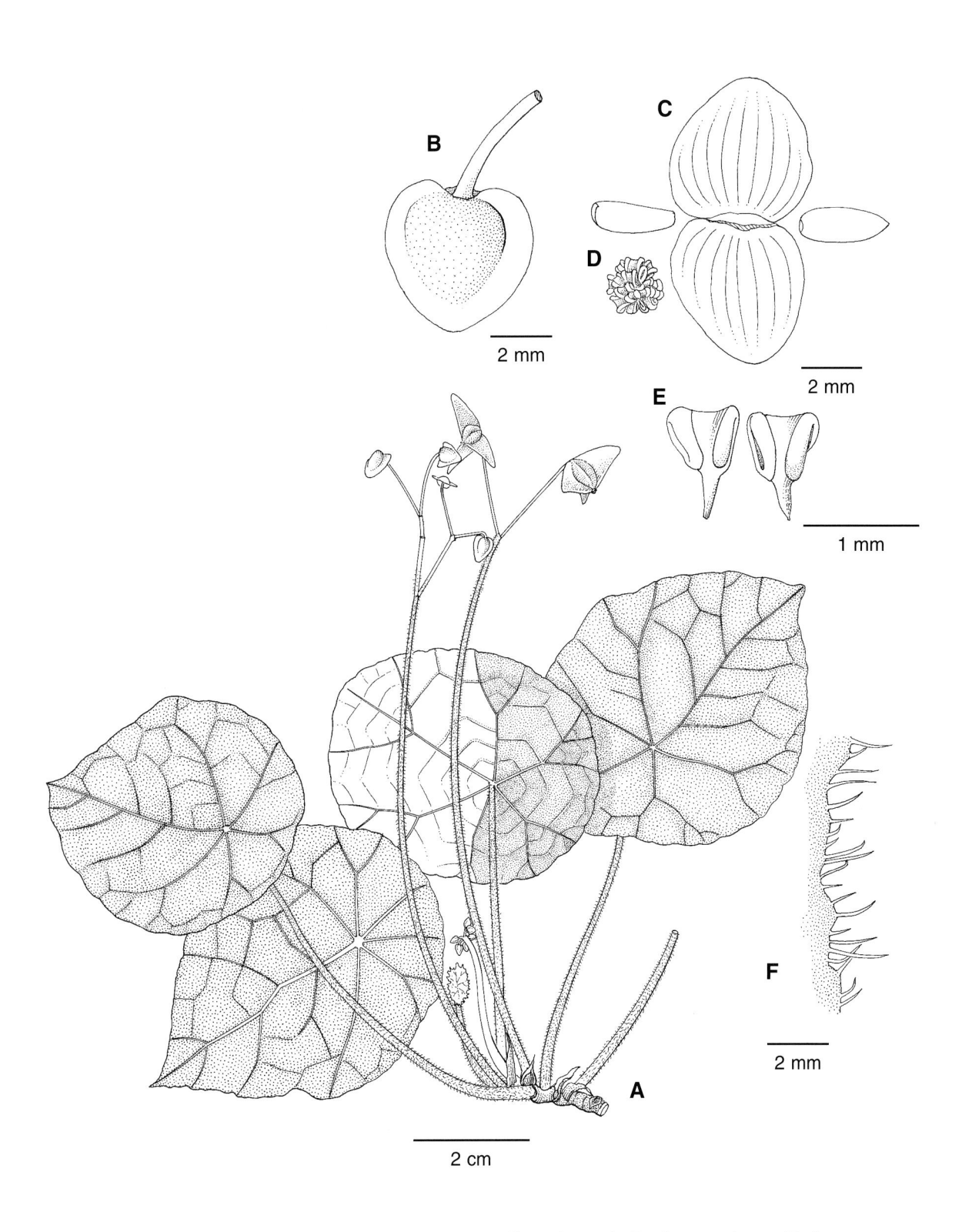

Begonia ignorata Irmsch. **A.** The plant. **B.** Male bud. **C.** Male tepals. **D.** Stamen cluster. **E.** Stamens. **F.** The leaf margin. (*RK 4922*)

BEGONIA IGNORATA

stalk reddish brown, 6–11(–20) cm long, round in cross-section, densely hairy, hairs *c.* 2 mm long; blade oblique or sometimes not oblique and the stalk is in line with the midrib, peltate, variegated with paler green veins against grey-green blade above and beneath or dark green or brownish green above and reddish beneath or dark purple-black or reddish purple above and reddish purple beneath, hairless above, hairy beneath, softly succulent, drying papery, slightly asymmetric, 4–7(–9) × (3–)4.5–5(–8) cm, broad side 2.5–5.5 cm wide, base rounded 0.75–3 cm long, margin not toothed but sometimes with rounded indentations (crenate), conspicuously fringed by long hairs 1—2 mm long, tip pointed or sometimes rounded; venation palmate-pinnate, 3 pairs with another 3 veins in peltate base, branching towards the margin, impressed or plane above, beneath slightly prominent and hairy. **Inflorescences** axillary, pale green or reddish-brown, sparsely hairy, longer than the leaves, 9–25 cm long with two main branches 1.5–4 cm long, stalk 8–22 cm long, male flowers many, female flowers 4–8, male flowers open first. Bract pair deflexed, obovate, 2–2.5 × 1.5–2 mm, margin not toothed, fringed by hairs, soon falling. **Male flowers** with pale green stalk 13–18 mm long; tepals 4, margin not toothed, tip rounded, outer two white with pinkish tinge towards base and outside pinkish with a few sparse hairs, broadly oval, 6–7 × 4.5–7 mm, inner two white, hairless, narrowly obovate, 5–6 × 2.75–3; stamens many, cluster globose, 2–2.5 mm across, not stalked or stalk to 0.5 mm long; filaments *c.* 0.25 mm long; anthers dull yellow, broadly obovoid, 0.75–1 mm long, tip notched, opening by slits. **Female flowers** with pale green stalk 5–6 mm long; ovary pale green, 6 × 13–14 mm, wings 3, equal, locules 3, placenta 1 per

The Fringed Limestone Begonia grows in profusion on the damp shaded base of limestone cliffs.

locule; tepals 2, white, slightly pink at base, hairless, rotund, margin not toothed, tip rounded, 6–7 × 6 mm; styles 3, styles and stigmas yellow, *c.* 4 mm long, stigmas spiral. **Fruits** dangling on a fine, hair-like stalk, (7–)10–12 mm long; capsule (5–)7–9 × 11–15 mm long, hairless, locules 3, wings 3, equal, triangular tapering to a rounded or slightly pointed tip, 3–6 mm wide, drying papery, splitting between locules and wings. **Seeds** barrel-shaped, *c.* 0.4 mm long, collar cells *c.* 4/5 seed length.

DISTRIBUTION. Endemic in Peninsular Malaysia—Kelantan, Pahang and Trengganu.

HABITAT. Restricted to limestone in the lowlands at about 100 m altitude. It is locally common on damp, lightly shaded rock faces below the tree canopy, on boulders and limestone-derived soil at the base of cliff faces and on guano in well-lit caves.

OTHER SPECIMENS. KELANTAN—Gua Batu Bali, *Kiew & Anthonysamy RK 3063* (SING); Gua Pehah Kerbau, *Abdullah Piee s.n.* 7 Aug 2002 (SING), *Kiew RK 5262* (SING), *Kiew & Anthonysamy RK 3045* (SING). PAHANG—Batu Luas, *Kiew RK 1201* (KLU, SING), *Kiew RK 1389* (SING); Bukit Batu, *Kiew RK 2372* (KEP, SING); Bukit Cheras, *Henderson SFN 25249* (SING), *Kiew RK 1556* (KEP, SING), *Kiew RK 4922* (SING); Kota Gelanggi, *Kiew RK 4919* (SING); Gunung Jebak Puyuh, *Kiew RK 2151* (KEP, SING); Gunung Senyum, *Anthonysamy SA 513* (SING), *Kiew RK 1588* (KEP, SING), *Kiew RK 5079* (SING); Taman Negara, *Chin 1304* (SING). TRENGGANU—Batu Biwa, *Kiew RK 2285* (KEP, SING).

NOTES. The Fringed Limestone Begonia is very distinctive in its soft peltate leaves with a hairy margin. Ridley originally described it as a hairy variety of *Begonia kingiana*, the first peltate species to be described from Peninsular Malaysia but *B. kingiana* is very different in having stiff, thickly fleshy leaves that are completely without hairs. In addition, *B. kingiana* has fruits with one wing longer than the other two.

Begonia ignorata shows a wide range of leaf shape on the different limestone hills. For example, on Kota Gelanggi and Batu Biwa the leaf blades are broadly oval in outline, on Gunung Senyum they are round with a wavy margin, on Bukit Cheras angular, and on limestone in Taman Negara the blades are longer, oval and sometimes straight-sided. However, there are no differences in other characters so this variation is not recognized at the subspecific level.

The Fringed Limestone Begonia regenerates abundantly from seed and many seedlings can be found. However, there is no doubt it is a perennial species due to its wiry rhizome. In addition, it regenerates naturally from leaf cuttings—when the leaf rots, the blade separates from the stalk and new plantlets develop from the severed veins.

4. THE TIGER BEGONIA
Begonia tigrina Kiew, sp. nov.
(Latin, tigris=tiger, referring to its growing in tiger country)

A *Begonia ignorata* Irmsch. tepalis masculis 2 (nec 4) et foliis margine glabris recedit. **Typus:** *Kiew RK 5271*, 19 February 2003, Gua Setir, Kelantan (holo SING, iso K, KEP).

Stem rhizomatous, succulent, densely hairy, unbranched, slender, up to 10 cm long with leaves in the top 3 cm, 3–4 mm thick; without a tuber. Stipules narrowly triangular, 8–10 × 3–4 mm, margin not toothed, fringed by long hairs, tip ending in a long hair, persistent. **Leaves** tufted, up to 3 mm apart; stalk reddish, sparsely hairy, 3–5(–8.5) cm long, grooved above; blade slightly oblique or not oblique and the stalk in line with the midrib, peltate, dull or plain green above or variegated and reddish brown between the veins, hairless above with microscopic scattered hairs beneath, thinly succulent in life, thinly papery when dried, broadly ovate, slightly asymmetric, 3–5 × 3.5–5.5 cm, broad side 2–3 cm wide, base rounded to truncate, sometimes slightly heart-shaped, 0.5–1 cm long, margin not toothed, with sparse, distant, short hairs to 0.5 mm long, tip pointed; venation palmate-pinnate, 2 pairs of veins at the base of the midrib and one pair along the midrib

Gunung Setir, home of the Tiger Begonia, *Begonia tigrina*.

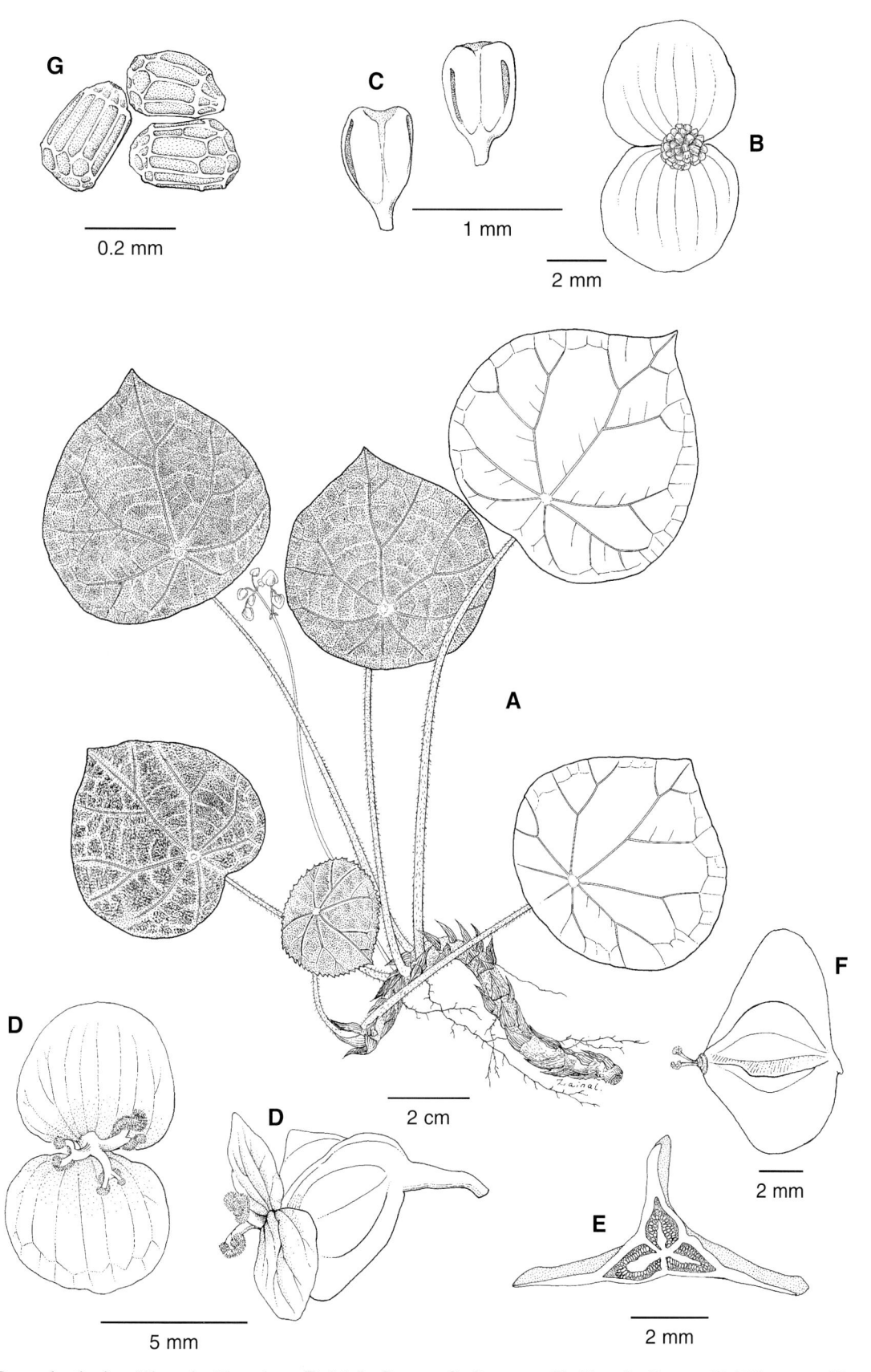

Begonia tigrina Kiew **A.** The plant. **B.** Male flower. **C.** Stamens. **D.** Female flower. **E.** T.S. ovary. **F.** Fruit. **G.** Seed. (*Piee s.n.*)

BEGONIA TIGRINA

The Tiger Begonia grows on shaded limestone boulders.

and another pair below the point of attachment with the stalk in the base, branching *c.* half-way to margin, veins plane above and beneath, beneath sparsely hairy. **Inflorescences** axillary, reddish, sparsely hairy, longer than the leaves, 6–16 cm long with two main branches 0.75–2.5 cm long, stalk 5–15 cm long, male flowers many, female flowers 4, male flowers open first. Bract pair narrowly obovate, 1–1.5 × *c.* 0.5 mm, hairless, margin not toothed, soon falling. **Male flowers** with a stalk 5–14 mm long; tepals 2, white, without hairs, rotund, 5–6 × 6–7 mm, margin not toothed, tip rounded; stamens many, cluster ovoid, 2–2.5 mm across, stalk *c.* 0.5 mm long; filaments *c.* 0.3 mm long; anthers yellow, broadly obovate, *c.* 0.5 mm long, tip slightly notched, opening by slits. **Female flowers** with a stalk 4–9 mm long; ovary white becoming green or reddish green, *c.* 4 mm long, wings 3, equal, locules 3, placenta one per locule; tepals 2, white, without hairs, rotund, 3–5 × 3.5–6 mm, margin not toothed, tip rounded; styles 3, styles and stigmas yellow, 1.5–2 mm long, stigmas U-shaped. **Fruits** dangling on a hair-like stalk 6–10 mm long; capsule 6 × 11–13 mm, hairless, locules 3, wings 3, equal, rounded, thinly papery, 3–4 mm wide, splitting between locules and wings. **Seeds** barrel-shaped, 0.25–0.3 mm long, collar cells *c.* half the seed length.

DISTRIBUTION. Endemic in Peninsular Malaysia—northwest Kelantan (Gua Maka and Gua Setir).

HABITAT. Restricted to limestone in the lowlands at *c.* 50 m altitude in rock crevices near the base of cliff faces or on limestone-derived soil at the base of the cliff face or in cave entrances, not common.

OTHER SPECIMENS. KELANTAN—Gua Maka, *Kiew & Anthonysamy RK 3021* (SING); Gua Setir, *Abdullah Piee s.n.* 8 August 2002 (SING).

NOTES. Gua Setir in the Jeli District was terrorized by man-eating tigers in 2002 when we collected the type specimen, hence its name. The area was so dangerous that we were accompanied by a local farmer with a gun. He told us that two of his cows had been killed by these tigers. Fortunately, we encountered only pug marks.

Peltate begonias are few in Peninsular Malaysia and three of the four peltate species are restricted to limestone. This new species is a dainty begonia with smooth leaves often prettily variegated. It is not common where it is found and at present it is known from only two hills.

With its ovary with three locules each with one placenta, it belongs to sect. *Reichenheimia*. Among species of this section, it most resembles the other peltate species, *Begonia ignorata*, but it is distinct from that species in its leaves that are wider than long (they are longer than wide in *B. ignorata*), and the leaf margin is not fringed by long hairs unlike *B. ignorata*, by its male flowers which have two hairless tepals (not four with the outer two sparsely hairy as *B. ignorata* has) and its small tepals in the female flower (3–5 mm long as opposed to 6–7 mm in *B. ignorata*).

Begonia tigrina. Female flowers.

5. THE SHILLING BEGONIA
Begonia sibthorpioides Ridl.
(Latin, *-iodes*=resembling, similar to *Sibthorpia europaea*, the Cornish pennywort)

Ridley, JFMSM 7 (1916) 42, FMP 1 (1922) 859; Irmscher, MIABH 8 (1929) 158. **Type:** *Robinson and Kloss 6047*, December 1915, Kedah Peak, Kedah (holo K).

Stem erect, red, densely hairy, unbranched or stem becoming prostrate and producing erect branches, 3–10 cm tall, succulent, extremely slender, less than 1 mm thick when dried; tuber cylindrical and up to 13 × 4 mm or globose and up to 8 mm diam., covered by a fawn brown corky layer. Hairs on all parts of the plant transparent, unbranched, *c.* 1 mm long, terminating in a rounded (glandular) cell. Stipules narrowly triangular, *c.* 1.5 × 1 mm, margin not toothed, densely hairy, margin fringed by long hairs, tip acute, persistent. **Leaves** sometimes opposite, distant and up to 15 mm apart; stalk 5–20 mm long, slender; blade not oblique, deep malachite green to brown green, reddish purple beneath, with scattered long hairs, thinly leathery when dried, rotund, symmetric, 7–15 mm diam., base slightly heart-shaped, basal lobes equal, *c.* 1–1.3 mm long, margin crenate, tip rounded; venation palmate with 2–3 pairs of veins, with one vein in each of the basal lobes, branching towards the margin, impressed above, beneath slightly prominent and hairy. **Inflorescences** terminal, red, hairy, longer than the leaves, 3–5.5 cm long, branches 3, male flowers 2–3, female flower 1. Bracts of two widely spaced pairs with prominent midribs, triangular, 1–1.5 × 0.75 mm, tip acute. **Male flowers** with a deep red stalk 1.5–7 mm long; tepals

The Shilling Begonia, *Begonia sibthorpioides*.

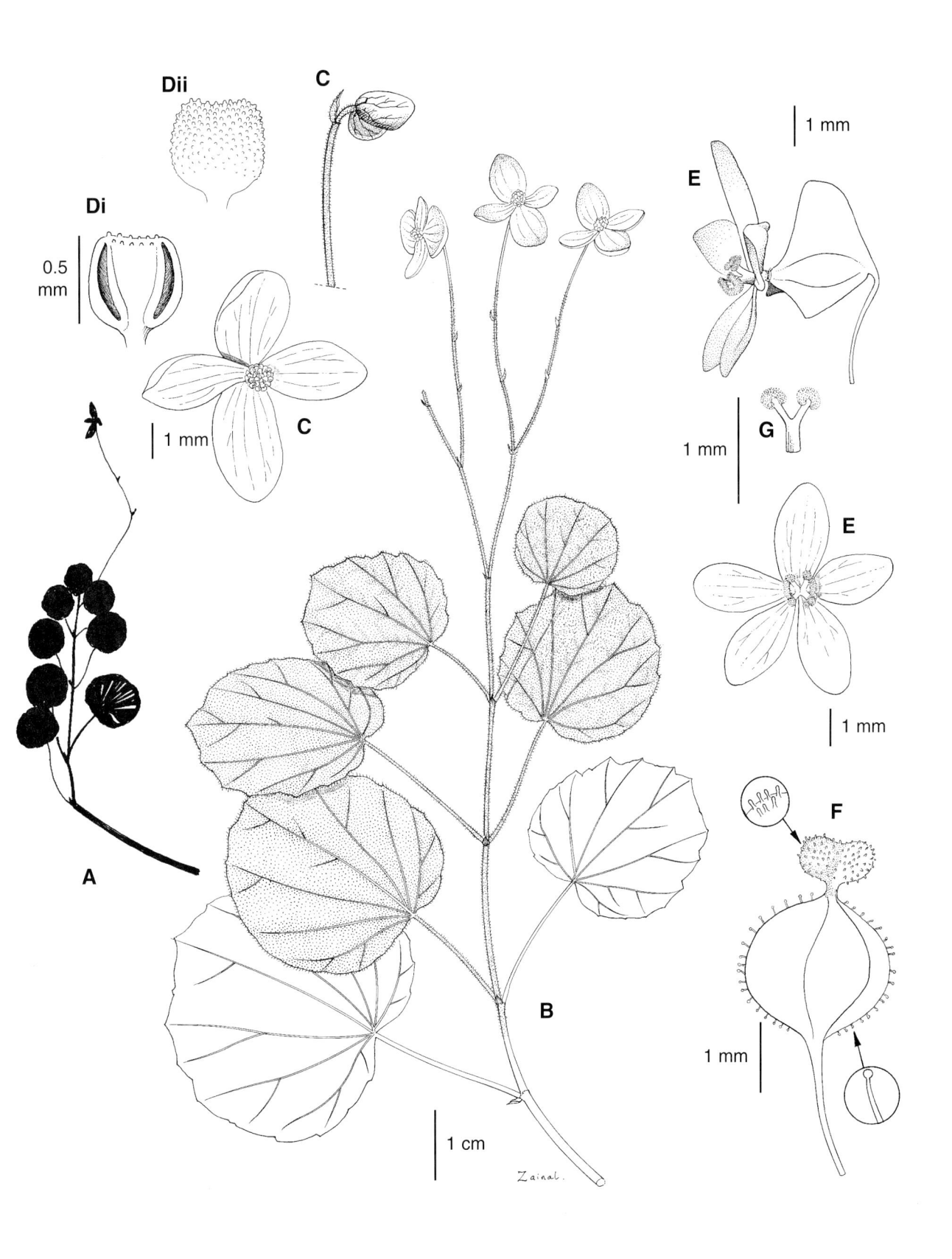

Begonia sibthorpioides Ridl. **A.** The plant (life-size). **B.** The plant. **C.** Male flower. **Di.** Inner face of the stamen. **Dii.** Outer face of the stamen. **E.** Female flowers. **F.** Ovary. **G.** Styles and stigmas. (*RK 5301*)

4, deep rosy pink fading to white, without hairs, margin not toothed, tip rounded, subequal. outer two oval, 3–4.5 × 2.5–3 mm, inner two 3–4 × 2–2.25 mm; stamens 12–15, cluster globose, *c.* 1.25 mm diam., stalk *c.* 1 mm long. filaments 0.15–0.25 mm long, anthers golden yellow, ovate to globose, *c.* 0.5 × 0.5 mm long, tip slightly notched, the top and back covered in minute, dense pimples, opening by slits on the inner face. **Female flowers** with a stalk 1.5–7 mm long; ovary green with red veins, 2–3.5 mm long, sparsely hairy, wings 3, unequal, the longer 1.5–2.2 mm wide, the shorter two scarcely developed, 0.3—0.5 mm wide, locules 2, placentas 2 per locule; tepals 5(–6), without hairs, broadly oval, margin not toothed, tip rounded, subequal, outermost 3.5–5.5 × 2–2.5 mm, innermost similar but smaller 2–5 × 1.5–2 mm; styles 2, styles and stigmas golden yellow, 1–1.5 mm long, stigmas U-shaped, positioned horizontally. **Fruit** with a stalk *c.* 4 mm long, capsule 2.5–3.3 × 5 mm, wings 3, unequal, the larger oblong rounded, 4–5 mm wide, the smaller two very short, *c.* 0.5 mm wide. **Seeds** not known.

DISTRIBUTION. Endemic in Peninsular Malaysia: Kedah—summit of Gunung Jerai (Kedah Peak) and Gunung Machinchang, Langkawi.

HABITAT. Growing on sandstone summits: on Gunung Jerai on a precipice below the summit at *c.* 1200 m altitude; Gunung Machinchang on mossy slopes at 700 m.

OTHER SPECIMENS. KEDAH: Gunung Jerai, *Haniff 626* (SING); Gunung Machinchang, Langkawi, *Saw FRI 44406* (KEP), *Kiew RK 5301* (KEP, SING).

NOTES. The Shilling Begonia has the smallest leaves of any begonia in the Peninsula, ranging in size from that of a 10 to 20 cent coin, hence the common name because coins are called shillings in Malaysia. It is a very pretty plant with tiny, dark green leaves and large, bright rosy pink flowers. At first sight, it looks like small plants of the Sparkling Begonia, *Begonia sinuata*, but it is different in its rounded leaf tip and the absence of the star-like hairs that are so characteristic of the Sparkling Begonia.

The Shilling Begonia grows on rocks near the summit of sandstone hills.

BEGONIA SIBTHORPIOIDES

Begonia sibthorpioides. (Left). Male flowers and fruits. (Right). Its rocky habitat.

The Shilling Begonia is extremely rare and very local. It had been collected twice from one precipitous rock face on Gunung Jerai but in spite of searching for it, it could not be refound there probably because the extensive infrastructure of the telecommunications station built on the summit has obliterated its habitat. In 2002, Saw Leng-Guan discovered a new population on the summit of Gunung Machinchang, Langkawi, where it grows on the same sandstone formation as that on Gunung Jerai.

It appeared so different from other begonias that Irmscher (1929) placed it in a section of its own, Sect. *Heeringia*, on account of its 'worm-like' tuber covered in golden hairs, the upper pair of leaves that are opposite (begonia leaves are almost always alternately or spirally arranged), its anthers that open by two pores and its erect capsule. However, the new material from Gunung Machinchang shows the tuber to be either 'worm-like' (cylindrical) or globose and that it is covered by a light brown corky layer rather than golden hairs. The capsule is not erect, although it is unusual in the larger wing apparently being held uppermost in contrast to begonias with splash cups where the larger wing is positioned downwards.

The most remarkable characters are seen in the stamen. The filament is extremely short and the small anther almost globose. It opens by two slits that extend along its entire length (not by pores) but, unlike any other Malaysian species, they are positioned on the inner face of the anther instead of on the side (laterally). Remarkable too are the dense, minute pimples on the top and back of the anther (on the side opposite to the one with the slits), another feature not seen in any other Malaysian begonia.

The U-shaped stigmas are unusual too in being positioned horizontally (usually they are erect) and the glandular cells are so large and dense that the stigma when receptive appears like a golden ball.

6. THE SPARKLING BEGONIA
Begonia sinuata Meisn.
(Latin, *sinuatus*=strongly waved, referring to leaf margin)

Meisner, Ber. Verhandl. Naturf. Ges. Basel 2 (1836) 42; C.B. Clarke in FBI 2 (1879) 650; King, JASB 71 (1902) 59; Ridley, FMP 1 (1922) 856; Irmscher, MIABH 8 (1929) 142; Henderson, MWF Dicot (1959) 163 & Fig 155; Burtt, Notes Roy. Bot. Gardn. Edinb. 32 (1973) 274. **Type:** *Wallich 3680*, Penang (holo K). **Synonym novae:** *B. clivalis* Ridl., JSBRAS 54 (1910) 43, *B. sinuata* var. *clivalis* (Ridl.) Irmsch., MIABH 8 (1929) 144 & Fig. 8. **Type:** *Ridley 13523*, Klang Gates (holo SING); var. *etamensis* Irmsch., MIABH 8 (1929) 142. **Type:** *Curtis 3098*, Penang (lecto SING–here designated); var. *langkawiensis* Irmsch., MIABH 8 (1929) 142. **Type:** *Curtis s.n.*, September 1890, Langkawi (lecto SING–here designated); var. *malaccensis* Irmscher, MIABH 8 (1929) 143. **Type:** *Maingay 675*, Malaya (B, L not seen); var. *monophylloides* Irmsch., MIABH 8 (1929) 143. **Type:** *Ridley 2643*, Tahan River (lecto SING–here designated); var. *penangensis* Irmsch., MIABH 8 (1929) 142. **Type:** *Forest Guard s.n.*, Batu Feringgi, Penang (lecto SING, here designated).

Stem glossy, succulent, with star-like hairs, erect, rarely branched, slender; tuber pale brown, globose, without hairs, 4–10 mm diam. Stipules narrowly triangular, covered with star-like hairs, *c.* 2 × 1 cm, margin not toothed, tip acute, soon falling. **Leaves** pendent; stalk greenish red to purple red, hairy, round in cross-section; blade not oblique, plain bright green and sparkling above,

The dark-leaved form of the Sparkling Begonia, *Begonia sinuata*.

(Top). The Sparkling Begonia covers rock faces. (Bottom left). Male flowers. (Bottom right). Female flowers.

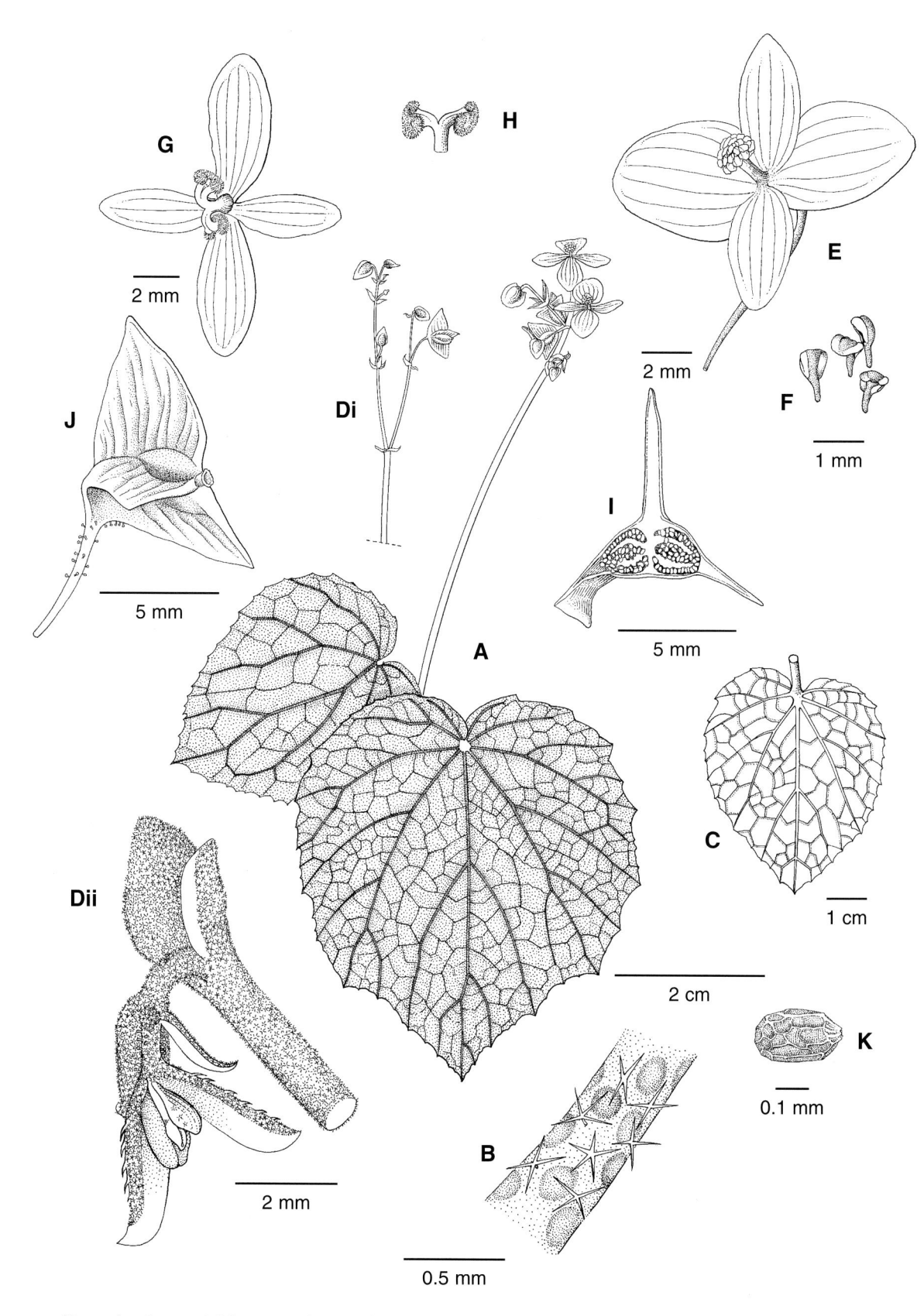

Begonia sinuata Meisn. var. ***sinuata*** **A.** The plant. **B.** Leaf stalk with star-like hairs. **C.** Underside of the leaf. **Di.** Inflorescence branch, **Dii.** Magnified to show star-like hairs. **E.** Male flower. **F.** Stamens. **G.** Female flower. **H.** Styles and stigmas. **I.** T.S. ovary. **J.** Fruit. **K.** Seed. (*RK 4945*)

thin in life, thinly papery when dried, broadly ovate, symmetric, base heart-shaped, with star-like hairs sparse above, dense on the margin and lower surface especially on the veins, tip pointed; venation palmate-pinnate, 4 pairs of veins with another pair in the basal lobes, branching towards the margin, deeply impressed and corrugated above, beneath prominent, the minor veins forming a conspicuous network. **Inflorescences** terminal, without hairs, longer than the leaves, 4.5–16.5 cm long with stalk 3.5–13 cm long and two main branches *c.* 3.5–6 cm long, male flowers 3–5, female flowers 2, male flowers open first. Bracts light pink or pale white, densely hairy, narrowly ovate, 2–12 × 1–8 mm, margin not toothed, persistent. **Male flowers** with a pale pink stalk 10–15 mm long; tepals 4, hairless, margin not toothed, tip rounded, sometimes slightly acute, outer two oval, 5–7 × 5–7 mm, inner two white, narrowly oval, 5–7 × 2–3 mm; stamens about 12, cluster globose, *c.* 1–2 mm across, stalk 0.75–1 mm, filaments *c.* 0.25 mm long, anthers lemon yellow, broadly obovate, *c.* 0.75 mm long, tip notched, opening by slits. **Female flowers** with a greenish stalk 3–8 mm long; ovary 2–5 × 4.5–7 mm, wings 3, equal, locules 2, placentas 2 per locule; tepals (4–)5, hairless, outer broadly oval, margin not toothed, tip rounded, 2–5 × 1.5–2.5 mm, inner similar in shape and size but slightly narrower; styles 2, styles and stigma golden yellow, 2–3 mm long, stigmas U-shaped. **Fruit** a pendent splash cup on a fine and hair-like stalk 5–9 mm long; capsule 6–9 × 11–20 mm, hairless, wings 3, unequal, thinly fibrous, larger conspicuously concave, 10–16 mm wide, smaller two curved, 3–9 mm wide, splitting irregularly between locules and wings. **Seeds** barrel-shaped, *c.* 0.25 mm long, collar cells *c.* half the seed length.

The Sparkling Begonia commonly grows on shaded vertical rock faces by forest streams.

BEGONIA SINUATA

NOTES. The Sparkling Begonia has a wide geographic distribution from Peninsular Thailand, east to Cambodia and Vietnam, and in Peninsular Malaysia south to Johore. It is also the most common begonia in the Peninsula, being frequently found on damp vertical rock faces often associated with small waterfalls.

It is a very distinctive begonia with a short fleshy stem, large first leaf and a few smaller ones above. The leaves are heart-shaped and symmetric with a pointed tip. The veins are deeply sunken so that the leaf surface is corrugated, the margin is crimson and jagged, and on the underside of the leaf the veins are deep crimson, the colour showing through on the upper surface. Besides having a scintillating surface, *Begonia sinuata* is unique among Malaysian species in having star-like hairs. These have a very short stalk and four to five radiating arms. They can just be seen with the naked eye and are particularly abundant on the leaf margin.

The Sparkling Begonia is seasonal in its growth and is the only strictly annual begonia in the Peninsula. It dies down towards the end of the year, regenerating profusely from seed but presumably also from its basal tuber and leaf bulbils. Up until about June, populations consist entirely of seedlings or small plants, which begin to flower at about 2.5 cm tall. From then onwards larger plants can be found. This accounts for the great range in size observed in herbarium specimens depending on when they were collected.

Plants start to flower when they are just a few centimetres tall and can grow to up to 29 cm tall with widely spaced leaves. However, the maximum size

Fruits of the Sparkling Begonia.

does vary between populations. Ridley's *Begonia clivalis* from Selangor was based on these large plants and similar populations of large plants are now known from Kedah and Trengganu. In contrast, leaf shape is quite uniform within a population on a single rock face, which tends to emphasize the differences between populations. Irmscher (1929) took account of this and recognized seven varieties in the Peninsula, most based on plants from a discrete area, although he noted that there were specimens that were intermediate between his varieties. Since his time, many more collections have been made over a much wider area, which blur the differences between most of his varieties, which can no longer be recognized as distinct. Two, however, var. *langkawiensis* and var. *monophylloides*, do appear distinct from herbarium specimens but field observations show that the first leaves of var. *langkawiensis* are no different from the general Malaysian population, it is only the larger leaves that have the more elongate leaf shape characteristic of this variety. Similarly, var. *monophylloides*, as its name suggests, has one large leaf. However, when John Dawn examined plants in the field, less than half the plants had just one leaf, most had two or more. Because of this variation within a population or within the lifetime of a single plant, neither of these varieties is accepted here.

BEGONIA SINUATA VAR. SINUATA

Having said that, one new variety is recognised here for two reasons—the first is that it is quite unlike other *B. sinuata* populations in having a deeply incised leaf, a character that is unique among Peninsular begonias. The second is that, because of its decorative leaf, it has potential as an ornamental plant so it is useful for it to have a botanical name.

While being typical of *B. sinuata* in its habit, in having star-like hairs, in range of leaf size, and the characters of flower and fruits, this new variety is distinct in its total lack of red pigmentation in the leaves, inflorescence and flowers; in its extremely thin leaves that are translucent in life; but more particularly in its leaf margin that is deeply incised.

Key to the Varieties

Leaf margin not incised, Leaf undersurface dark green, reddish or rosy purple .... var. *sinuata*
Leaf margin incised, incisions 3–22 mm deep, leaf undersurface pale green and translucent .... var. *pantiensis*

var. sinuata

Stem pale crimson to reddish brown, sometimes greenish white, (1.5–)2–8(–29) cm tall, flowering at 2.5 cm tall, *c.* 2–5 mm thick. **Leaves** (1–)2–3(–4), distant and 1–3.5 cm apart; stalk (0.5–)2–6.5 cm long; blade dark green or completely reddish, sometimes rosy purple beneath, 2–12 × 2.5–14 cm, basal lobe 0.5–1(–2) cm long, margin shallowly scalloped and finely toothed, veins green or crimson, the colour showing through on upper surface. **Inflorescence** white at base, crimson towards top, tepals of the male flowers white on the inner surface, rosy pink outside; ovary of female flower crimson and tepals pale rosy pink.

DISTRIBUTION. Southern Thailand, Cambodia, Vietnam and Peninsular Malaysia.

HABITAT. Deeply shaded, vertical rock faces (on granite, sandstone or quartzite but not on limestone) or large boulders often close to streams and small waterfalls, rarely (in Langkawi) by the seashore; lowland usually below 500 m altitude but up to about 1175 m on Gunung Jerai, Kedah.

OTHER SPECIMENS. JOHORE—Gunung Panti, *Kiew RK 623* (SING), Kota Tinggi, *Corner SFN 31437* (SING), *Teruya 876* (SING), Mawai, Rubber Estate *Teruya 1080* (SING); KEDAH—Bukit Panchur, *Spare SFN 36306* (SING), Bukit Perangan, *Chan FRI 021744* (KEP), Kedah Peak, *Allen s.n.* (SING), *Haniff 4194* (SING), *Ridley s.n.* (SING), *Spare SFN 36312* (SING), Langkawi, *Anthonysamy SA 949* (KEP), *Corner s.n.* (SING), *Corner SFN 37887* (SING), *Haniff & Nur 7077* (SING), *Kiew RK 4943* (SING), *Kiew RK 4945* (SING), *Ridley s.n.* (SING), *Stone 14237* (KLU), *Tay 136* (SING), Lata Mengkuang, *Kiew RK 5161* (SING), Lata Peringing, *Kiew RK 5165* (SING), Rawei River *Ridley s.n.* (SING), Somme Estate, *Spare 3078* (SING); KELANTAN—Gunung Ayam, *Saw & Kamarudin FRI 37609* (KEP), Kuala Aring, *Kiew KBH 24* (SING), Machang, *Damanhuri & Khairuddin FRI 36004* (KEP), PAHANG—Gunung Tahan, *Corner s.n.* (SING), *Kiah SFN 31714* (SING), *Kiew RK 1224* (SING), *Yong s.n.* (KEP), *Wray & Robinson 5539* (SING), Ulu Kinchin, *Kiew KBH 86-25* (KEP); PENANG—Penang Hill, *Burkill 1524* (SING), *Burkill 2590* (SING), *Chan s.n.* (SING), *Curtis 390* (SING), *Curtis 3168* (SING), *Henderson SFN 21356* (SING), *Kiew RK 1602* (SING), *Nauen SFN 38065* (SING), *Ridley 9229* (SING) *Shimizu et al 12980* (SING), *Shimizu et al 13176* (SING), *Sinclair SFN 39103* (SING), Pulau Jerajah *Curtis 481* (SING), Telok Bahang, *Wong FRI 35259* (KEP), SELANGOR—Genting Simpah, *Strugnell 13057* (KEP), Klang Gates Ridge, *Kiew RK1084* (SING), *RK 1254* (SING), *Sinclair SFN 40132*

BEGONIA SINUATA VAR. PANTIENSIS

(KEP, SING), Semenyih, *Henderson 8273* (SING), Ulu Ampang Waterfall, *Shah 1* (KEP), TRENGGANU—Batu Biwa, *Kiew RK 2347* (SING), Bukit Kajang, *Corner s.n.* (SING), Jeram Gajah, *Corner 25844* (SING), Jeram Tanduk, *Kiew RK 5080* (SING), Sekayu, *Anthonysamy SA 649* (KEP), *Hume 150* (KLU, SING), *Saw & Kamarudin FRI 37632* (KEP), Sg. Kemia, *Chua et al FRI 40560* (KEP), Ulu Brang, *Moysey & Kiah SFN 33859* (SING), Ulu Setui, *Anthonysamy SA 667* (KEP).

The typical form of the Sparkling Begonia.

THE MAPLE-LEAF BEGONIA
var. pantiensis Kiew, var. nov.
(Latin, -ensis=originating from)

A var. *sinuata* foliis incisis statim dignoscenda. **Typus:** *Kiew RK 5113*, 25 September 2000, Gunung Panti, Johore (holo SING; iso KEP, K, L).

Stem pale green, to 14 cm tall and 3–4 mm thick, flowering at 2.5 cm tall. **Leaves** 3–4, pale green, distant *c.* 0.5–2.5 cm apart; stalk 2–5.5 cm long, blade pale green, 4–12.5 × 4.5–13 cm, basal lobe 0.5–1.75 cm long, thin and translucent in life, extremely thinly papery when dried, margin deeply incised, incisions 3–5 mm deep in small leaves, (7–)12(–22) mm deep in larger leaves, veins pale green above and beneath. **Inflorescences** pale green; tepals of the male and female flowers completely white, ovary pale green.

DISTRIBUTION. Endemic in Peninsular Malaysia: Gunung Panti, Johore.

HABITAT. In deep shade, on large granite boulders in lowland forest at *c.* 390–450 m altitude.

OTHER SPECIMENS. JOHORE—Gunung Panti, *Kiew RK 3780* (SING), *Saw FRI 37733* (KEP), *Weber 840723-2/2* (KEP); *Teruya 483* (Kota Tinggi) (SING).

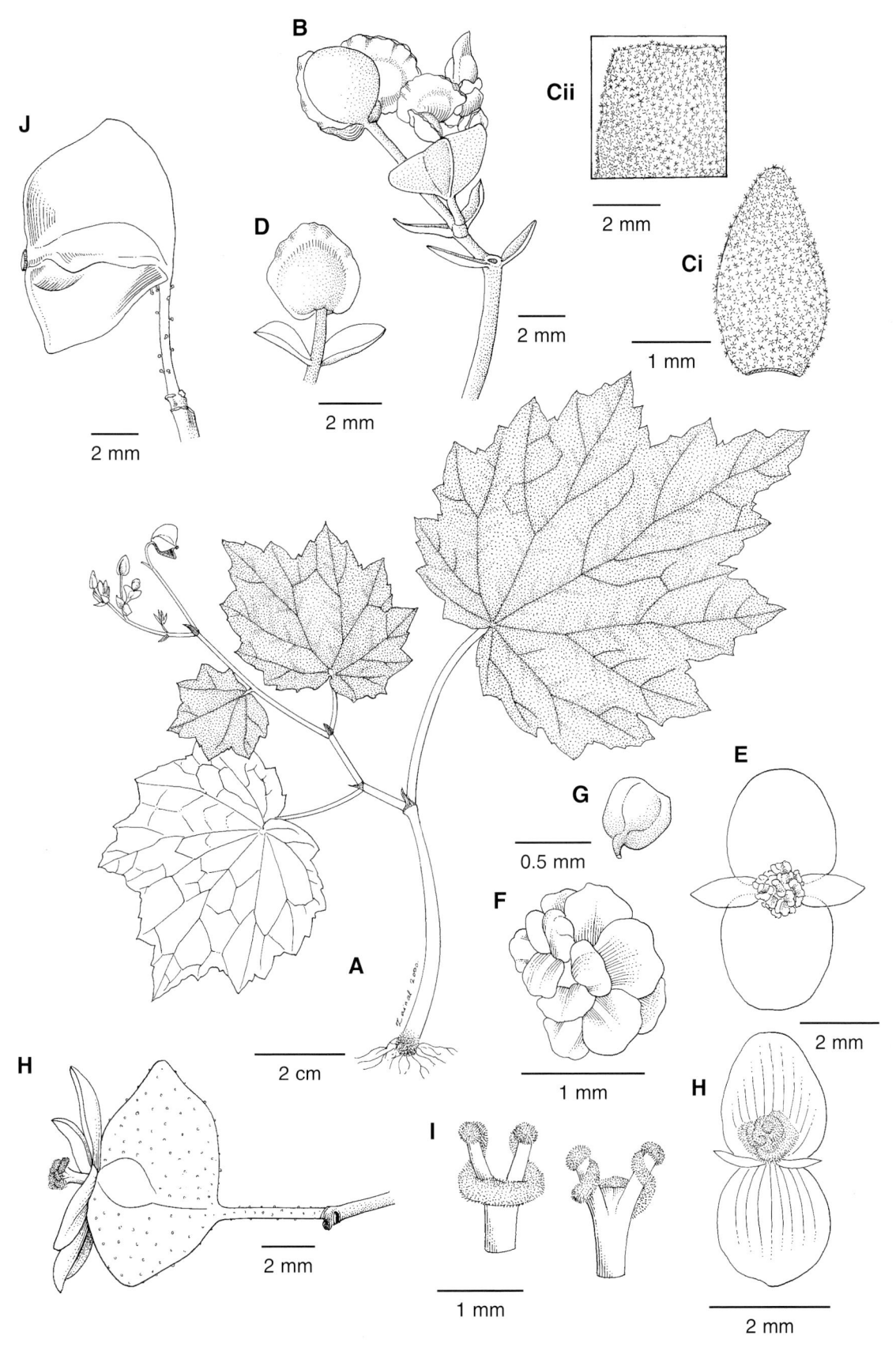

Begonia sinuata var. ***pantiensis*** Kiew **A.** The plant. **B.** Inflorescence branch. **Ci.** Bract. **Cii.** Magnified to show star-like hairs. **D.** Male bud. **E.** Male flower. **F.** Stamen cluster. **G.** Stamen. **H.** Female flower. **I.** Styles and stigmas. **J.** Fruit. (*RK 5780*)

BEGONIA SINUATA VAR. PANTIENSIS

The Maple-leaf Begonia, *Begonia sinuata* var. *pantiensis*.

BEGONIA SINUATA VAR. PANTIENSIS

(Left). The Maple-leaf Begonia with male flowers at the top of the inflorescence. (Inset top). Fruit. (Inset bottom). Female flower.

7. THE BURMESE BEGONIA
Begonia martabanica A. DC. var. **pseudoclivalis** Irmsch.
(Greek, pseudo=false, referring to this species looking like *Begonia clivalis*)

Irmscher, MIABH 8 (1929) 145 & Fig 9. **Type:** *Ridley 14606*, July 1909, Temangoh [Temangor], Perak (holo SING; iso K).

Stem succulent with short, straight, brown hairs, erect, sometimes branched at base, 6–9 cm tall; tuber globose, *c.* 7 mm across. Stipules narrowly triangular, up to 1.5 mm long, margin not toothed, tip pointed, soon falling. **Leaves** 3–5 per plant, distant and 3–6 cm apart; stalk up to 6 cm long with short brown hairs; blade not oblique, plain green, thinly papery when dried, upper surface with microscopic straight hairs, broadly ovate, almost symmetric, 2.5–5 × 3.25–6.5 cm long, broad side 1.8–3.5 cm wide, base truncate sometimes slightly heart-shaped, basal lobes equal 2–6 mm long, margin minutely and distantly toothed, tip acute; venation palmate-pinnate with 2 pairs of veins at the base, branching *c.* halfway to the margin, and one pair near the top of the midrib and another pair in the basal lobes, beneath veins prominent and sparsely brown hairy. **Inflorescences** terminal, minutely hairy, 5–11 cm long, branches 1–2.5 cm long, stalk 2.5–10 cm, male flowers many, female flower one on a branch from the base of the inflorescence; male flowers are still produced after the fruit has ripened. Bract pair *c.* 2 mm long, narrowly triangular. **Male flowers** with a stalk up to 17 mm long; tepals 4, margin not toothed, tip rounded, outer two oval, 2.75–3 × 2–2.5 mm, inner two obovate, 2 × 1 mm; stamens many, joined at the base in a stalk 1 mm long; filaments shorter than the anthers; anthers narrowly obovate, *c.* 0.5 mm long, tip notched, opening by slits. **Female flower** not known. **Fruit** pendent on a fine, hair-like stalk up to 16 mm long; capsule *c.* 7 × 10 mm, hairless, locules 2, placentas 2 per locule, wings 3, unequal, rectangular, thinly fibrous, larger 5 mm wide, the two smaller *c.* 2 mm wide. **Seeds** not known.

DISTRIBUTION. *Begonia martabanica* var. *martabanica* is found in Myanmar and Thailand. *B. martabanica* var. *pseudoclivalis* is endemic in Peninsular Malaysia.

HABITAT. Ridley (1911) reported it from 'sandy banks'.

OTHER SPECIMEN. PERAK—Temangoh *Ridley s.n.* July 1909 (SING).

NOTES. Variety *pseudoclivalis* is different from *Begonia martabanica* from Myanmar and Thailand in having more leaves (up to five as compared to two or three in var. *martabanica*). However, var. *pseudoclivalis* begins to flower at 6 cm tall when it has three leaves. Irmscher recorded three other differences between the two varieties: greater size, many-flowered inflorescence and smaller flowers. Since then, *B. martabanica* has been collected several times from Thailand including larger plants up to 14 cm tall (but still with three leaves). It is likely that when the Thai begonias are better known, the distinction between the two varieties will disappear. The Malaysian variety has not been recollected and its flowers and fruits are still poorly known. Until this species is better known, I have kept the Malaysian specimens as a separate variety.

Ridley (1911) listed the type specimen of var. *pseudoclivalis* under his *B. clivalis*, but apparently changed his mind as he did not list it under this species in his flora account (Ridley, 1922). (*B. clivalis*, which has star-like hairs, is a synonym of *B. sinuata*). Variety *pseudoclivalis* does indeed look like tall plants of *B. sinuata*, but is distinct in its straight hairs whereas *B. sinuata* has star-like hairs.

Begonia martabanica, like *B. integrifolia*, is an example of an Asian begonia that reaches its southernmost distribution in northern Malaysia (Perak). It has not been refound since Ridley collected it in 1909. It is possible that its original habitat is now covered by the Temangor Dam and that this variety is extinct.

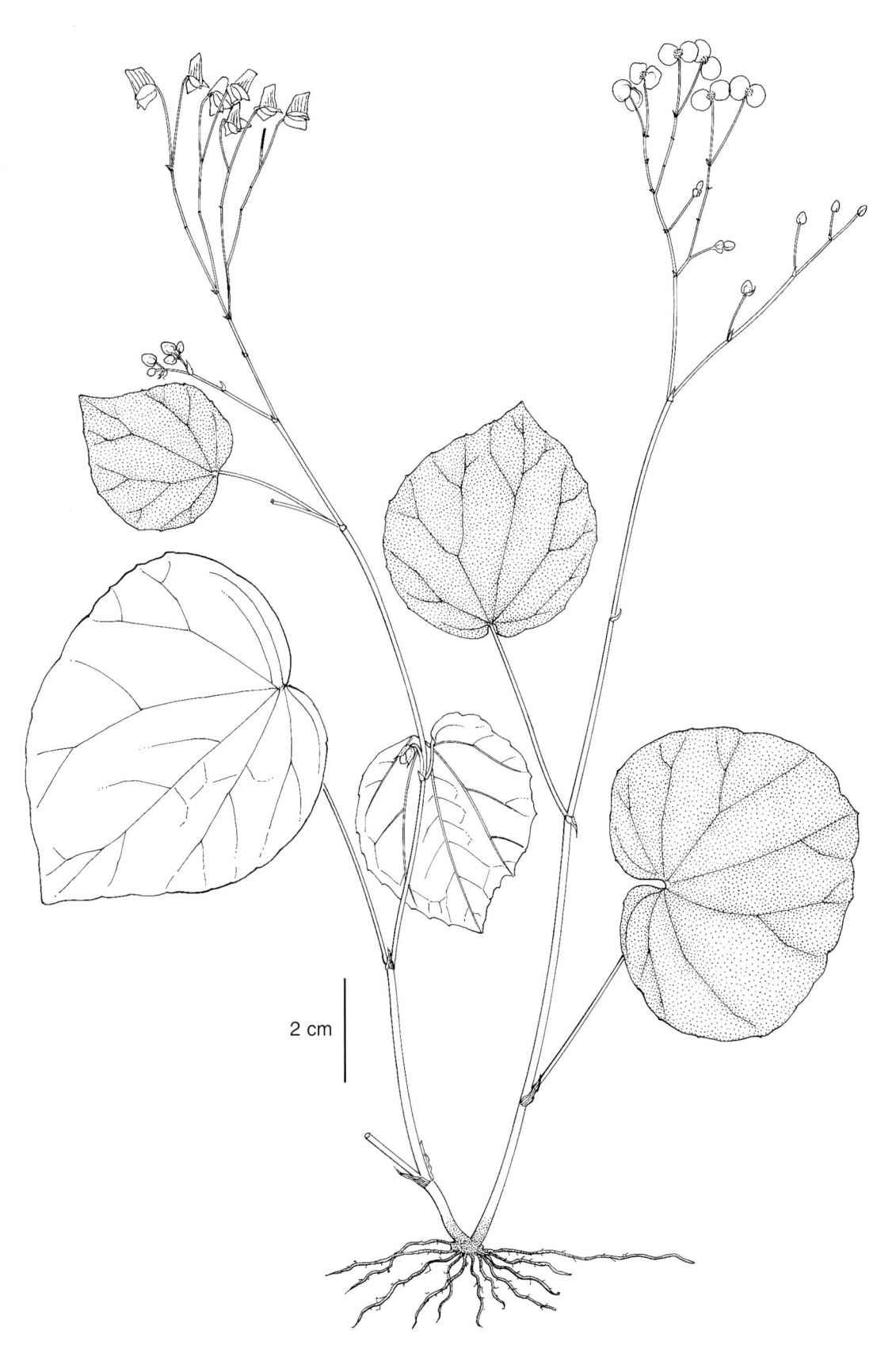

Begonia martabanica A.DC. var. ***pseudoclivalis*** Irmsch. Plant with male flowers and fruits. (*Ridley 14606*).

8. THE LUSH-LEAVED BEGONIA
Begonia carnosula Ridl.
(Latin, *carnosula*=slightly fleshy, referring to the leaf)

Ridley, JFMSM 4 (1909) 20, FMP 1 (1922) 857; Irmscher, MIABH 8 (1929) 146. **Type:** *Ridley 14124*, November 1908, Telom, Pahang (lecto SING, designated by Irmscher).

Stem pale green, semi-translucent, succulent, mealy or sparsely hairy, erect, falling and rooting, larger plants branched from first or second leaf axil, up to 7–12 cm long and *c.* 7 mm thick, flowering when *c.* 3 cm tall at the two-leaf stage; tuber globose, 5–7 mm diam. Stipules white or greyish, matted with hairs, narrowly triangular, 8–17 × 2–7 mm, margin not toothed, tip acute, persistent. **Leaves** up to 5(–8) per plant, distant and up to 2.5 cm apart, lowest leaf appressed to the ground, upper held horizontally; stalk fleshy, at first pale green and sparsely hairy, becoming grey and scurfy, 2–14 cm long, grooved above; blade oblique, plain glossy metallic green above with bluish hue or mid-green depending on the angle to the light, pale green beneath, with scattered hairs *c.* 0.5–0.75 mm long above and beneath, blade slightly raised between veins, thin in life, thinly papery when dried, broadly ovate, asymmetric, (5.5–)9(–18) × (6–)9–13(–17.5) cm, broad

The Lush-leaved Begonia, *Begonia carnosula*.

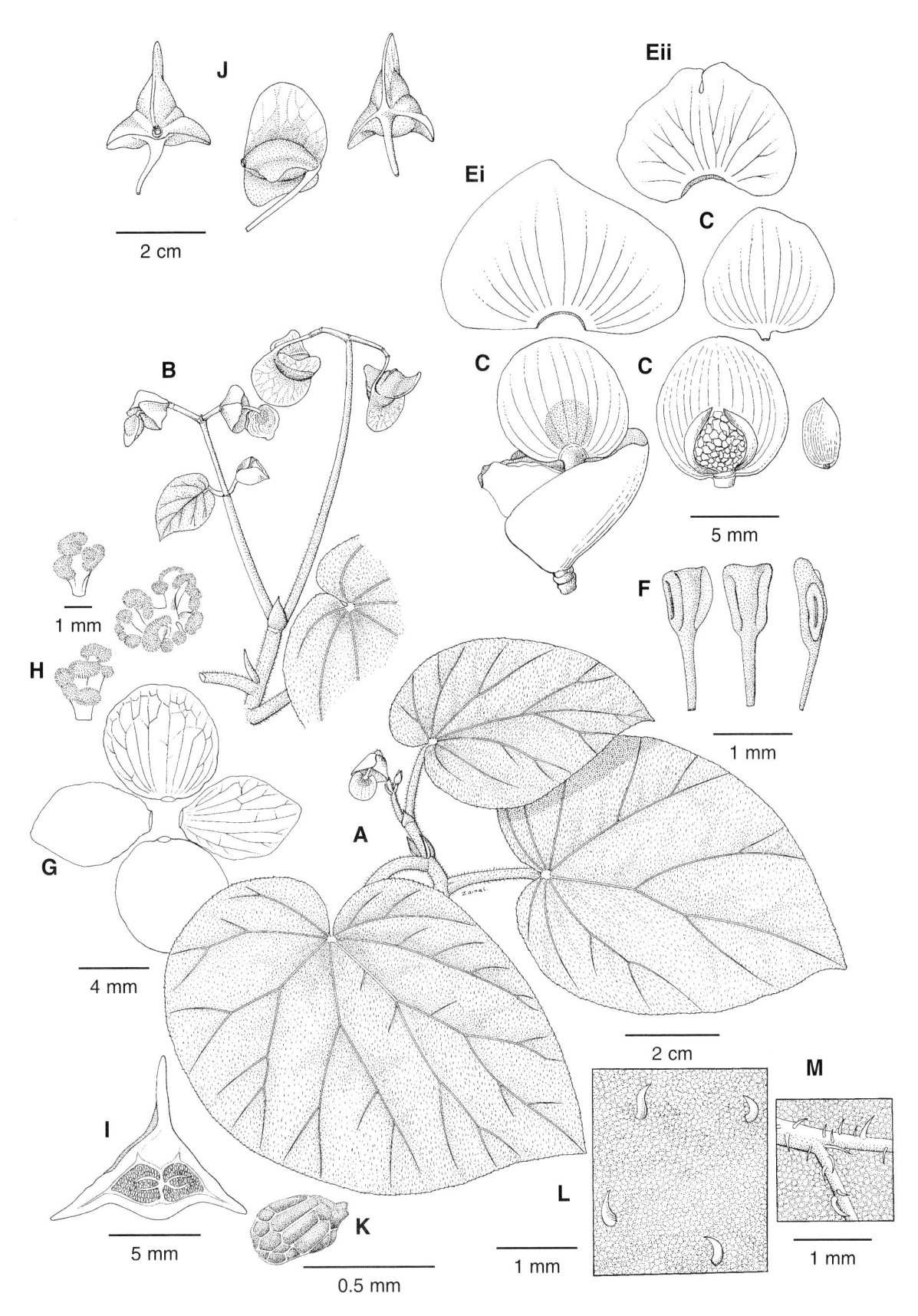

Begonia carnosula Ridl. **A.** The plant. **B.** Inflorescence. **C.** Male bud. **D.** Bract. **Ei.** Outer male tepal. **Eii.** Inner male tepal. **F.** Stamens. **G.** Female flower. **H.** Styles and stigmas. **I.** T.S. ovary. **J.** Fruit. **K.** Seed. **L.** The upper leaf surface. **M.** The lower leaf surface. (*RK 4742*)

BEGONIA CARNOSULA

Male flowers of the Lush-leaved Begonia.

side 3.5–8 cm wide, base slightly unequal, rounded and slightly overlapping or deeply heart-shaped, basal lobes 2–3.5 cm long, margin minutely toothed and fringed with hairs, tip broadly acute or rounded; venation palmate, 2–3 pairs, branching *c.* halfway to the margin, and one pair in the basal lobes, raised above, beneath prominent, same colour as blade, densely hairy. **Inflorescences** terminal, pale green with scattered hairs, succulent, erect, longer than the leaves, twice branched, branches 5–8 mm long, stalk 2–6.5 cm, lengthening to 9–17.5 cm in fruit, male flowers 5, female flowers 2, male flowers open first. Bract pair enclosing the developing inflorescence, pale green or white, broadly ovate, 10–15 × 14–17 mm, margin not toothed, tip pointed or slightly notched, soon falling. **Male flowers** with a white stalk 10–15 mm long; tepals 4, completely white, hairless, margin not toothed, tip rounded, outer two broadly ovate, 13–16 × 9–12 mm, inner two narrowly oval, 7–13 × 3–5 mm; stamens many, cluster globose, 2–4 mm across, stalk 0.5–1 mm long; filaments 0.5–1.5 mm long; anthers golden yellow, oblong, 0.5–1 mm long, tip slightly notched, opening by slits. **Female flowers** with a stalk 4 mm long; ovary pale green with a red stripe between the locules and wings, 7–11 mm long, wings 3, unequal, locules 2, placentas 2 per locule; tepals 4–5, white, hairless, broadly oval, margin not toothed, tip rounded, 7–8 × 5–7 mm; styles 2, styles and stigmas golden yellow or slightly greenish yellow, 2–3 mm long, stigmas spiral. **Fruit** a splash-cup pendent on a stiff stalk 4–8 mm long; capsule 7–12 × 10–19 mm, hairless, locules 2, wings 3, thinly fibrous, concave, unequal, largest wing up to 8–12 mm wide, tip rounded, smaller two 4–7 mm wide, tip pointed, splitting irregularly between locules and wings. **Seeds** barrel-shaped, *c.* 0.3 mm long, collar cells half to three quarters of the seed length.

DISTRIBUTION. Endemic in Peninsular Malaysia—Perak: foothills of Cameron Highlands between 350–750 m altitude.

HABITAT. Earth banks on hillsides, locally common.

BEGONIA CARNOSULA

OTHER SPECIMENS. PERAK: Tapah, Jor, *Burkill SFN 14203* (SING), *Burkill SFN 14246* (SING); lower stretches of road to Cameron Highlands, *Kiew RK 3243* (K, KEP, SING), *RK 4742* (KEP, SING).

NOTES. The Lush-leaved Begonia belongs to the *Begonia integrifolia* group but differs from the others in its rounder, less asymmetric, non-variegated leaves with the veins radiating from the base, and in the lower leaves that have stalks longer than the blades.

Ridley described the plant as acaulescent (stemless) with a creeping rhizome but in fact it has a short, succulent, erect stem, which readily roots on contact with the soil. He also described the inflorescences as axillary when in fact they are terminal. This may have led Doorenbos *et al.* (1998) to place *B. carnosula* in sect. *Diploclinium*. This is clearly in error as the Lush-leaved Begonia is neither rhizomatous nor has a 3-loculate fruit. It belongs in sect. *Parvibegonia*.

The changing colour of the leaves from iridescent green to blue depends on the angle of the light, but is not as bright as in the Peacock Begonia, *B. pavonina*.

(Left). The Lush-leaved Begonia grows on steep earth banks in hill forest. (Top right). Female flowers. (Bottom right). Male flowers and fruits.

9. THE SPOTTED BEGONIA *ASAM RIANG*
Begonia integrifolia Dalzell
(Latin, *integrifolia*=entire-leaved)

Dalzell in Hook. *f.* J. Bot. Kew Gard. Misc. 3 (1851) 230; Dalzell & Gibson, Bombay Flora (1861) 104; C.B. Clarke in FBI 2 (1879) 648. **Type:** *Dalzell s.n.*, Bombay (lecto K). **Synonym:** *Begonia guttata* Wall. *ex* A.DC, Prodr. 15,1 (1864) 352; C.B. Clarke in FBI 2 (1879) 648; King, JASB 71 (1902) 61; Ridley, FMP 1 (1922) 858; Irmscher, MIABH 8 (1929) 153; Henderson, MWF Dicot (1959) 166 & Fig 159. **Type:** *Wallich Cat. No. 3671A*, Penang (holo K). **Synonym nova:** *B. guttata* forma *elongata* Irmsch., MIABH 8 (1929) 154. **Type:** *Curtis s.n.*, Aug 1896, Penang (lecto SING, here designated). **Synonym nova:** *B. debilis* King, JASB 71 (1902) 60; Ridley, FMP 1 (1902) 858; Irmscher, MIABH 8 (1929) 157. **Type:** *King's Collector 8289*, October 1883, Gunung Pondok, Perak (lecto K, here designated). **Synonym nova:** *B. leucantha* Ridl., JSBRAS 57 (1911) 49, FMP 1 (1922) 857; Irmscher, MIABH 8 (1929) 157. **Type:** *Ridley 14731*, July 1909, Temengoh, Perak (lecto SING, here designated, iso BM, K). **Synonym nova:** *B. curtisii* Ridl. JSBRAS 59 (1911) 106, FMP 1 (1922) 856; Irmscher, MIABH 8 (1929) 149. **Type:** *Curtis 3234*, Nov 1896, Kasum, Thailand (SING). **Synonym nova:** *B. haniffii* Burkill, JSBRAS 79 (1918) 104; Ridley, FMP 1 (1922) 859); Irmscher, MIABH 8 (1929) 151. **Type:** *Burkill 1494*, 1917, cultivated Penang Botanic Garden from a plant collected by Md Haniff from Langkawi, Kedah (SING lecto, here designated; iso K).

Stem translucent red or pale green, weak, succulent, erect, little branched, readily rooting in contact with soil, up to 25 cm tall, occasionally to 43 cm, flowering at 1.5 cm tall, 3–9 mm thick; densely or sparsely hairy, hairs erect, short, white, unicellular and hooked; tuber globose, 7–12 mm diam. Stipules translucent white, brown, pale green or red, triangular, minutely hairy outside, 5–15 × 2.5–9 mm, margin not toothed, tip pointed, ending in a hair, persistent. **Leaves** 2–3(–6), distant

Begonia integrifolia. (Left). Male and female flowers. (Right). Male flowers.

The Spotted Begonia, *Begonia integrifolia*. (Painting by Wendy Gibbs)

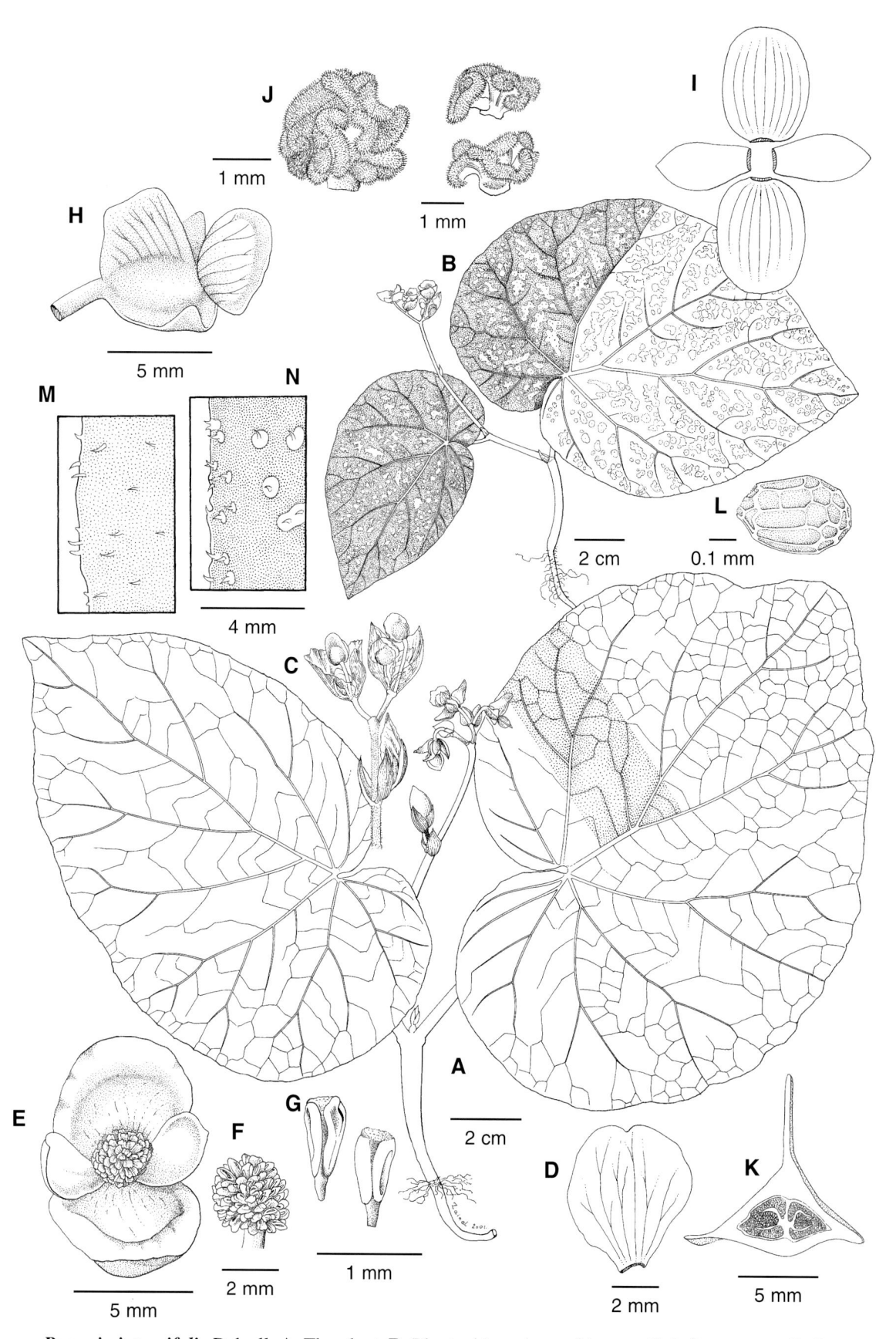

Begonia integrifolia Dalzell. **A.** The plant. **B.** Plant with variegated leaves. **C.** Inflorescence. **D.** Bract. **E.** Male flower. **F.** Stamen cluster. **G.** Stamens. **H.** Female buds. **I.** Female tepals. **J.** Styles and stigmas. **K.** T.S. ovary. **L.** Seed. **M.** The upper surface of a plain leaf. **N.** Upper surface of a variegated leaf. (*RK 4940*)

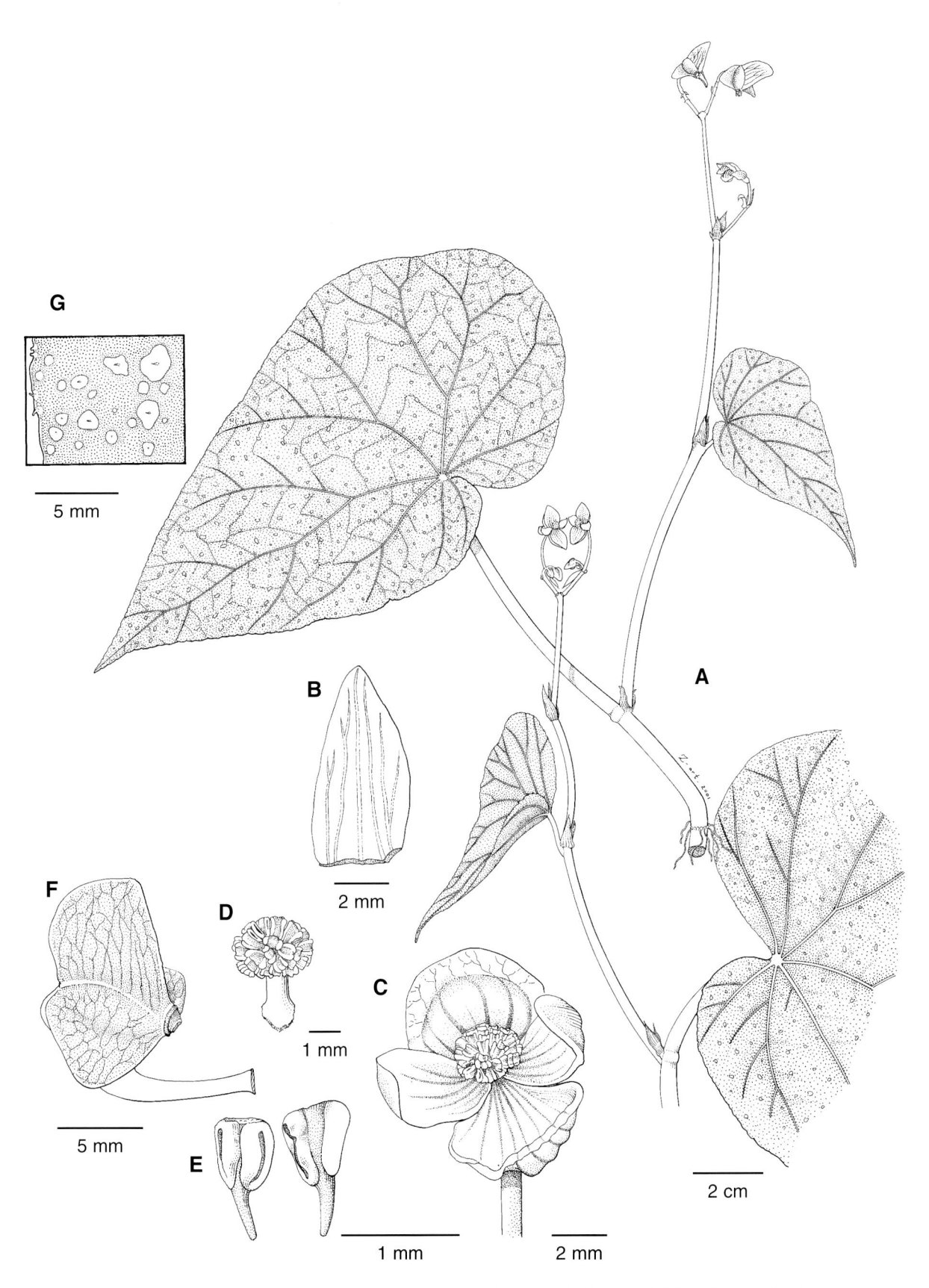

Begonia integrifolia Dalzell. **A.** The plant with spotted leaves. **B.** Bract. **C.** Male flower. **D.** Stamen cluster. **E.** Stamens. **F.** Fruit. **G.** The upper leaf surface. (*RK 4936*)

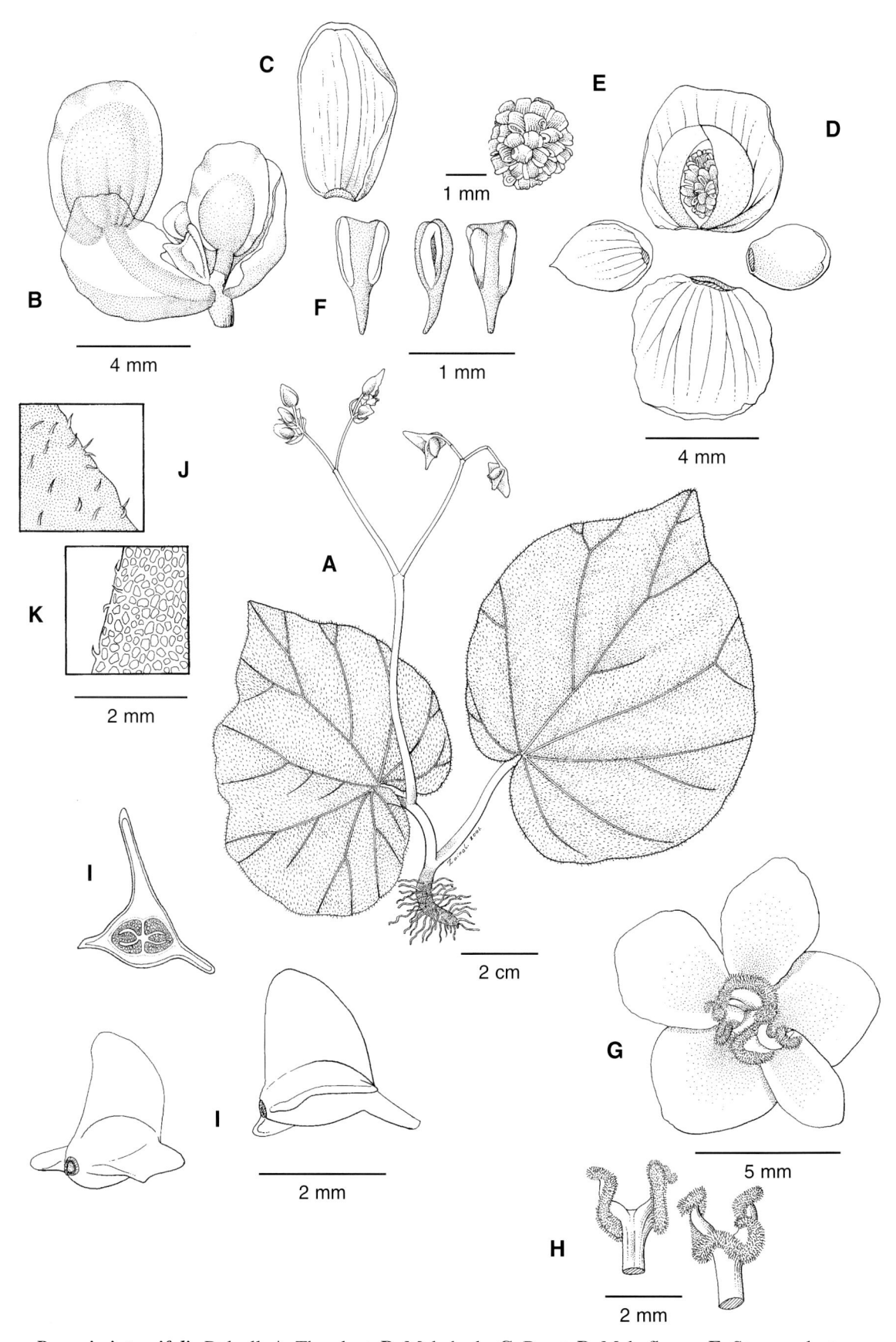

Begonia integrifolia Dalzell. **A.** The plant. **B.** Male buds. **C.** Bract. **D.** Male flower. **E.** Stamen cluster. **F.** Stamens. **G.** Female flower. **H.** Styles and stigmas. **I.** Fruit. **J.** The upper leaf surface. **K.** The lower leaf surface. (*RK 5168*)

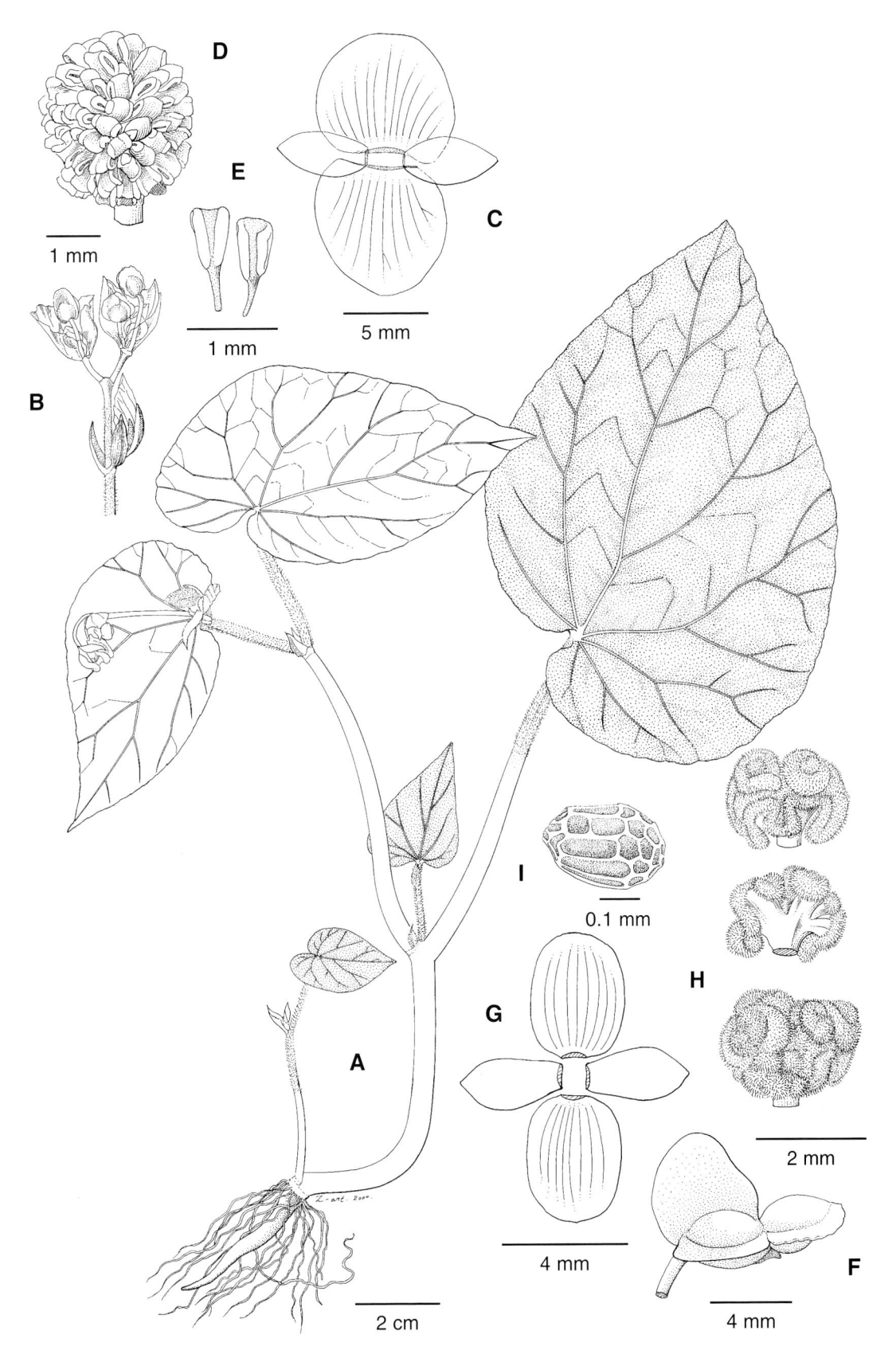

Begonia integrifolia Dalzell. **A.** The plant. **B.** Inflorescence. **C.** Male tepals; **D.** Stamen cluster. **E.** Stamens; **F.** Female bud. **G.** Female tepals; **H.** Styles and stigmas. **I.** Seed. (*RK 4944*)

TAN JIEW HOE

and up to 2–13 cm apart, held flat against the soil surface; sometimes with up to 3 fawn, hairy, globose bulbils (1 large and 2 small) in leaf axil, up to 7 × 7 mm; stalk translucent pale red or whitish green, densely or sparsely hairy, 1–11(–22.5) cm long, 4–7 mm thick, flat or grooved above; blade dull plain pale green and beneath whitish or mid-green and pale reddish beneath, sometimes with bluish sheen above, rarely brown to black above and deep rosy purple beneath, sometimes with many tiny silver spots or with silver spots *c.* 4 mm across with a hair in the centre, hairs on both surfaces but sparse above; thin and soft in life, thinly papery when dried, obliquely broadly ovate, asymmetric, (4–)11(–22) × (4–)8(–18) cm wide, broad side (1.5–)4.5(–10) cm wide, base unequally heart-shaped, not overlapping, basal lobe (1–)3(–6) cm long, margin not toothed with sparse hairs or minutely toothed, sometimes undulate, tip blunt or rounded, rarely acute; venation palmate-pinnate, 4–6 pairs of veins, branching three times before reaching the margin, with another 1–2 in the basal lobe, plane or slightly impressed above, beneath slightly prominent, whitish green or reddish, often hairy. **Inflorescences** terminal, rarely with an additional one from the lowest leaf axil, pale green or red, without hairs, longer than the leaves, branched, (1.5–)7–11(–19.5) cm long with up to 5 branches *c.* 1–1.5 cm long, stalk *c.* 1.5 cm long, male flowers *c.* 9, female flowers 2, male flowers open first. Bract pair enclosing the developing inflorescence, translucent, white or pale green with a reddish tinge, broadly oval, *c.* 13 × 5 mm, soon falling. **Male flowers** with pinkish white stalk up to 20 mm long; tepals 4, white, pale pink

(Opposite). Variation between different populations of the Spotted Begonia for leaf shape, spot size and density and whether the leaf is purple underneath.

BEGONIA INTEGRIFOLIA

The plain form of the Spotted Begonia, *Begonia integrifolia*.

towards base, sometimes pale yellowish-green, hairless, margin not toothed, tip rounded, outer two rotund, 5–12 × 5–9 mm, inner two narrowly oval, 5–10 × 2–3 mm; stamens many, cluster globose, 2 × 2 mm, stalk 1 mm long; filaments 0.5–0.75 mm long; anthers pale to golden yellow, obovate, *c.* 1 mm long, tip rounded, opening by slits. **Female flowers** with stalks *c.* 8 mm long; ovary green, sometimes with red veins, 4–5 mm long, wings 3, unequal, longest *c.* 7 mm wide, shorter two *c.* 3 mm wide, locules 2, placentas 2 per locule; tepals 5, white or very pale pink, hairless, margin not toothed, tip rounded, outer two rotund, *c.* 5 × 6 mm, innermost oval, smaller 2–4 × 1–2 mm; styles 2, styles and stigmas golden yellow, 3–4 mm long, stigmas spiral. **Fruit** dangling on a thin stalk *c.* 10 mm long; capsule 7–12 × 10–18 mm, hairless, wings 3, unequal, thinly fibrous, longer wing rounded, 4–10(–15) mm wide, shorter two pointed, *c.* 3–4 mm wide, splitting irregularly between the locule and the wing. **Seeds** barrel-shaped, *c.* 0.25–0.3 mm long, collar cells *c.* half the seed length.

DISTRIBUTION. India, Myanmar, Thailand, Laos, Vietnam and Peninsular Malaysia as far south as Gopeng, Perak, and Kuala Lipis, Pahang; common on the west coast, rather rare east of the Main Range.

HABITAT. Usually on bare earth slopes among rocks, frequently on limestone rocks or on the damp base of limestone cliff faces, also on shale cliffs or damp rotting logs in lowland forest from sea level (Langkawi) up to 650 m altitude.

OTHER SPECIMENS. KEDAH—Baling, *Kiew RK 5168* (SING); Gua Manik, *Kiew RK 5152* (SING); Gunung Jerai, *Weber s.n.* 1986 (SING); Pulau Langkawi, *Anthonysamy SA 956* (KEP), *SA 960* (KEP, SING), *SA 978* (KEP), *SA 1004* (SING), *SA 1025* (SING); *Corner s.n.* 1941 (SING); *Curtis s.n.* 1890 (SING); *Holttum SFN 15127* (SING); *Kiew RK 4944* (SING); *Keng 23* (SINU), *66* (SING, SINU), *81* (SINU); *Md Nor 31383* (KLU); Soepadmo & Mahmud 1213 (KLU); *Stone 10953* (KLU), *14280* (KEP); *Tay 115* (SING), *119* (SING), *156* (SING); Pulau Dayang Bunting, *Corner s.n.* 1941 (SING); *Kiew RK 4940* (SING); *Stone 14197* (KLU); Rimba Teloi FR, *Weber s.n.* 1986 (SING); Pulau Langgon, *Boey 546* (KLU), *Kiew RK 4936* (SING); Pulau Timun, *Stone 10983* (KLU); KELANTAN —Bertam, *UNESCO Limestone Exhibition 26* (SING); Gua Panjang, *UNESCO Limestone Exhibition 584* (SING); Kuala Aring, *Hamid H 23* (SING), *H 24* (SING), *H25* (SING), *H 26* (SING). PAHANG—Kuala Lipis, *Kiew RK 3887* (SING); Gua Tipus, *Henderson SFN 19408* (SING). PENANG—*Birch 1897* (SING); *Curtis 3097* (SING), *3179* (SING); *Md Haniff SFN 6127* (SING), *SFN 6124* (SING), *Md Haniff s.n.* 1919 (SING); PERAK Belum N-E Highland, *Davison B2* (SING); Gunung Bujang Melaka, *Anthonysamy SA 824* (SING); PERLIS—Bukit Ketari, *Kiew RK 3616* (KEP), Bukit Pinang, *Kiew RK 3666* (SING); Bukit Rongkit, *Kiew RK 3698* (SING); Gua Burma, *Kiew RK 3650* (KEP), *RK 3658* (KEP), Wang Besar, *Kiew RK 3605 (KEP), RK 3606A* (KEP), *RK 3606B* (KEP); Wang Kelian, *Jaman RJ 5497* (UKMB), *RJ 5514* (UKMB), *RJ 5585* (UKMB); *Kiew RK 5147* (SING).

NOTES. The Spotted Begonia shows a great range in size—it begins to flower when only 1.5 cm tall with two leaves each 4 cm long, but in damp and sheltered conditions it grows into a very lush plant up to about 40 cm tall with up to six well-spaced leaves, the largest up to 22.5 cm long. Its stem, which remains succulent, gains support by growing against earth banks or rocks. The leaf stalk shows a great range in length even on the same plant—those of the lowermost leaves are three to four times longer than those of the uppermost. Leaf colour is also variable ranging from bright green to mid-green to distinctly bluish green to purplish brown (liver-coloured) to black. Some populations include plants with either spotted or plain leaves, while other populations are all plain-leaved; some plants have pale green undersides to the leaf, others are reddish green underneath. These colour forms look so different that they appear to be different species!

BEGONIA INTEGRIFOLIA

In Malaysia, this species was known as *Begonia guttata*. However, Clarke (1879) had noted that *B. integrifolia* and *B. guttata* were probably one species. In 1921, Gagnepain in the flora of Indo-China treated *B. guttata* as a variety of *B. integrifolia*. The major difference between the two species was whether the leaf margin was fringed by hairs or was hairless. However, examination of various populations in the field shows that this is not a fundamental difference but a difference only of degree. All have at least a few short hooked hairs on the margin. In addition, in some populations the undersurface of the young leaves appears mealy from the dense hairs but this wears off with age. I, therefore, agree with Clarke that *B. guttata* is the same as *B. integrifolia*.

King described a new species, *B. debilis,* which he distinguished from *B. guttata* by its being caulescent with a tuberous rootstock. *Begonia guttata* he described as being stemless with a creeping rhizome. However, the type of *B. guttata* from Penang is an example of a well-grown plant with a long stem with a tuber at the base. King's *B. debilis* is clearly the same as this, i.e. it is the same as *B. integrifolia*.

Irmscher's *B. guttata* forma *elongata* is, as he suggested, merely a form adapted to local conditions. The stem is spindly (38 cm long) but is no taller than well-grown plants in other populations. In view of the variation shown by this species, it is futile to single out particularly large or small plants and give them taxonomic status.

Another three species also belong to *B. integrifolia*. Ridley described *B. leucantha* from northern Perak, *B. curtisii* from southern Thailand, and Burkill described *B. haniffii* from Langkawi, which Ridley (1902) considered the same as his *B. curtisii* from Thailand. None is distinct in any way from *B. integrifolia*. At the time they were described, the variation of *B. guttata/B. integrifolia* was not understood.

Further confusion is caused by Doorenbos *et al.* (1998) who placed *B. integrifolia* in sect. *Platycentrum* (which has axillary inflorescences) and *B. guttata* in sect. *Parvibegonia* (which has terminal inflorescences). This is clearly in error because, as early as 1876, Clarke had noted the inflorescence of *B. integrifolia* 'as though terminal'. Examination of type specimens show both *B. integrifolia* and *B. guttata* to have terminal inflorescences. *Begonia integrifolia* therefore belongs in sect. *Parvibegonia*.

There is a group of four begonias in Peninsular Malaysia closely related to *B. integrifolia* that all have succulent, low stems with a small tuber at the base, thin leaves (spotted in some species), terminal inflorescences and two-loculate fruits with three unequal wings, the largest wing not being more fibrous than the two shorter ones. In contrast to the widespread *B. integrifolia*, the others have restricted distributions and three are endemic in Peninsular Malaysia. They are *B. carnosula* from the foothills of Cameron Highlands, *B. phoeniogramma* from Selangor, and *B. variabilis* from Temangor, Perak. The fourth species, *B. elisabethae*, grows in Kedah (on Langkawi and the mainland) and Peninsular Thailand. Among these, *B. integrifolia* is most closely similar to *B. phoeniogramma* and is distinguished from the latter primarily by its plain as opposed to red-striped tepals. (A few populations of *B. integrifolia* on Langkawi have tepals with the veins slightly red on the outer surface; the tepals of *B. phoeniogramma* have veins that are deep red on the inner as well as the outer surface and, in addition, are ribbed on the outer surface). Their distributions do not overlap—*B. integrifolia* reaches its southern limit in south Perak; *B. phoeniogramma* is confined to Selangor.

The Spotted Begonia is apparently a short-lived species dying down in the dry season, then sprouting again from the tuber. It also regenerates readily from seed; small plants and seedlings can always be found.

10. THE RED STRIPED BEGONIA
Begonia phoeniogramma Ridl.
(Latin, *phoeniceus*=scarlet, *grammatus*=striped, referring to tepals)

Ridley, JSBRAS 75 (1917) 35, FMP 1 (1922) 858; Irmscher, MIABH 8 (1929) 155; Henderson, MWF Dicot (1959) 167 & Fig 160; *B. paupercula non* King, Ridley, JSBRAS 54 (1909) 42. **Type:** *Ridley 13430*, August 1908, Batu Caves (lecto K *ex* SING).

Stem pale translucent reddish brown, succulent, without hairs, erect, little branched, slender, 13–27 cm tall, flowering at 3 cm tall, 4–7 mm thick; tuber small, up to 12 × 14 mm. Stipules pale reddish brown with red veins, hairless, ovate, 5–6 × 3–4 mm, margin not toothed, tip pointed, soon falling. **Leaves** *c.* 6, distant and up to 9.5 cm apart; stalk greenish red, hairless, 4–11 cm long, grooved above; blade oblique, plain pale green above and beneath or plain dark green with a metallic bluish sheen and magenta beneath or variegated with grey spots on pale green leaves or grey-green silvery spots on dark green leaves, thinly succulent in life, thinly papery when dried, glossy, ovate, strongly asymmetric, 8.5–19 × 6.5–12 cm, broad side 3.5–4(–9) cm wide, base rounded, sometimes overlapping, basal lobe 3.5–6 cm long, margin undulate, tip acute to elongate; venation palmate-pinnate, 1–2 pairs of veins at the base and 2–3 pairs along the midrib, branching

The Red Striped Begonia grows not only on limestone rocks but also on granite rocks (above) and on earth banks.

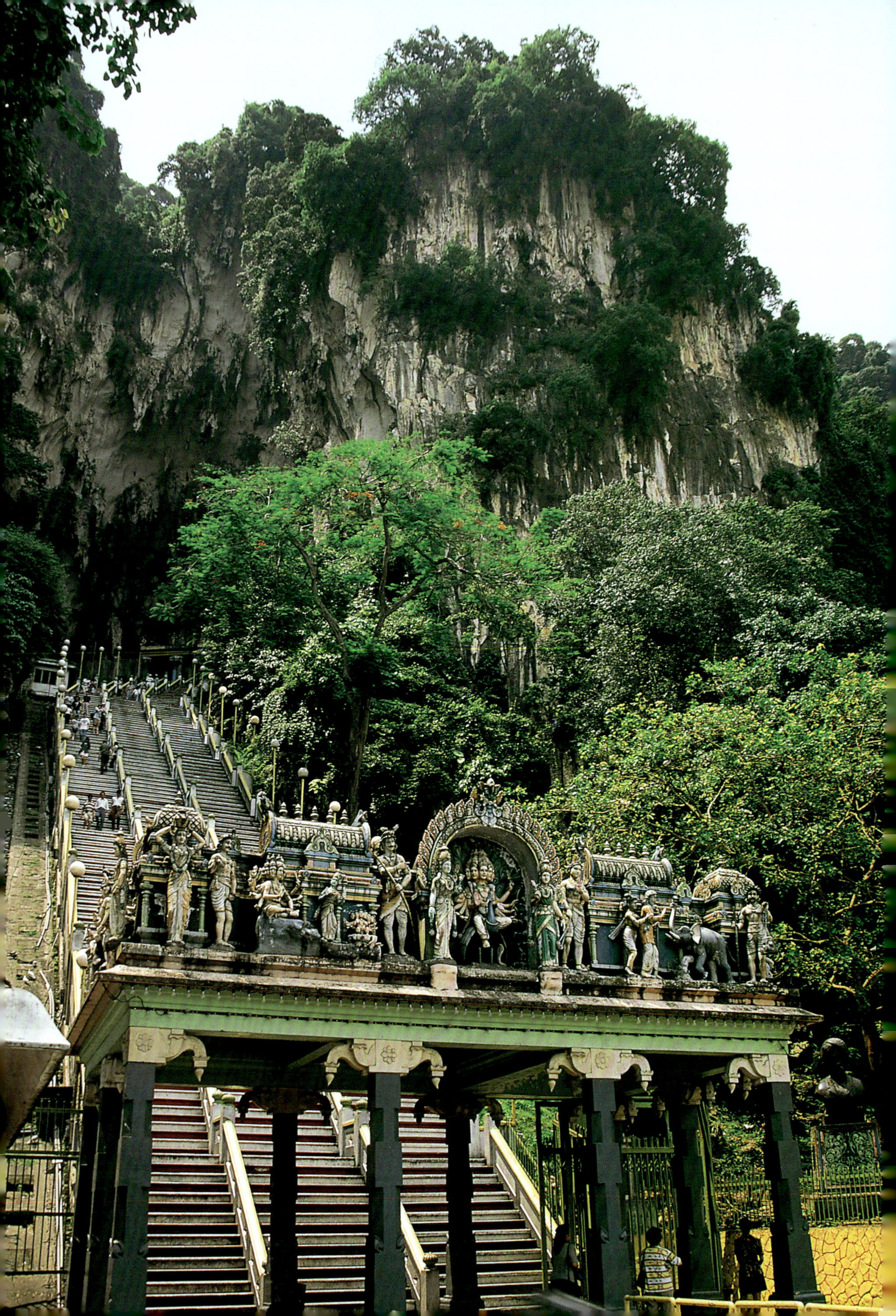

BEGONIA PHOENIOGRAMMA

more than halfway to margin, with another 1–2 veins in the basal lobe, veins plane above, beneath prominent, hairless and the same colour as blade. **Inflorescences** terminal, pale reddish brown, without hairs, erect, little branched, (2.5–)5–9 cm long with two main branches 1.5–2.5 mm long, stalk 1–8 cm long, lengthening to 13.5 cm in fruit, male flowers many, female flowers 2, male flowers open first. Bract pair greenish red with red stripes, ovate, *c.* 10 × 3 mm, margin not toothed, soon falling. **Male flowers** with pale red hairless stalk *c.* 11 mm long; tepals 4, white with red veins raised as ribs on the outer surface, hairless, margin not toothed, outer two obovate, 9–11 × 6–8 mm, tip acute, inner two narrowly obovate, 8–10 × 4–5 mm, tip rounded; stamens many, cluster globose, 2.5–3 mm across, stalk *c.* 1 mm long; filaments *c.* 0.75 mm; anthers pale yellow, obovate, *c.* 1 mm long, tip slightly notched, opening by slits. **Female flowers** with pale reddish brown, hairless stalk, *c.* 4 mm long; ovary green with reddish veins on the wings, 5–8 mm long, wings 3, unequal, locules 2, placentas 2 per locule; tepals 5–6, pale pink with raised red stripes, hairless, ovate, margin not toothed, tip rounded, outer tepals 4–6 × 4.5 mm, innermost smaller *c.* 4 × 2.5 mm; styles 2, styles and stigmas yellow, *c.* 2.5 mm long, stigmas spiral. **Fruit** a splash cup pendent on a fine and hair-like stalk 8–11 mm long, capsule 10–13 × 21–22 mm, hairless, locules 2, wings unequal, tips pointed, drying almost papery, larger wing 10–17 mm wide; smaller two 3–6 mm wide, splitting irregularly between locules and wings. **Seeds** barrel-shaped, *c.* 0.25 mm long, collar cells *c.* ³/₄ seed length.

Distribution. Endemic in Peninsular Malaysia—Selangor.

Habitat. In the lowlands and foothills west of the Main Range to *c.* 350 m altitude in light shade in forests on steep earth slopes or on granite rocks or on limestone (Batu Caves).

The red striped flowers of *Begonia phoeniogramma*. (Left & centre). Male flowers and buds. (Right). Female flowers. (Opposite). The population of the Red Striped Begonia on Batu Caves is threatened by disturbance.

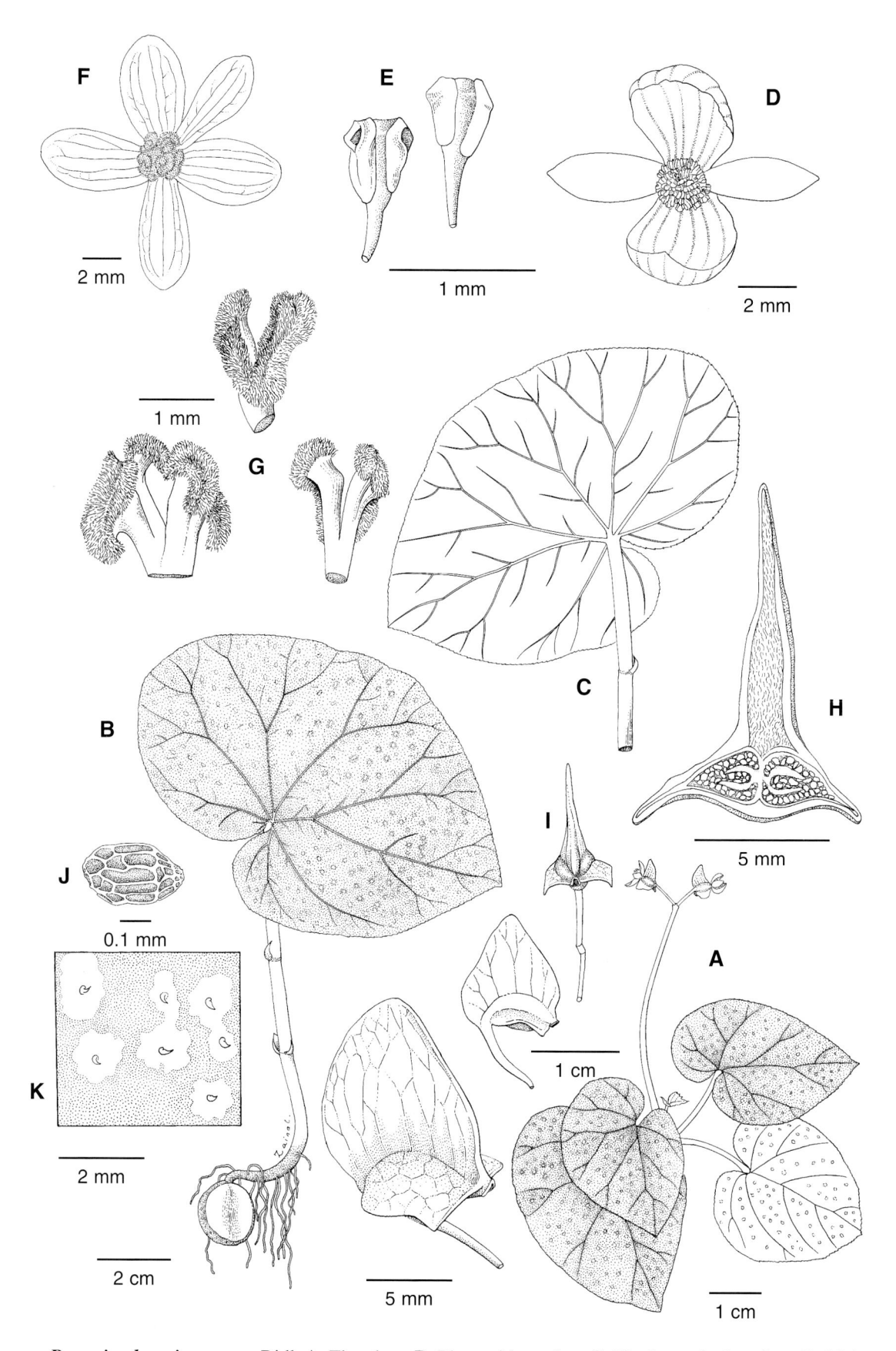

Begonia phoeniogramma Ridl. **A.** The plant. **B.** Plant with a tuber. **C.** The lower leaf surface. **D.** Male flower. **E.** Stamens. **F.** Female flower. **G.** Styles and stigmas. **H.** T.S. ovary. **I.** Fruit. **J.** Seed. **K.** The upper leaf surface. (*RK 3252*)

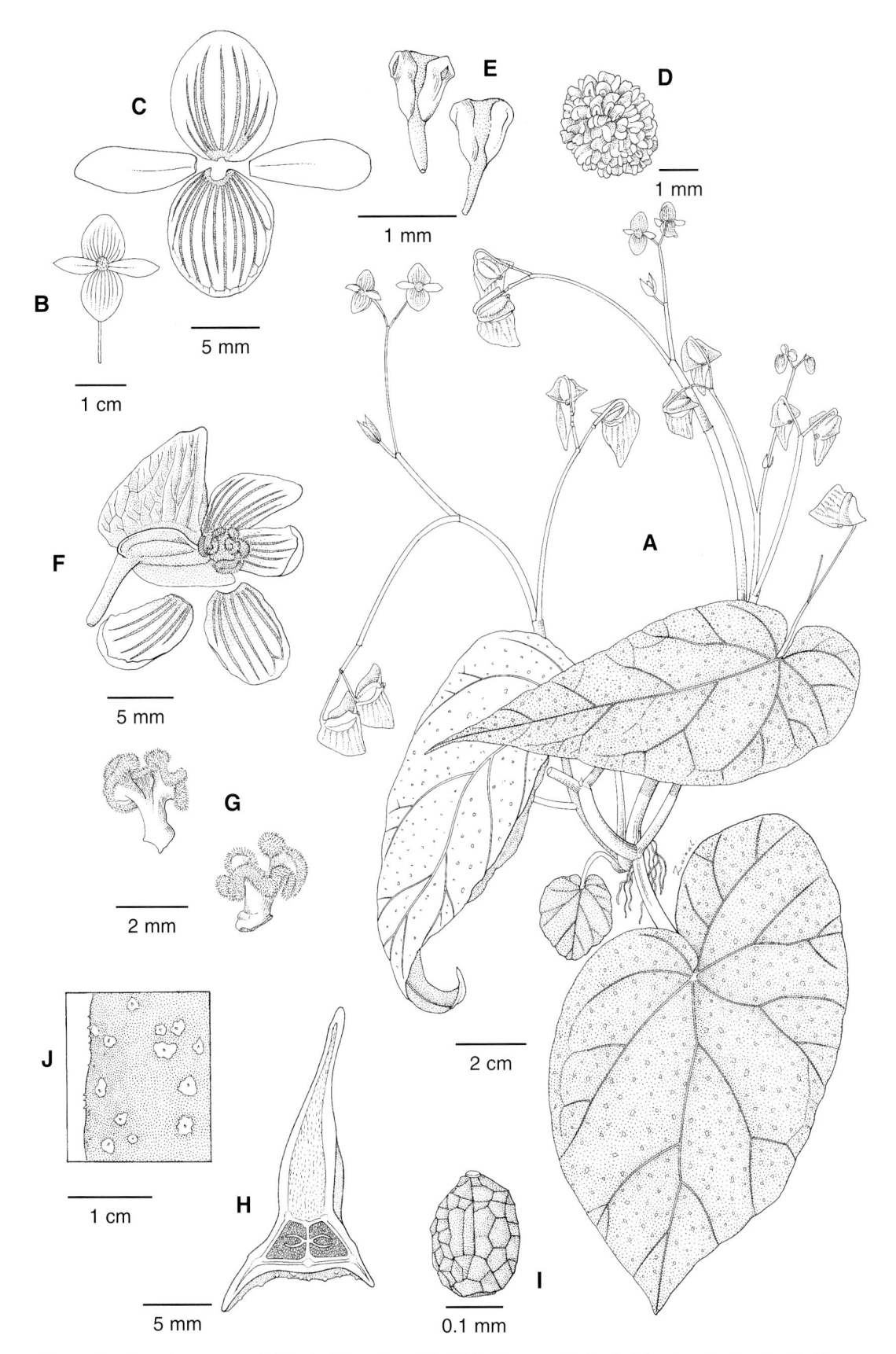

Begonia phoeniogramma Ridl. **A.** The plant. **B.** Male flower. **C.** Red striped male tepals. **D.** Stamen cluster. **E.** Stamens. **F.** Female flower. **G.** Styles and stigmas. **H.** T.S. ovary. **I.** Seed. **J.** The upper leaf surface. (*RK 5105*)

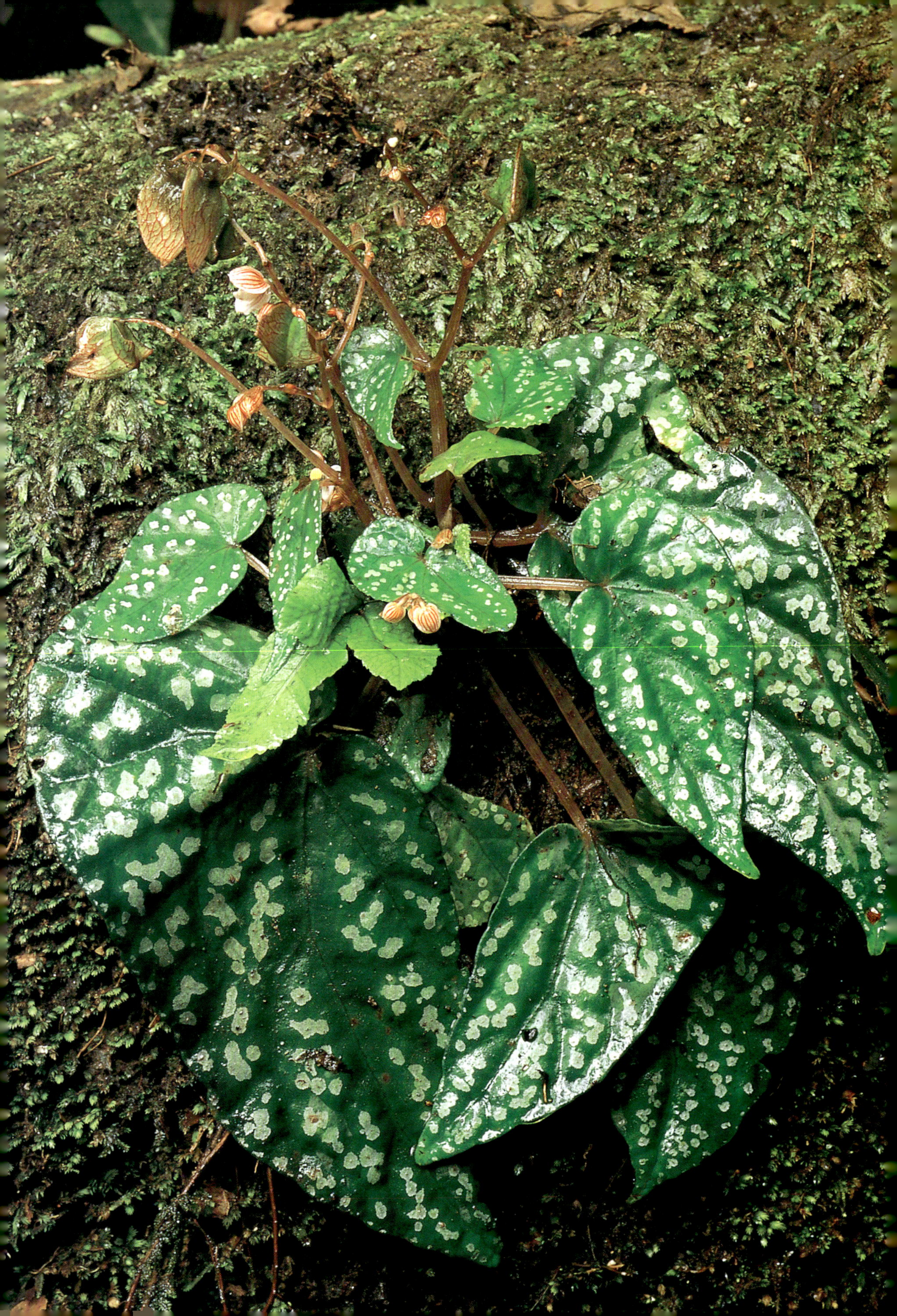

OTHER SPECIMENS. SELANGOR—Ampang Catchment Area, *Anthonysamy SA 432* (KEP), *Kiew RK 1306* (KEP); Batu Caves, *Burkill SFN 6362* (SING) *SFN 6366* (SING), *SFN 6367* (SING), *Kiew RK 1257* (KEP), *RK 1258* (KEP), *RK 1259* (SING), *RK 1340* (SING), *RK 3252* (SING); *Md Nur SFN 8972* (SING); *Ridley 2882* (SING), *8281* (SING), *s.n.* 1897; *Sinclair SFN 40067* (KEP, SING), *SFN 10720* (SING); *Teruya 502* (SING), *3212* (SING); *Wyatt-Smith KEP 79241* (KEP), *KEP 85206* (KEP); Bukit Lagong FR, *Blanc & Vethevelu FRI 32945* (KEP), *Kochummen KEP 79034* (KEP), *Sands 3510* (KEP), *Saw FRI 34268* (KEP), *Stevens 31* (KEP), *Vethevelu & Sands FRI 32934* (KEP); Gabai, *Kiew RK 5106* (SING); Genting Bidai, *Ridley 7289* (SING); Gombak Valley, Kiew RK 1098 (SING), *RK 1196* (KEP); Lolo *Halijah s.n.* (SING), *Kiew RK 5105* (SING); Sungai Degek, *Yong s.n.* (SING), Sungai Lallang, *Kiew RK 3800* (SING), Sungai Buloh, *Kiew RK 1610* (KEP).

NOTES. Apart from the striking ribbed, red-veined tepals, this species is hardly different from *Begonia integrifolia*. (There are several populations of *B. integrifolia* in Pulau Langkawi that have pink flowers with redder veins but the veins are not distinctly raised as they are in *B. phoeniogramma*). Indeed, Ridley even confused the two when identifying specimens. Like *B. integrifolia*, the Red Striped Begonia shows the same variability in size, and leaf shape, variegation and colour. Other slight differences from *B. integrifolia* include the tepals of the male flower that are relatively narrower and the more pointed fruit wings. The two species are, however, geographically separated: *B. phoeniogramma* is found in Selangor as far north as Batu Caves, while *B. integrifolia* is found as far south as Gopeng, Perak.

Irmscher used wing size to separate them—whether the largest wing was twice as wide as long in *B. phoeniogramma* or about as wide or shorter than long in *B. integrifolia*. In fact, both species show the same range of size, most wings being slightly shorter or the same size and only occasionally are plants of both species found with wings twice as wide as long. In addition, this feature is difficult to see in herbarium specimens as the wings are thin and shrivel.

Begonia variabilis also has red-veined tepals but is very different in leaf shape so would not be confused with this species.

(Left & opposite). The Red Striped Begonia, *Begonia phoeniogramma*.

11. THE VARIABLE BEGONIA
Begonia variabilis Ridl.
(Latin, *variabilis*=variable, referring to leaf shape)

Ridley, JSBRAS 57 (1911) 50, FMP 1 (1922) 858; Irmscher, MIABH 8 (1929) 147 & Fig. 10. **Type:** *Ridley 14730*, July 1904, Ulu Temango (lecto SING, here designated).

Stem pale red, succulent, without hairs, erect, often branched, falling and rooting at the nodes, slender, 2–26 cm tall, flowering at 2 cm tall, 3–7 mm thick; tuber globose to corm-shaped, 7 × 15 mm to 25 × 12.5 mm. Stipules pale green with red lines, hairless, narrowly triangular, 6–11 × 2.5–4 mm, margin not toothed, tip elongate, ending in a hair, soon falling. **Leaves** 2–3(–8); distant and 6–11 cm apart; stalk pale red, without hairs, slender, 3.5–10.5 cm long, grooved above; blade very oblique, plain pale green above and beneath or dark green above and purple beneath, frequently variegated with small light green or silvery spots or blotches between the veins, sometimes appearing velvety above, minutely and sparsely hairy when young becoming hairless with age, soft and thin in life, thinly papery when dried, glossy above, strongly asymmetric, base deeply heart-shaped, lobes not overlapping, margin wavy, not toothed, tip narrowing to an elongated tip, lower leaves ovate, 10.5–16.5(–25) × 7.5–11 cm, basal lobe rounded 4.5–6(–8.5) cm long, broad side 5–7 cm wide, upper leaves often narrowly ovate and elongate, 7–14 × 2.5–7 cm, basal lobe 2.5–4.5 cm long, broad side 2.5–4.5 cm wide; venation palmate-pinnate, (2–)4–6 pairs of veins with another 2–3 pairs in basal lobe, branching towards the margin, veins impressed above, beneath prominent

(Above). The Variable Begonia, *Begonia variabilis*, grows both on limestone and on earth banks. (Opposite). As its name suggests, this begonia has very variable leaves for shape, colour and the pattern of the spots.

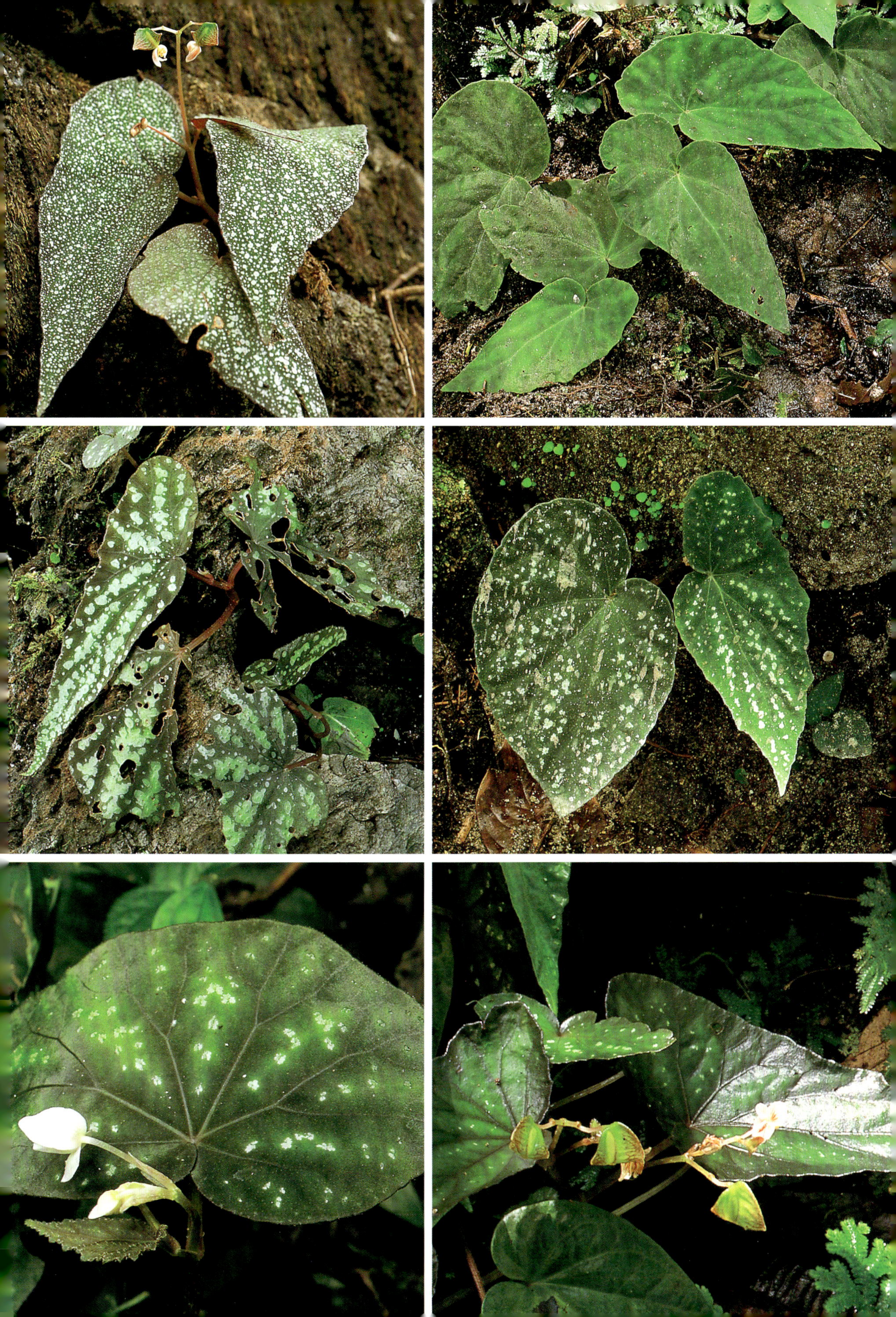

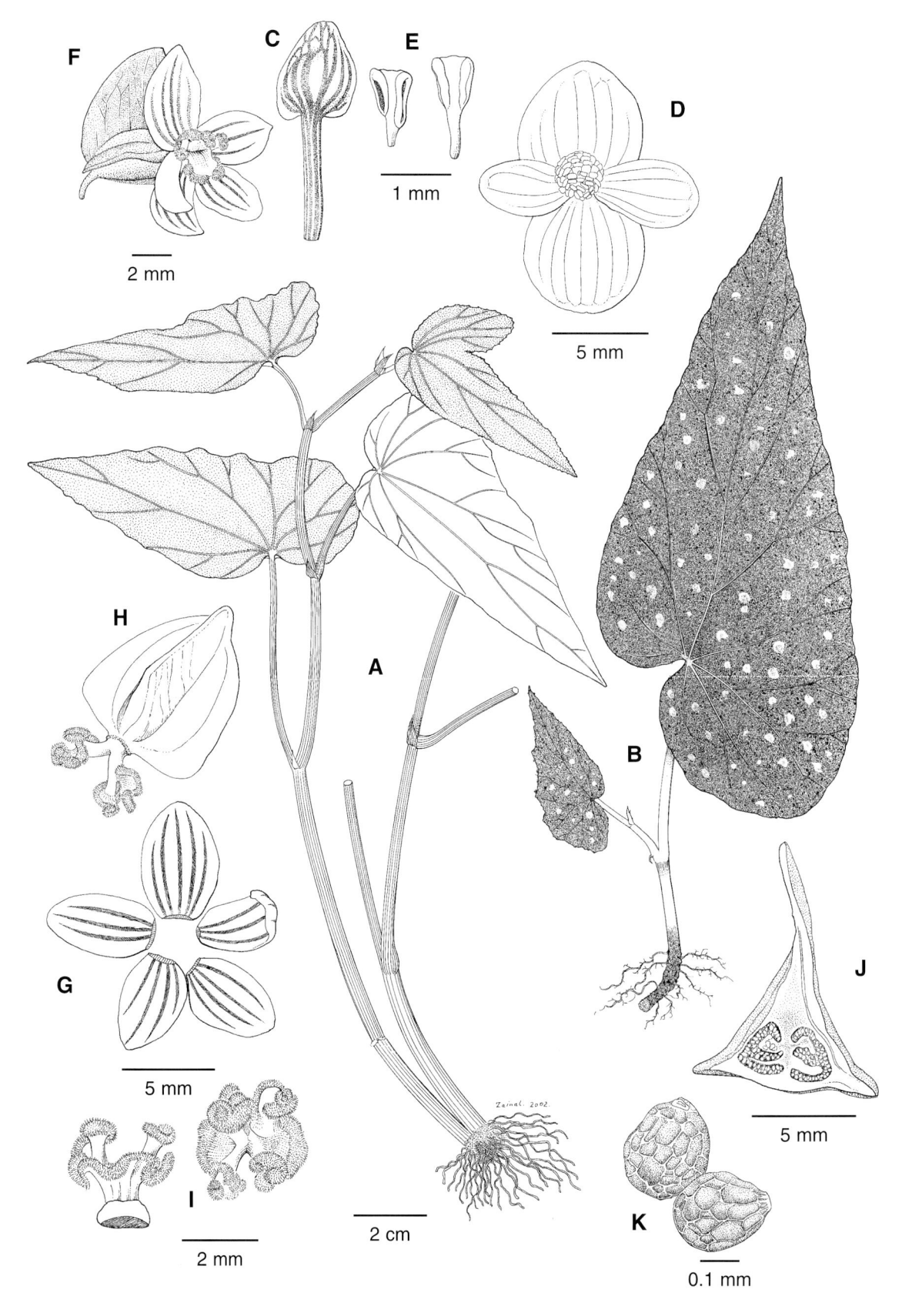

Begonia variabilis Ridl. **A.** The plant. **B.** Variegated leaf. **C.** Male bud. **D.** Male flower. **E.** Stamens. **F.** Female flower. **G.** Female tepals. **H.** Ovary. **I.** Styles and stigmas. **J.** T.S. ovary. K. seed. (*RK 5097*)

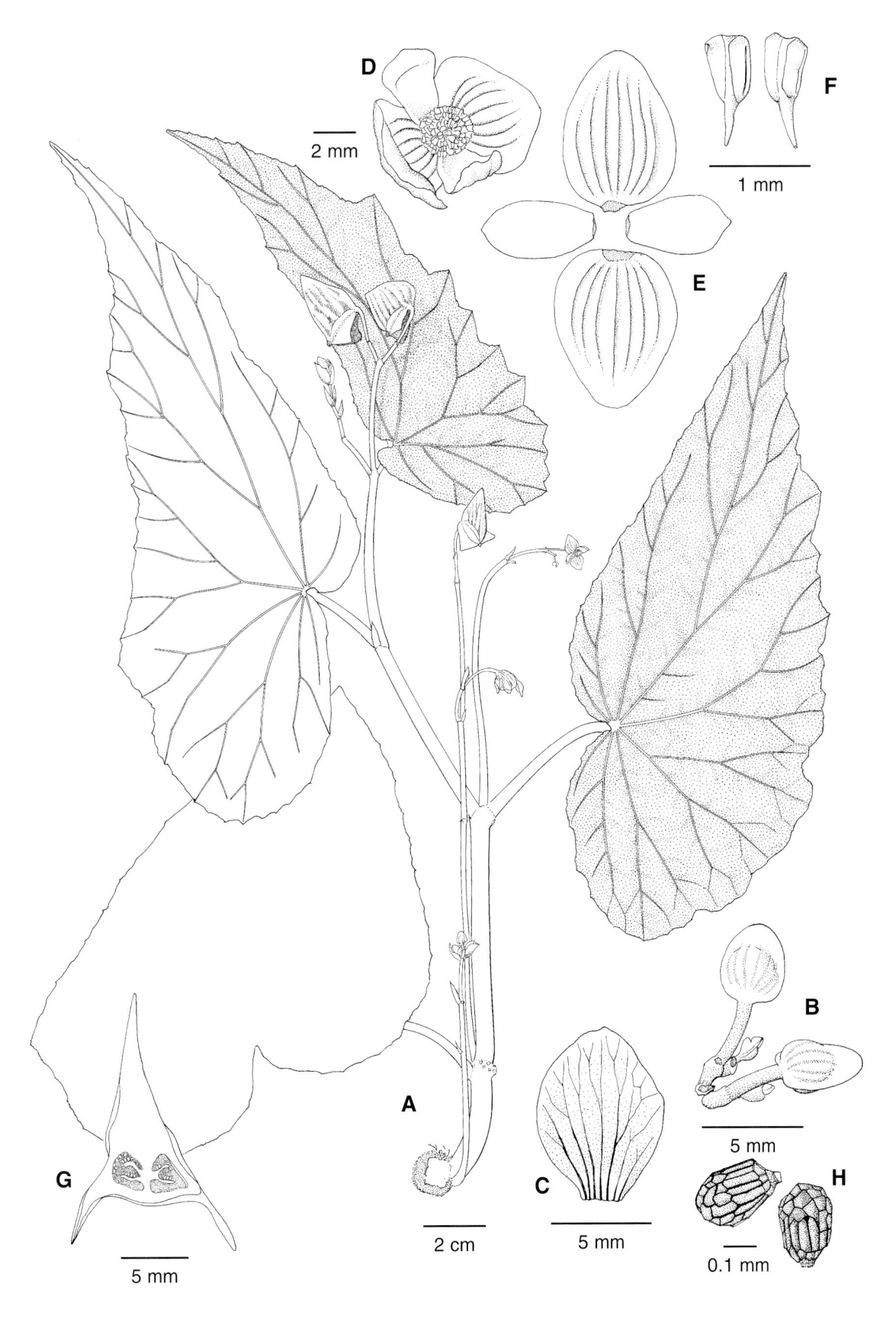

Begonia variabilis Ridl. **A.** The plant with a tuber. **B.** Male bud. **C.** Bract. **D.** Male flower. **E.** Red striped male tepals. **F.** Stamens. **G.** T.S. ovary. **H.** Seed. (*RK 5099*).

BEGONIA VARIABILIS

Begonia variabilis. (Top left). Male flowers. (Bottom left). Fruits. (Right top & bottom). Female flowers.

and reddish. **Inflorescences** at first terminal, then from lower leaf axils, pale red, hairless, shorter than the leaves, little branched, 2–10 cm with stalk 1–9 cm long, branches 0.75–1 cm long; male flowers many, female flowers 2, male flowers open first. Bract pair enclosing the developing inflorescence, pale green with red lines, hairless, ovate, 7–10 × 4–8 mm, margin not toothed, soon falling. **Male flowers** with a reddish stalk 7–14 mm long; tepals 4, outer two white with red veins, hairless, broadly ovate, 5–15 × 4.5–12 mm, margin not toothed, tip rounded, inner two pure white, narrowly obovate, 5–12 × 2.5–6 mm; stamens many, cluster globose, 2–6 mm across, stalk *c.* 2 mm long; filaments 0.5–2 mm long; anthers pale yellow, narrowly obovoid, 1–2 mm long, apex rounded to slightly pointed, opening by slits. **Female flower** with stalk *c.* 9 mm long; ovary green, 5–6 mm long, wings 3, unequal, longer wing with red veins, locules 2, placentas 2 per locule; tepals 5, outer two with red stripes, inner ones pure white, without hairs, narrowly oval, 5–7 × 1.5–4 mm, margin not toothed, tip acute; styles 2, styles and stigmas pale yellowish green, 2–4 mm long, stigmas spiral. **Fruit** a splash cup pendent on a thin, stiff stalk 6–15 mm long; capsule 4.5–13 × 9–17 mm long, hairless, locules 2, wings 3, thinly fibrous, unequal, larger wing broadly triangular, 10–11 mm wide, smaller two narrow, 3.5–5 mm long, splitting irregularly between locules and wings. **Seeds** barrel-shaped, *c.* 0.25 mm long, collar cells *c.* half seed length.

BEGONIA VARIABILIS

DISTRIBUTION. Endemic in Peninsular Malaysia: Upper Perak and Kelantan.

HABITAT. Lowlands at 100–150 m altitude, in deep shade on banks or granite rocks in lowland dipterocarp forest, not common, or in deep cracks or gullies on limestone in the Temangor Dam and on Gunung Reng.

OTHER SPECIMENS. KELANTAN—East-West Highway, *Kiew RK 5274* (SING); Gunung Reng, *Davison s.n.* (SING), *Kiew RK 5097* (SING), *Kiew & Anthonysamy RK 3010* (SING). PERAK—Belum F.R., *Hanim s.n.* (SING), *Rahimatsah A2* (SING), *Rahimatsah A3* (SING), *Rahimatsah A4* (SING); Sg Halong, *Turner & Yong 13* (SING), Sg. Sera, *Saw FRI 39969* (KEP); Grik, Kuala Kendrong, *Burkill SFN 12447* (SING), *Poore 402* (KLU); Temango, *Ridley s.n.* July 1909 (SING); Temangor Dam, Batu Putih, *Kiew RK 5099* (SING), *Rahimatsah B3* (SING), *Rahimatsah B4* (SING).

NOTES. The Variable Begonia is very distinctive, often with very elongated leaves with a very large basal lobe that is a third or even half as long as the leaf blade. Its scientific name is appropriate as it is remarkable for the variation in leaf shape on a single plant. The upper leaves can be extremely narrow for their length compared with the lower ones that are almost twice as wide as the upper leaves.

On limestone, plants have a very pronounced tuber, which suggests that they die down in dry periods. They begin to flower when just 2 cm tall with two leaves.

It belongs to the *Begonia integrifolia* group and, like *B. integrifolia* and *B. phoeniogramma*, the Variable Begonia shows variation in leaf coloration and pattern (whether spotted or purple underneath or plain green). It is also similar to these two species in growing not only on limestone, but also on granite rocks and in some cases on earth banks. It was not previously reported from limestone.

Like *B. phoeniogramma*, the tepals are striped with conspicuous longitudinal red lines, but these are not ribbed as they are in *B. phoeniogramma*.

Begonia variabilis grows on shaded gullies and on damp shaded boulders on limestone in the Temangor Dam.

12. ELISABETH'S BEGONIA
Begonia elisabethae, Kiew sp. nov.
(Elisabeth Eber-Chan, collector of this species)

Begoniam variabilem Ridl. aemulans, sed foliis ad apices radicantibus et tepalis masculis pilosis differt. **Typus:** *E. Eber-Chan 0124* October 1996, Langkawi (holo SING).

Stem translucent, weak and succulent, erect, unbranched or once branched near base, weak and falling and rooting on contact with soil, slender, up to 17 cm tall, (2–)5–7 mm thick, hairs rough, sparse; without a tuber. Stipules pale green with rough, sparse hairs, narrowly triangular, 8–10 × *c.* 4 mm, margin not toothed, soon falling. **Leaves** 2–4, distant, 3.5–4.5(–9) cm apart, held horizontally with the tip arching and touching the soil where it roots and eventually forms a new plantlet; stalk succulent with rough, sparse hairs, 3.5–10.5(–16.5) cm long, 2–6 mm thick, round in cross-section; blade oblique, glossy, plain pale green or blue-green in deep shape or densely speckled with small white spots *c.* 1.5–3 mm wide, each with a hair in its centre, paler beneath, in life thin, thinly papery when dried, ovate-lanceolate, asymmetric, 13–24.5 × 6–7 cm, broad side 4–8.5 cm wide, base very unequal, not overlapping, basal lobe 3–8.5 cm long, margin slightly scalloped and minutely toothed, tip attenuate; venation palmate-pinnate, 3 pairs of veins branching *c.* one third of way to margin with another 2 veins in the basal lobe, plane above, prominent beneath, without hairs. **Inflorescences** terminal, hairy, erect, shorter than the leaves, up to 11 cm long with *c.* 4 branches 1.5–3 cm long, stalk 4–5 cm long, male flowers many, female flowers few, male flowers open first. Bract pair leafy and enclosing the developing inflorescence, green with red veins, broadly ovate, 13–15 × 11–15 mm, tip broadly rounded, soon falling. **Male flowers** with a white to pale pink, hairy stalk 10–19 mm long; tepals 4, white, margin not toothed, tip rounded, outer tepals obovate, 8–15 × 6–11 mm, outer surface sometimes pale rosy pink, densely hairy, hairs glassy, uniseriate slightly hooked at tip; inner two narrowly obovate, 7–12 × 2–6 mm, without hairs; stamens many, cluster globose, *c.* 3 mm across, stalk *c.* 1 mm long; filaments *c.* 0.3 mm long; anthers golden yellow, broadly ovate, *c.* 1 mm long, tip notched, opening by slits. **Female flowers** with a white stalk 4–6 mm long; ovary *c.* 7.5 mm long, wings 3, unequal, longer wing pale green *c.* 5–6 mm wide, tip broadly rounded, shorter wings white 4–5 mm wide, locules 2, placentas 2 per locule; tepals 5–6, pure white, hairless, margin not toothed, outer two broadly ovate, *c.* 10 × 7 mm,

(Left). Female flowers of *Begonia elisabethae*. (Right). Plantlets growing from the leaf margin, a unique feature among Malaysian begonias. *Begonia elisabethae*. (Opposite). Foliage of *Begonia elisabethae*.

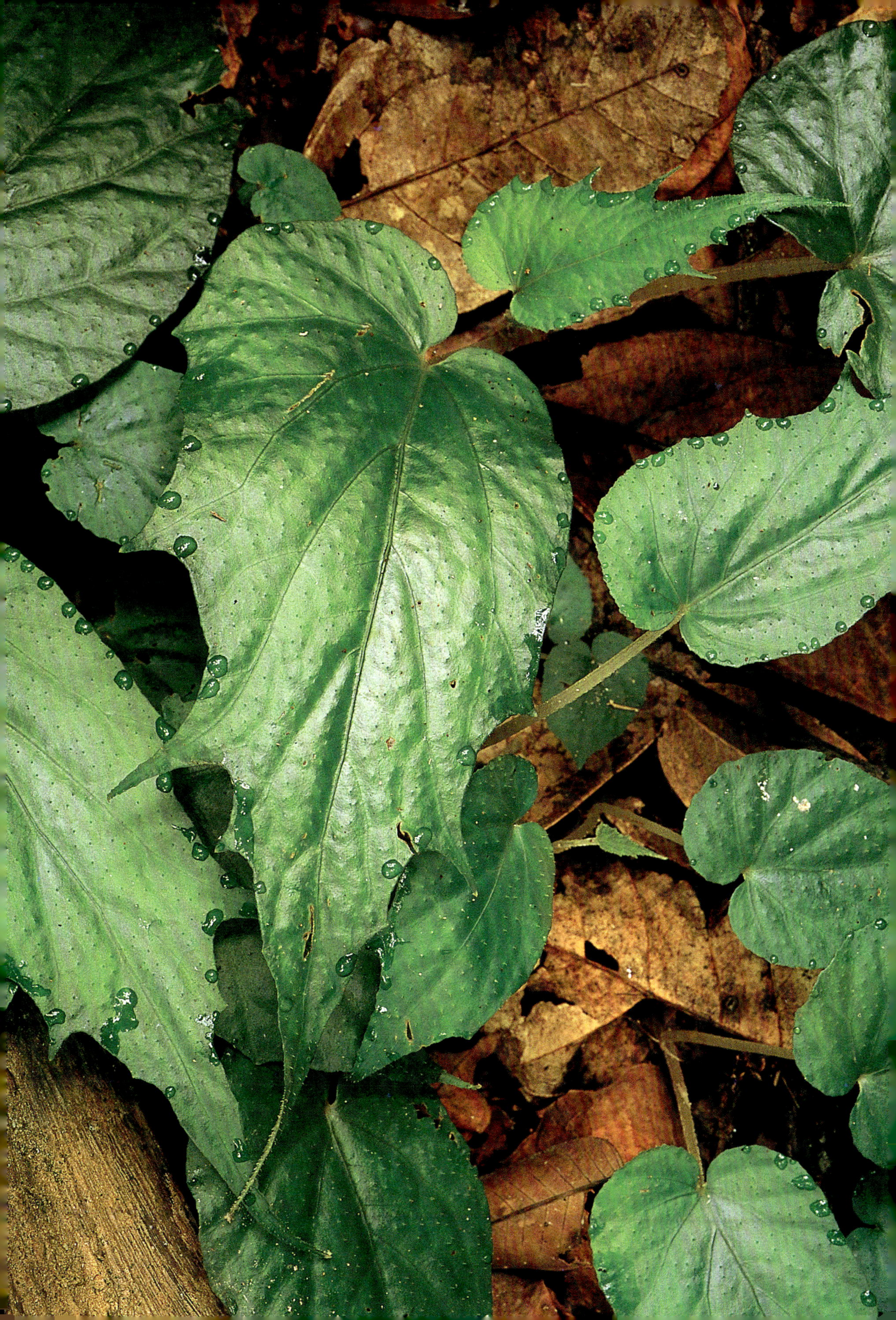

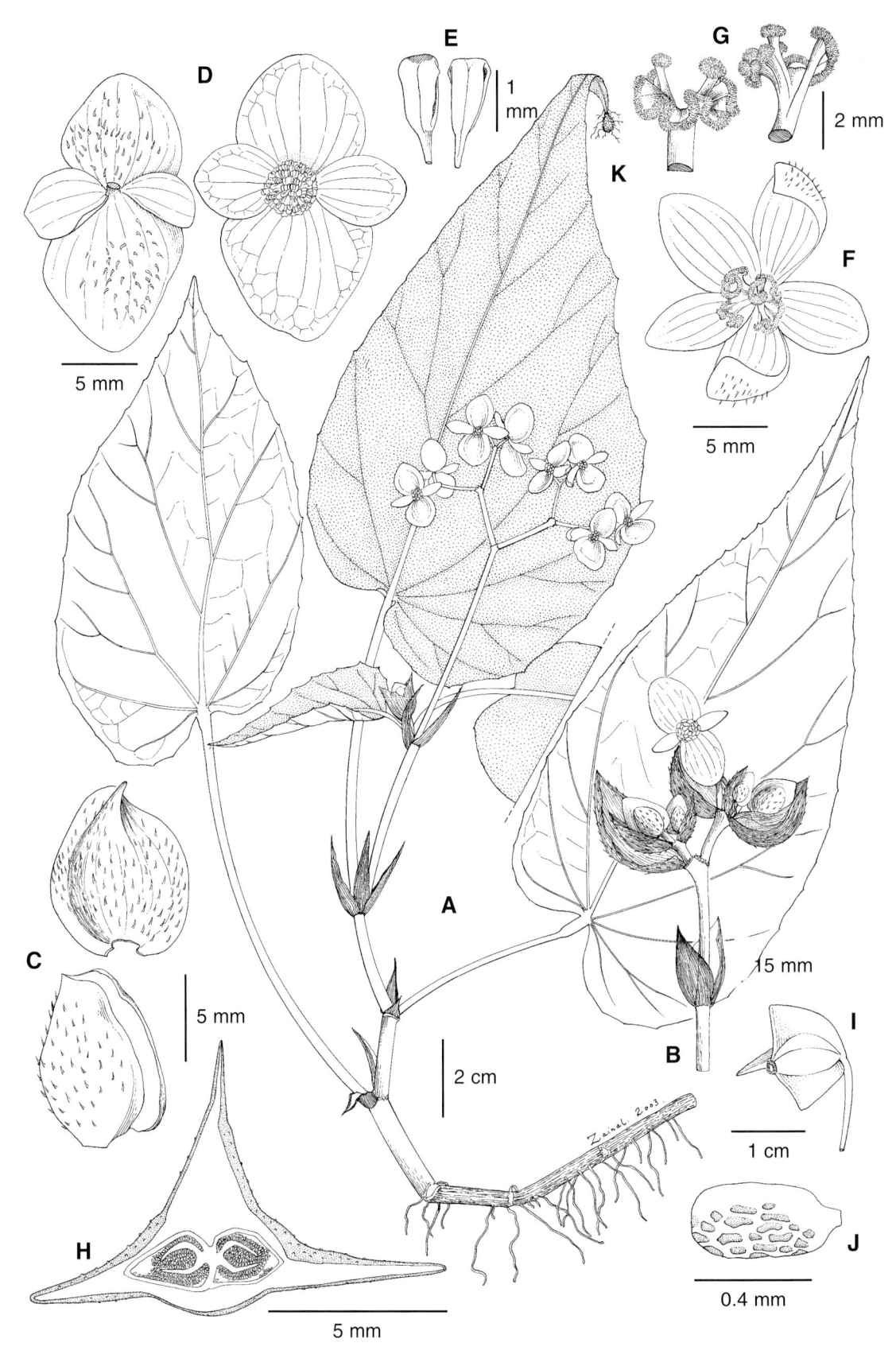

Begonia elisabethae Kiew. **A.** The plant. **B.** Young inflorescence. **C.** Bracts. **D.** Male flower. **E.** Stamens. **F.** Female flower. **G.** Styles and stigmas. **H.** T.S. ovary. **I.** Fruit. **J.** Seed. **K.** Leaf tip forming a plantlet. (*Eber-Chan 0124*)

decreasing in size to innermost tepal ovate, *c.* 10 × 4.5 mm; styles 2, styles and stigmas golden yellow, *c.* 3 mm long, stigmas spiral. **Fruits** pendent on stiff, slender stalks 4–10 mm long; capsule 7–10 × 14–15 mm, hairless, wings 3, broadly rounded, thin and papery, unequal, longer wing 8–9 mm wide, shorter two 3–6 mm wide, splitting irregularly between the locules and wings. **Seeds** barrel-shaped, *c.* 0.4 mm long, collar cells *c.* half the seed length.

DISTRIBUTION. Peninsular Thailand and Peninsular Malaysia (Langkawi and Bukit Weng, Kedah).

HABITAT. In Langkawi, it grows close to the seashore on sandy banks of a shallow stream and on low granite rocks in semi-swampy areas. Elsewhere it grows on granite rocks in moist places by streams or waterfalls. In Thailand it is recorded from limestone.

OTHER SPECIMEN. KEDAH—Bukit Weng *Weber s.n.* (SING).

NOTES. The very elongated leaf tip that arches and roots at the tip is unique among Malaysian begonias. It recalls the growth habit of the walking fern, *Bolbitis heteroclita* (Pr.) Ching, that grows in similar wet places. In addition, if the leaf rests on the wet soil, plantlets develop at the point where the vein reaches the leaf margin. Observations from growing this begonia suggest that the plant is very vulnerable to drying out but can persist and regrow from these plantlets.

Among Malaysian begonias, it belongs to the *Begonia integrifolia* group and is most similar to *B. variabilis* in its elongated leaf-shape and few-flowered inflorescences, but it is different from the Variable Begonia in its larger leaves (those of *B. variabilis* grow up to 15.5 × 9 cm) and from all of the group in the hairy outer tepals of the male flower and its propensity to form plantlets from the elongated arching leaf tip.

Elisabeth's Begonia is basically a Thai species, which just gets into the extreme north west of Peninsular Malaysia, where it is known from just two populations. In both cases, the populations are recorded as small and local, which makes this species extremely vulnerable to habitat disturbance, in the case of Langkawi by beach resort development. However, it is not in danger of extinction as it is more common in Peninsular Thailand.

Begonia elisabethae belongs to Sect. *Parvibegonia* because it has an erect stem, terminal inflorescences, male flowers with four tepals and fruits with two locules, each locule with two placentas. The begonia is named for Elisabeth Eber-Chan, a keen gardener, who first discovered this species and brought it to the author's attention.

(Right). Male flowers of *Begonia elisabethae*. Note how the leaf tip curves downwards. When it touches the soil, it roots and produces a plantlet.

13. THE TAIPING BEGONIA
Begonia thaipingensis King
(Latin, *-ensis* indicates origin as it was first discovered in Taiping)

King, JASB 71 (1902) 61: Ridley, FMP 1 (1922) 857; Irmscher, MIABH 8 (1929) 145. **Type:** *King's Collector 8511*. Thaiping (lecto K, here designated).

Stem creeping and rooting at nodes (stoloniferous), succulent, reddish with downy soft white hairs, little branched, nodes not swollen, slender, up to 20 cm long, *c.* 3 mm thick; without a tuber. Stipules dark red, hairy, triangular, *c.* 3 × 2 mm, margin not toothed, tip acute, soon falling. **Leaves** distant, (0.5–)1–4 cm apart, held horizontally on erect leaf stalks; stalks red, hairy, 2.5–7.5 cm long, *c.* 2.5 mm thick, grooved above; blade oblique, velvety, plain pale or mid-green, blue-green, bluish-black or golden-green above, beneath pale green or magenta or purplish, succulent in life, slightly leathery when dried, kidney-shaped or rounded, slightly asymmetric, 3–6 × 4–6.5 cm, broad side 2.5–4 cm wide, base heart-shaped and overlapping, basal lobes subequal, 0.75–1.5 cm long, margin slightly toothed, each tooth ending in a hair, tip slightly acute or rounded; venation palmate, 2(–3) pairs of veins, branching *c.* halfway to margin, veins almost plane above, beneath slightly prominent, hairy, and the same colour as blade. **Inflorescences** terminal, sometimes axillary, green or reddish, hairy, very slender, longer than the leaves, 12.5–27 cm long, much branched, branches *c.* 2.5–3 cm long, stalk 9–19 cm long, male flowers many, female flowers 2 per branch, female flowers opening first. Bract pair narrowly triangular, hairy, 3–4 × 1.5–2 mm,

The Taiping Begonia, *Begonia thaipingensis*.

The Taiping Begonia, *Begonia thaipingensis*. (Painting by Wendy Gibbs)

BEGONIA THAIPINGENSIS

Begonia thaipingensis. (Top). The blue-leaved form. (Centre and bottom right). Female flowers. (Bottom left). Male flowers.

104

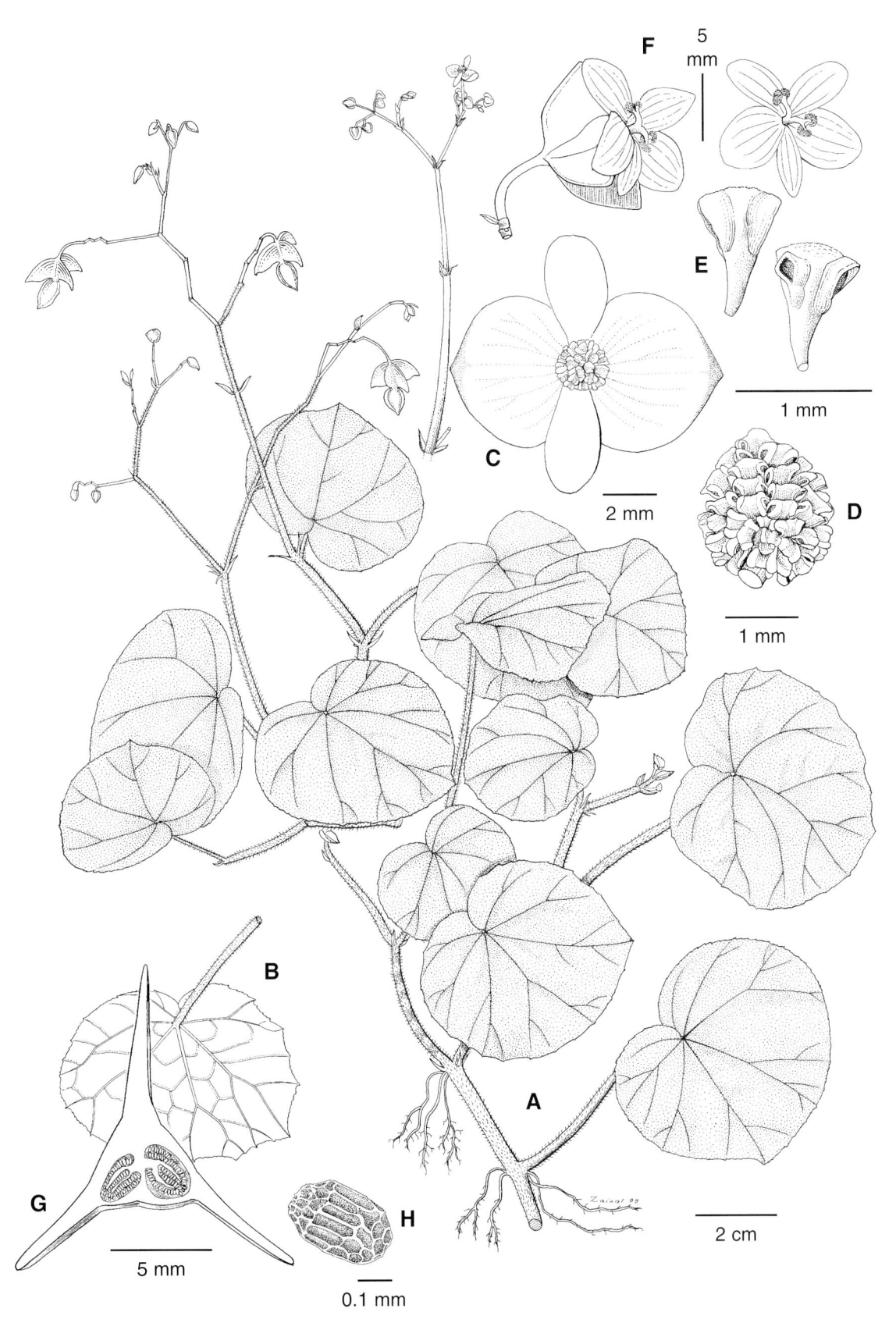

Begonia thaipingensis King **A.** The plant. **B.** Underside of the leaf. **C.** Male flower. **D.** Stamen cluster. **E.** Stamens. **F.** Female flower. **G.** T.S. ovary. **H.** Seed. (*RK 4705*)

margin not toothed, soon falling. **Male flowers** with a white or pale pink stalk, 5–7 mm long; tepals 4, white or pale pink with deeper red tip, without hairs, margin not toothed, tip rounded, outer two oval, 7–10 × 6–7 mm, inner two narrowly oval, 5–9 × 2–3 mm; stamens many, cluster globose, 1.5–2 mm across, stalk 1–2.5 mm long; filaments *c.* 0.3 mm long; anthers lemon yellow, narrowly obovate, *c.* 0.5 mm long, tip slightly notched, opening by slits. **Female flowers** with white or pale green stalk 5–6 mm long; ovary pale rosy red, 6–8 × 8–12 mm, wings 3, almost equal, locules 2, placentas 2 per locule; tepals 5, white or pale pink, without hairs, margin not toothed, tip rounded, outer broadly ovate, 5–8 mm × 4–5 mm, inner narrowly oval, 5–7 × 1.5–2 mm; styles 2, styles and stigmas yellow, *c.* 2.5 mm long, stigmas spiral. **Fruit** a splash cup pendent on a fine stalk 4–12 mm long; capsule 5–11 × 10–15 mm, hairless, locules 3, wings 3, thinly fibrous, unequal, larger wing 8–12 mm wide, tip curved and concave, smaller two curved 2–7 mm wide with pointed tip, splitting between the locules and the wings. **Seeds** barrel-shaped, *c.* 0.3 mm long, collar cells *c.* one third of the seed length.

DISTRIBUTION. Endemic in Peninsular Malaysia on the Main Range, from Maxwell Hill, Taiping, and Gunung Bubu in the north to Genting Peras on the Selangor/Negri Sembilan border in the south.

HABITAT. It grows in deep shade on earth banks and slopes in lowland and hill forest up to *c.* 1000 m altitude.

OTHER SPECIMENS. PAHANG—Genting Highlands, *Kiew RK 1547* (SING), *Stone 13769* (KLU), *Weber RK 1322* (KEP); PAHANG—Gunung Rajah, *Chua et al. FRI 40502* (KEP); PERAK—Gunung Bubu, *Chew CWL 1193* (KEP, SING), *Hou 611* (KEP); SELANGOR—Genting Pera, *Kiew RK 1613* (SING), Kanching VJR, *Kiew RK 1662* (SING), *Kiew RK 4705* (SING), *Md Nur SFN 34327* (SING); Kuala Kubu Baru-Gap Road, *Kiew RK 5230* (SING), *Ridley 8591* (SING).

NOTES. The Taiping Begonia shows a great range of leaf colour in part dependent on whether the underside is green, red or purple and in part because its velvety surface gives the leaf an iridescent sheen so that on turning the leaf, the colour changes from blue to green, blue to golden fawn or blue to black. The form with golden leaves was common in the area where the Awana Resort has been built at Genting Highlands, but it can no longer be found there.

It is unusual among Malaysian begonias in its slender, long creeping stem with widely spaced leaves. (Most begonias with horizontal stems are rhizomatous and cling to rocks and their leaves are close together and tufted.) The Taiping Begonia has leaves raised above the leaf litter on long stalks that are often longer than the leaf blades. Only *Begonia corneri* has a similar habit.

Young fruits of *Begonia thaipingensis*.

14. THE BERRY BEGONIA
Begonia longifolia Blume
(Latin, long-leaved)

Blume, Catalogus. (1823) 102; Tebbitt, Brittonia. 55 (2003) 25. **Type:** *Blume 740* Java, Salak (holo B). **Synonym:** *B. roxburghii sensu* Ridl., JFMSM 4 (1909) 20 *non* A.DC. *Begonia tricornis* Ridl., JSBRAS 75 (1917) 35, FMP 1 (1922) 854; Irmscher, MIABH 8 (1929) 109; Henderson, MWF Dicot (1959) 161 & Fig 153. **Type:** *Ridley 14123*, Telom (holo SING, iso K).

Stem erect, cane-like, brownish-red flecked with white, glossy, woody, without hairs, unbranched, nodes swollen, top arching with leaves held horizontally, sometimes falling and rooting, stout, up to 175 cm tall, flowering at *c.* 40 cm tall, *c.* 2 cm thick at base; without a tuber. Stipules pale green, hairless, narrowly triangular, 10–17 × 2–3 mm, margin not toothed, tip narrowing into a hair 2–3 mm long, soon falling. **Leaves** distant, 3–14 cm apart; stalk pale green, 5.5–10.5 cm long, grooved above; blade oblique, dull plain mid-green with a few sparse short hairs on the upper surface, thinly leathery in life, papery when dried, oblong-lanceolate, asymmetric, 10–29.5 × 4–13 cm, broad side 3.5–7.5 cm wide, basal lobe rounded 3–6 cm long, margin minutely toothed, tip elongate; venation pinnate, 5–6 pairs of veins along the midrib and another 2–3 veins in the basal lobe, branching towards the margin, impressed above, beneath prominent, the same colour as the blade, hairless. **Inflorescences** axillary, few-flowered, once branched, sometimes with two per axil, shorter than the leaves, reflexed, 9–13 mm long, elongating to 14–17 mm in fruit, stalk green, without hairs,

(Above). The Berry Begonia, *Begonia longifolia*. (Following page). The Berry Begonia grows in the understorey of mountain forests.

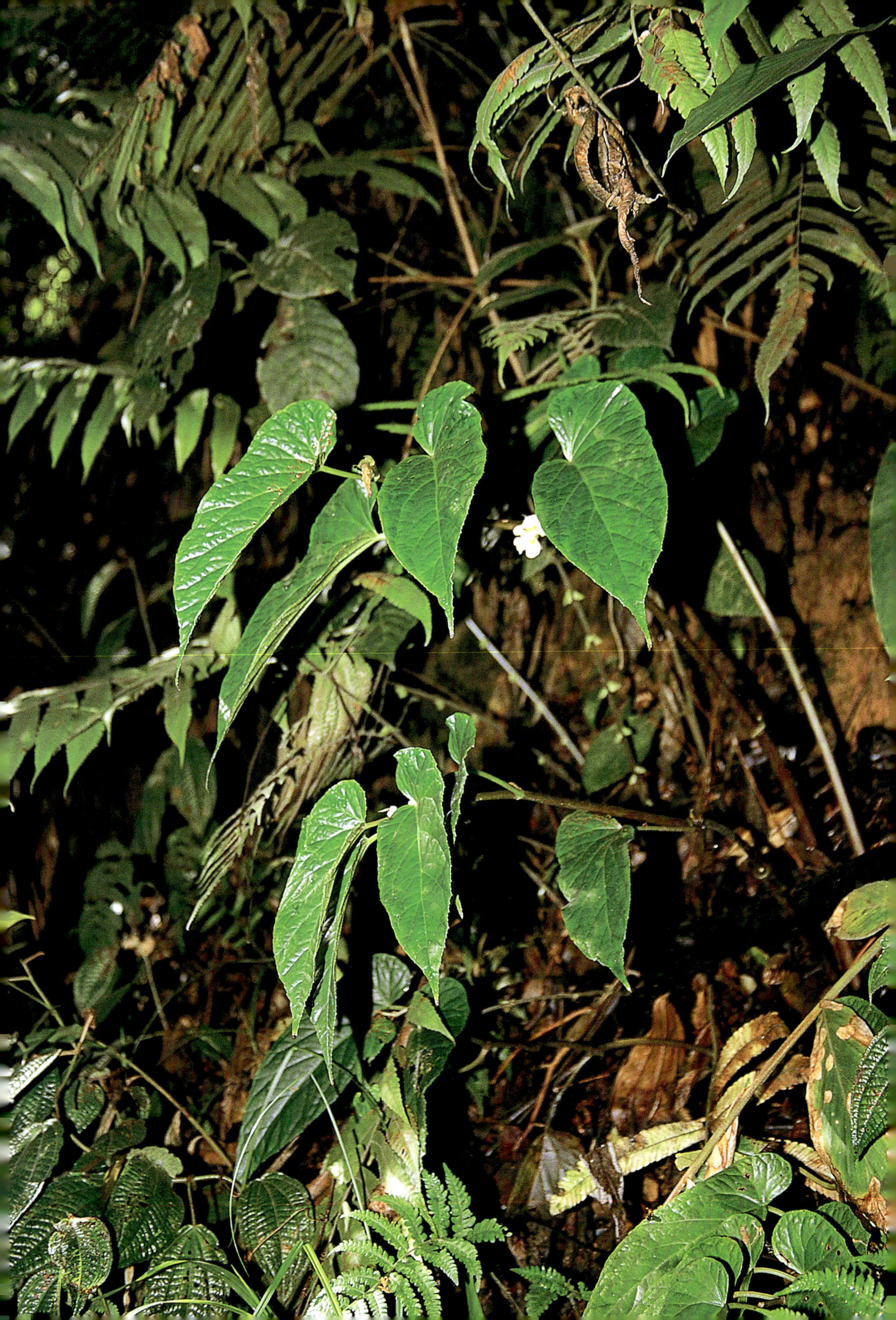

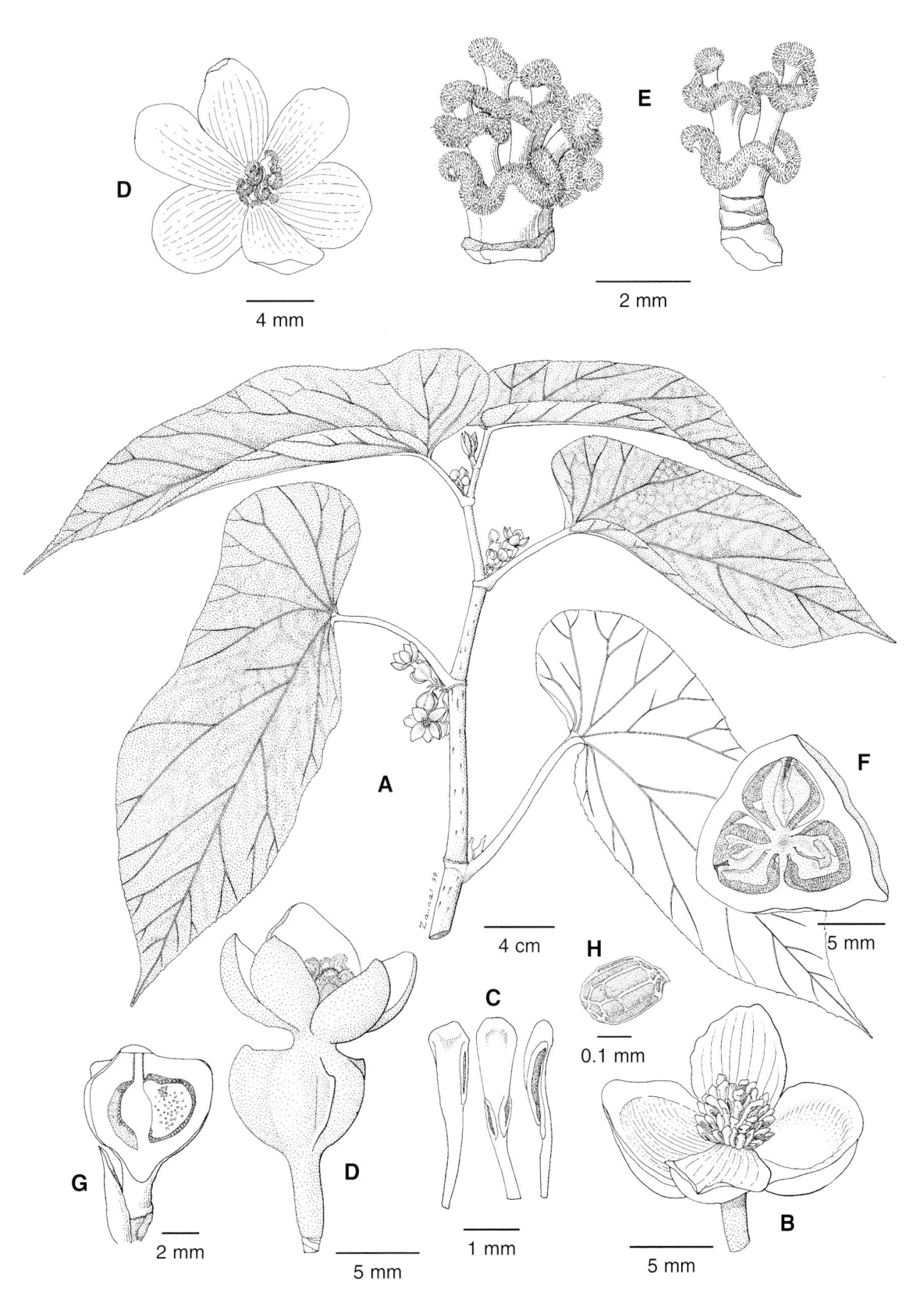

Begonia longifolia Blume. **A.** The plant. **B.** Male flower. **C.** Stamens. **D.** Female flower. **E.** Styles and stigmas. **F.** T.S. ovary. **G.** L.S. Ovary. **H.** Seed. (*RK 4733*)

BEGONIA LONGIFOLIA

5–10 mm long, male flowers 3, female flowers 4, male flowers open first. Bract pair pale or whitish green, narrowly triangular, 6–12 × 2–4 mm, margin not toothed, tip narrowing and ending in a hair, persistent. **Male flowers** with a pale green stalk, 6–25 mm long; tepals 4, pure waxy white, without hairs, rotund, margin not toothed, tip rounded, outer two *c.* 11 × 11 mm, inner two similar in shape, 10–11 × 7–9 mm; stamens many, cluster globose, *c.* 5 mm across, stalk 1.5 mm long; filaments *c.* 1 mm long; anthers pale yellow, narrowly obovate, *c.* 2 mm long, tip notched, opening by slits. **Female flowers** with a pale green stalk 5–7 mm long; ovary white becoming green, thick and fleshy, 3-angled and top-shaped, 8–11 × 10–13 mm, locules 3 (rarely 2), placentas 2 per locule; tepals (4–)6, pure white, hairless, broadly oval, margin not toothed, tip rounded, 11–12 × 8–9 mm; styles 3, styles and stigmas greenish yellow, 4–5 mm long, stigmas spiral. **Fruits** pendent on stiff fleshy stalk 7–10 mm long, berry green when ripe, fleshy, 14–17 × 12–20 mm, globose elongated at top into a fleshy beak *c.* 4–6 mm long, hairless, 3-lobed, each lobe with a fleshy ridge *c.* 1 mm thick with white or pale green warts, without wings, locules (2–)3, not splitting, stigmas persisting. **Seeds** barrel-shaped, 0.25–0.3 mm long, collar cells almost as long as or $^3/_4$ the seed length.

The Berry Begonia, *Begonia longifolia*. (Painting by Wendy Gibbs)

BEGONIA LONGIFOLIA

DISTRIBUTION. NE India, Bhutan, southern China, Taiwan, Myanmar, northern Thailand, N & C Vietnam, Peninsular Malaysia, Sumatra, Java, Bali and Sulawesi. In Peninsular Malaysia on the Main Range from southern Perak (Jor) to Selangor (Genting Bidai), and on Pulau Tioman, Pahang.

HABITAT. On shaded earth banks or slopes; on the Main Range in hill and lower montane forest at about 1000–1200 m altitude and on Pulau Tioman at *c*. 500 m.

OTHER SPECIMENS. PAHANG—Cameron Highlands *Ja'amat 27597* (KEP), *Kiew RK 172* (KEP), *RK 3244* (KEP), *RK 4733* (SING), *Md Nur SFN 32961* (KEP, SING), *Saw 29341* (KEP); Fraser's Hill *Burkill & Holttum 8669* (SING); Genting Highlands *Anthonysamy SA 738* (KEP), *Stone 15428* (KLU); Pulau Tioman *Kiew 5107* (SING), *Weber 890726-513* (SING) PERAK—Jor *Md Haniff SFN 14233* (SING); Sungai Groh *Ng FRI 1583* (KEP); Ulu Selim *Kiew 5116* (SING); SELANGOR—Genting Bidai *Ridley s.n.* (SING); Genting Simpah *Burkill 9989* (SING); Ulu Gombak *Anthonysamy SA 432* (KEP), *Kiew RK 1299* (SING), *RK 1552* (KEP).

NOTES. Tebbitt (2003) considers that the begonia Ridley called *Begonia tricornis* is the same species as the one Blume had described earlier from Java as *B. longifolia* and indeed I can find no character to separate them.

This is the only begonia in Peninsular Malaysia that has a berry for a fruit, all the rest have dry capsules that split to release the seeds. However, its seeds are identical to those from capsules, i.e., they have a sculptured seed coat, indicating that the fleshy berry type of fruit has evolved from dry capsules. Tebbitt (2003) attributes its wide distribution to the more efficient animal dispersal compared with wind dispersal of dust seeds from dry capsules. However, the dispersal agent of the Berry Begonia is not known.

Male flowers (above) and a female flower (below) of the Berry Begonia.

Tebbitt (2003) also suggested that the Berry Begonia had originated in the mountainous region stretching from the Himalayas to the north of Vietnam and spread south as far as Sulawesi. Many other mountain herbs, the temperate element of Malaysian flora, show this same distribution pattern, for example, species of *Anemone* (Ranunuculaeae), *Crawfurdia* (Gentianaceae), *Sarcopyramis* (Melastomataceae) and *Viola* (Violaceae). In spite of many similarities between the montane floras of Sumatra, Java and Peninsular Malaysia, this is the only Malaysian begonia that is distributed in mountains in Java and Sumatra.

Its distribution in the Peninsula, however, is rather peculiar as it is found in a restricted area on the Main Range and then at a distance on Pulau Tioman and there at a lower elevation. There is one other herb, *Loxonia hirsuta* Jack (Gesneriaceae) that shows this disjunct distribution, being found in the Peninsula only on Pulau Tioman whereas it is common in Sumatra and Java and rare in Borneo and the Anambas islands. The other begonia on Pulau Tioman, *B. rheifolia*, is found on the mainland but only on the East Coast.

The leaves are eaten by *orang asli* (aboriginal people) in Jor, Perak (*Md Haniff SFN 14233*).

15. HOLTTUM'S BEGONIA
Begonia holttumii Irmsch.
(R.E. Holttum, Assistant Director and Director of Botanic Gardens Singapore;
Professor Botany, University of Malaya, 1922–1955)

Irmscher, MIABH 8 (1929) 113; Henderson, MWF Dicot (1959) 167 & Fig 161. **Type:** *Curtis 1262*, October 1889, Batu Etam, Penang (lecto SING, here designated; isolecto KEP)). **Synonym:** *B. isoptera sensu* Ridl., FMP 1 (1922) 855 *pro pte. non* Dryand.

Stem green, brownish above node, minutely hairy when young, cane-like, erect, branched from basal rhizome, nodes swollen, stout and woody, up to *c.* 130 cm tall and 7–8 mm thick; without a tuber. Stipules minutely hairy, narrowly triangular, 10–20 × 4–5 mm, margin not toothed, tip elongate ending in a long hair, soon falling. **Leaves** distant up to (2.5–)6(–11) cm apart, held horizontally; stalk green, brownish towards base, minutely hairy, 4–12 cm long, grooved above; blade oblique, dull plain green on both surfaces, thin and soft in life, microscopically hairy above, papery when dried, broadly ovate, strongly asymmetric, (9–)18(–21) × (6–)9(–13) cm, broad side 4–9 cm wide, basal lobe broadly rounded, 1.5–5 cm long, margin toothed at the vein endings with minute teeth in between, tip narrowly elongated; venation palmate-pinnate, 4–6 pairs of veins with

Begonia holttumii is one of the wild cane-like begonias that grow in lowland forest.

BEGONIA HOLTTUMII

Begonia holttumii. (Left). Female flower. (Right). Fruits.

another 2 veins in basal lobe, branching *c.* 2/3 of way to margin, veins plane above, prominent and when young minutely hairy beneath. **Inflorescences** terminal and axillary, pale green, shorter than the leaves, with the main axis up to 12 cm with the stalk 1.5–5.5 cm long, with up to 3 pairs of female flowers at base and many fine branches with many male flowers above, female flowers open first. Bracts leafy, ovate, 2.5–11.5 × *c.* 1 mm, margin not toothed, soon falling. **Male flowers** with greenish white stalk 2.5–12 mm long; tepals 2, white or greenish white sometimes slightly pink outside, minutely hairy outside, broadly elliptic, 3–11 × 2.5–7 mm, margin not toothed, tip rounded; stamens many, cluster conical, *c.* 4 × 2.5 mm, not stalked; filaments *c.* 1 mm; anthers pale yellow, obovate, 0.7–1.2 mm, tip notched, opening by slits. **Female flowers** with a stalk *c.* 3 mm long; ovary pale green, oblong, 5–17 mm long, minutely hairy, wings 3, equal, *c.* 4 mm wide, locules 3, placentas 2 per locule; tepals 5, pure white, minutely hairy outside, margin not toothed, tip rounded, outermost obovate, 4–15 × 3–7 mm, innermost narrowly elliptic, 3–5 × 2–3 mm; styles 3, styles and stigma pale yellow, 4–5 mm long, stigmas spiral. **Fruit** with a stiff stalk 4–8 mm long, bent downwards with the fruit held below the leaves; capsule broadly oblong, 15–25 × 12–20 mm, hairless, locules 3, wings 3, equal, 2–3 mm wide, oblong, tip rounded, thinly fibrous, splitting between the locules and wings. **Seeds** barrel-shaped, *c.* 0.25 mm long, collar cells *c.* half the length of seed.

DISTRIBUTION. Endemic in Peninsular Malaysia—most common in the west (as far north as Penang) and south, also in Trengganu, but not yet recorded from upper Perak and Kedah, Kelantan and west Pahang.

HABITAT. In deep shade in lowland rain forest often by streams, not found above about 350 m altitude.

OTHER SPECIMENS. JOHORE—Gunung Panti, *Corner SFN 30750* (SING), *Md Nr SFN 20047* (KLU); Gunung Pulai, *Sinclair SFN 39555* (SING), *Weber 840702 –2/4* (KEP); Ulu Maduk, *Holttum SFN 10635* (SING); MALACCA—Sungai Udang, *Goodenough 1703* (SING); NEGRI SEMBILAN—Gunung Angsi, *Holttum SFN 9892* (SING), *Md Nur SFN 11592* (SING); *Jelebu*

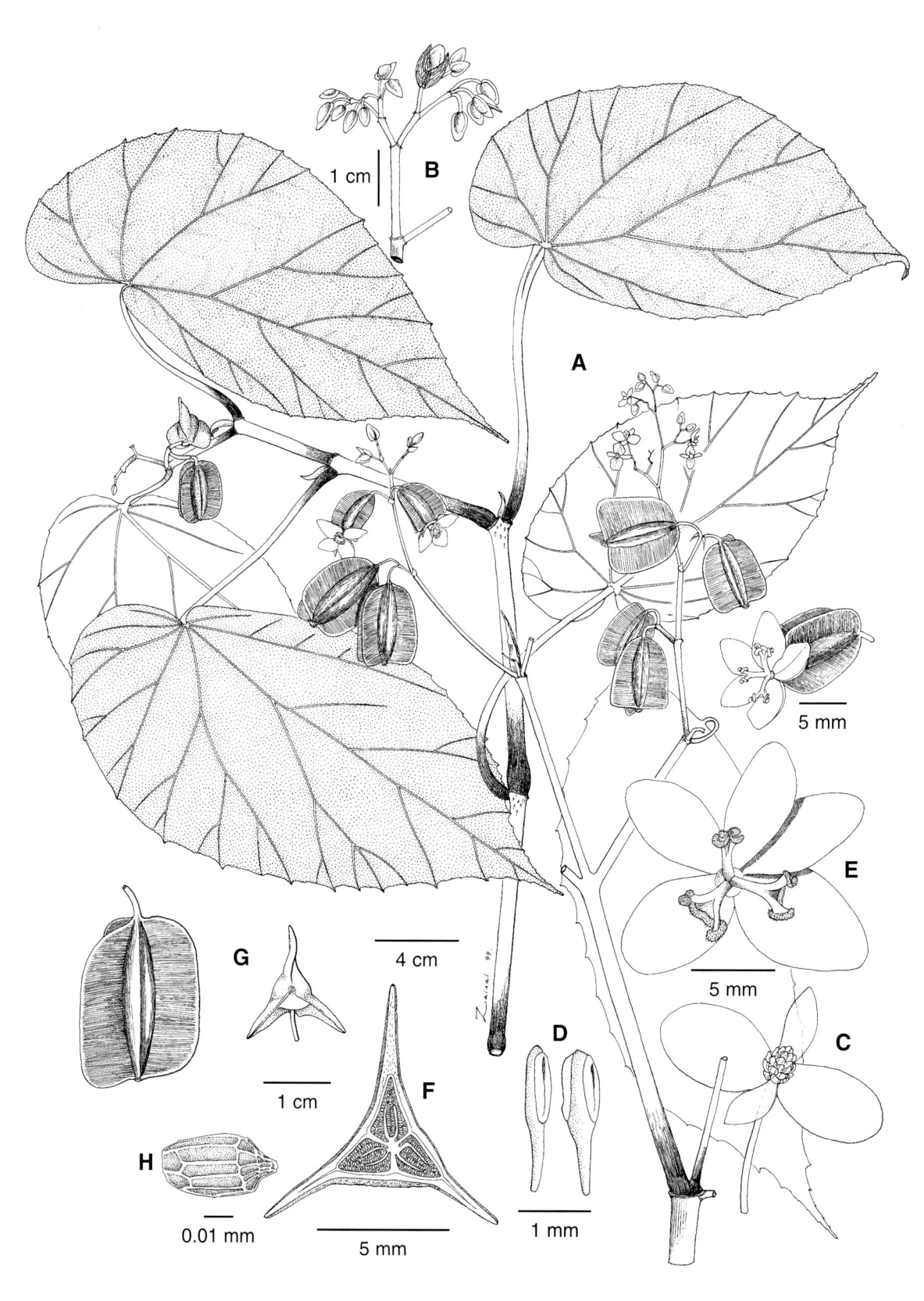

Begonia holttumii Irmsch. **A.** The plant. **B.** Inflorescence branch with male buds. **C.** Male flower. **D.** stamens. **E.** Female flower. **F.** T.S. ovary. **G.** Fruit. **H.** Seed. (*RK 5085*)

Everett KEP 104918 (KEP, SING); Jeram Toi, *Anthonysamy SA 688* (KEP), Pasoh, *Ja'amat 46084* (KEP), Perhentian Tinggi, *Ridley 10026* (SING), Tampin, *Md Nur SFN 2505* (SING), PAHANG—Bentong, *Addison SFN 37213* (SING), *Mohd Shah 197* (SING); PENANG—*Curtis 1738* (SING), *Curtis s.n.* (SING), *Kiew RK 1672* (KEP), *Ridley 7094* (SING) PERAK—*Wray s.n.* (KLU), Bukit Kepayung, *Ridley s.n.* (SING); Jor, *Burkill SFN 14207* (SING); Kledang Saiong FR, *Kiew RK 2564* (KEP), Telok Anson, *Md Haniff 14176* (SING); SELANGOR—Bukit Lagong FR, *Abd Hamid 37587* (KEP), Genting Bidai, *Ridley s.n.* (SING), Klang Gates Ridge, *Henderson 7291* (KLU); Semenyih, *Hume 7932* (SING); Sungai Lalang FR, *Symington 22604* (KEP), Sungai Pisang, *Kiew RK 3249* (KEP), Ulu Gombak, *Jumaat 42997* (KEP), *Ridley 13152* (KLU), *Saw FRI 37695* (KEP); TRENGGANU—Besut, *Chua FRI 40563* (KEP), Bukit Kajang, *Corner SFN 30206* (SING), *Kiew RK 2681* (KEP); Jeram Tanduk, *Kiew RK 5085* (SING).

NOTES. This is a relatively widespread begonia and like the Common Malayan Begonia, *Begonia wrayi*, is a cane-like begonia found in lowland forest. *Begonia holttumii* has a more southerly distribution, while *B. wrayi* is more common in the north. Their distribution overlaps in central Pahang, southern Selangor and northern Negri Sembilan. These two species are easily told apart by the length of the leaf stalk (long in *B. holttumii*, very short in *B. wrayi*), the very large basal leaf lobe of *B. holttumii* leaf blade and its inflorescence that has pairs of female flowers. The inflorescence of *B. holttumii* is unique among the Peninsular species in having a long unbranched axis.

Holttum's Begonia, *Begonia holttumii*.

16. THE BERUMBAN BEGONIA
Begonia isopteroidea King
(Latin, *-oides*=resembling, similar to *Begonia isoptera*, a Javanese species)

King, JASB 71 (1902) 59; Ridley, FMP 1 (1922) 856; Irmscher, MIABH 8 (1929) 112. **Type:** *Wray 1548*, October 1884, Gunung Berumban [Brumber], Pahang (lecto K, here designated).

Stem cane-like, erect, *c.* 1 m tall, woody, ribbed when dry, little branched, nodes swollen, without hairs; without a tuber. Stipules without hairs, narrowly oval with a distinct midrib, 15–20 × 5–6 mm, margin not toothed, tip pointed, ending in a hair, persistent. **Leaves** distant and up to 3–3.75 cm apart; stalk crimson, hairless, 7–10 cm long, grooved above; blade very oblique, plain green, thin when dried, elliptic-ovate, strongly asymmetric, 8–11 × 3.5–5 cm, broad side 2.25–3 cm wide, basal lobe well-developed and rounded, 2.5–3 cm long, margin bluntly toothed, tip narrowly elongate; venation palmate-pinnate, 4 pairs of veins with another 3 veins in basal lobe, branching *c.* halfway to margin, in dried leaf veins plane above and beneath, without hairs. **Inflorescences** axillary, without hairs, erect, few-flowered, shorter than the leaves, *c.* 1.5 cm long lengthening to 2.25 cm in fruit, male flowers *c.* 2, female flower 1, male flowers open first. Bract pair narrowly oval, *c.* 18 × 4 mm, tip pointed, margin not toothed, soon falling. **Male flowers** with stalk *c.* 5 mm long; tepals 4, very pale pink, hairless, margin not toothed, tip rounded, outer two broadly oval 15–20 × 10–12 mm, inner two similar in shape but smaller *c.* 12 mm long; stamens many, cluster globose, *c.* 6 × 5 mm, stalk *c.* 1 mm; filaments *c.* 2 mm long; anthers narrowly oblong, *c.* 1 mm long, tip notched. **Female flowers** with stalk 2 mm long; ovary *c.* 2 mm long, wings 3, subequal; tepals 4, very pale pink, hairless, rotund, margin not toothed, tip rounded, 'nearly as large as male flowers' (King); styles and stigma *c.* 5 mm long, persistent in fruit. **Fruit** pendent on a fine stalk *c.* 30 mm long, capsule *c.* 20–28 × 20–22 mm, hairless, locules 3, wings subequal, larger *c.* 9 mm wide, smaller *c.* 7 mm wide, wing narrowly oblong, papery, splitting between the locules and wings. **Seed** not known.

DISTRIBUTION. Endemic in Peninsular Malaysia: Pahang—Gunung Berumban.

HABITAT. In montane forest at about 1350 m altitude.

NOTES. This species is known only from the collection made by Wray in 1884 when he attempted to reach the summit of Gunung Berumban in what is now called Cameron Highlands. He approached Gunung Berumban from the southeast, a route not taken by others since, which may explain why this begonia has not been recollected.

Like *Begonia holttumii* and *B. wrayi* that also grow at Cameron Highlands, it has cane-like stems and dry capsules with three wings of almost the same size. However, it is very different from these in its ribbed stem, persistent stipules, very oblique leaves (the leaf bade is almost at right angles to the leaf stalk), shorter few-flowered inflorescences and much larger flowers. *Begonia venusta*, which also grows on Gunung Berumban, has equally large flowers and few-flowered inflorescences, but it is a semi-erect begonia with broad, ovate leaves and its inflorescences are long-stalked, and it has a fruit with one wing that is much larger and very fibrous. There is therefore no doubt that this is a distinct species.

King (1902), followed by Ridley (1922), stated that the fruits had four styles and locules but this was corrected by Irmscher (1929) to three styles and locules.

Begonia isopteroidea King. The plant with fruits. (*Wray 1548*)

17. THE COMMON MALAYAN BEGONIA *RIANG BATU*
Begonia wrayi Hemsl.
(L. Wray Jr. Curator of Perak State Museum, 1883–1908)

Hemsley, J. Bot. 25 (1887) 203; Kiew, Mal. Nat. J. 47 (1994) 311. **Type:** *Wray 55*, Ulu Kenering, Perak (holo K, iso SING). **Synonym:** *B. isoptera sensu* King, JASB 71 (1902) 58 *pro pte. non* Dryand.; Ridl., FMP 1 (1902) 855 *pro pte.*; *B. pseudoisoptera* Irmscher, MIABH 8 (1929) 112; Henderson, MWF Dicot (1959) 162 & Fig 154. **Type:** *Ridley 2246*, 1891, Tahan River, Pahang (lecto SING, here designated; iso BM).

Stem translucent, red or greenish red, darker above nodes, cane-like, erect, at first unbranched and the top part arching over, then branching to form a bushy canopy, young stem with short brown hairs, nodes slightly swollen, rooting from lower nodes or nodes of fallen stems, up to 130 cm tall, flowering at 50 cm tall, woody, 8–10 mm thick at base; without a tuber. Stipules pale green, hairless, narrowly triangular, 13–20 × 1.5–4.5 mm, narrowed to a long hair, margin not toothed, soon falling. **Leaves** distant and up to 6 cm apart; stalk translucent, greenish red with short brown hairs, 8–12(–20) mm long, grooved above; blade not oblique, yellow green when young, becoming plain, dark green and velvety above, sometimes coppery red or reddish purple beneath, hairless above, thin in life, papery when dried, narrowly obovate-oblong, slightly asymmetric, 8–18 × 3–7.5 cm, broad side 2.25–4.5 cm wide, base rounded, basal lobe 0.5–3 cm long, margin crimson, scalloped and toothed at the vein endings and minutely toothed between, tip elongated, 1–3 cm

The Common Malayan Begonia, *Begonia wrayi*.

BEGONIA WRAYI

Common Malayan Begonia, *Begonia wrayi*. (Painting by Wendy Gibbs)

long; venation palmate-pinnate with 1 pair at the base, 3–4 pairs along the midrib branching *c.* a third of the way to the margin and 1 vein in the basal lobe, veins impressed above and the blade corrugated, beneath prominent, deep purplish brown, the colour of the veins showing through the upper surface, with short brown hairs when young. **Inflorescences** terminal, shorter than the leaves, pale yellow green, hairless, 4.5–9 cm long with stalk *c.* 2.5 cm long with 1–3 female flowers at base and above with many male flowers on thin branches *c.* 5–25 mm long, female flowers open first. Bracts foliaceous, 5–6.2 × 3–3.5 mm, margin not toothed, soon falling. **Male flowers** with stalk 5–9.5 mm long; tepals 2, white or greenish white, reddish brown at the tip, without hairs, broadly oval, 3.5–6.5 × 1.5–5 mm, margin not toothed, tip rounded; stamens many, cluster ovoid, *c.* 3 × 2 mm, not stalked; filaments *c.* 0.75 mm long; anthers yellow, narrowly oval, *c.* 1–1.2 mm long, tip notched, opening by slits. **Female flower** with a stalk 7.5 mm long; ovary pale green, broadly oblong, 12–15 × 7–13 mm long, wings 3, equal *c.* 1.2 mm wide, locules 3, placentas 2 per locule; tepals (4–)5(–6), very pale green or greenish pink, minutely hairy outside, elliptic, margin with 3–7 pale purple jagged teeth toward the tip, tip pointed, outermost (6–)15–16 × (2.5–)7–9 mm, innermost obovate, slightly smaller, up to 15 × 7 mm; styles 3, styles and stigmas pale yellow, 1.5–2 mm long, stigmas spiral. **Fruit** with a stiff stalk 8–15 mm long, bent back so the fruit occupies a horizontal position under the leaves, capsule 20–23 × 14–20 mm, hairless, locules 2, wings 3, equal, 2–7 mm wide, triangular with rounded tip, thinly fibrous, splitting

Begonia wrayi. (Top right). Male flowers. (Top right). Fruits. (Bottom). Female flowers. Note the toothed tepals of the female flowers.

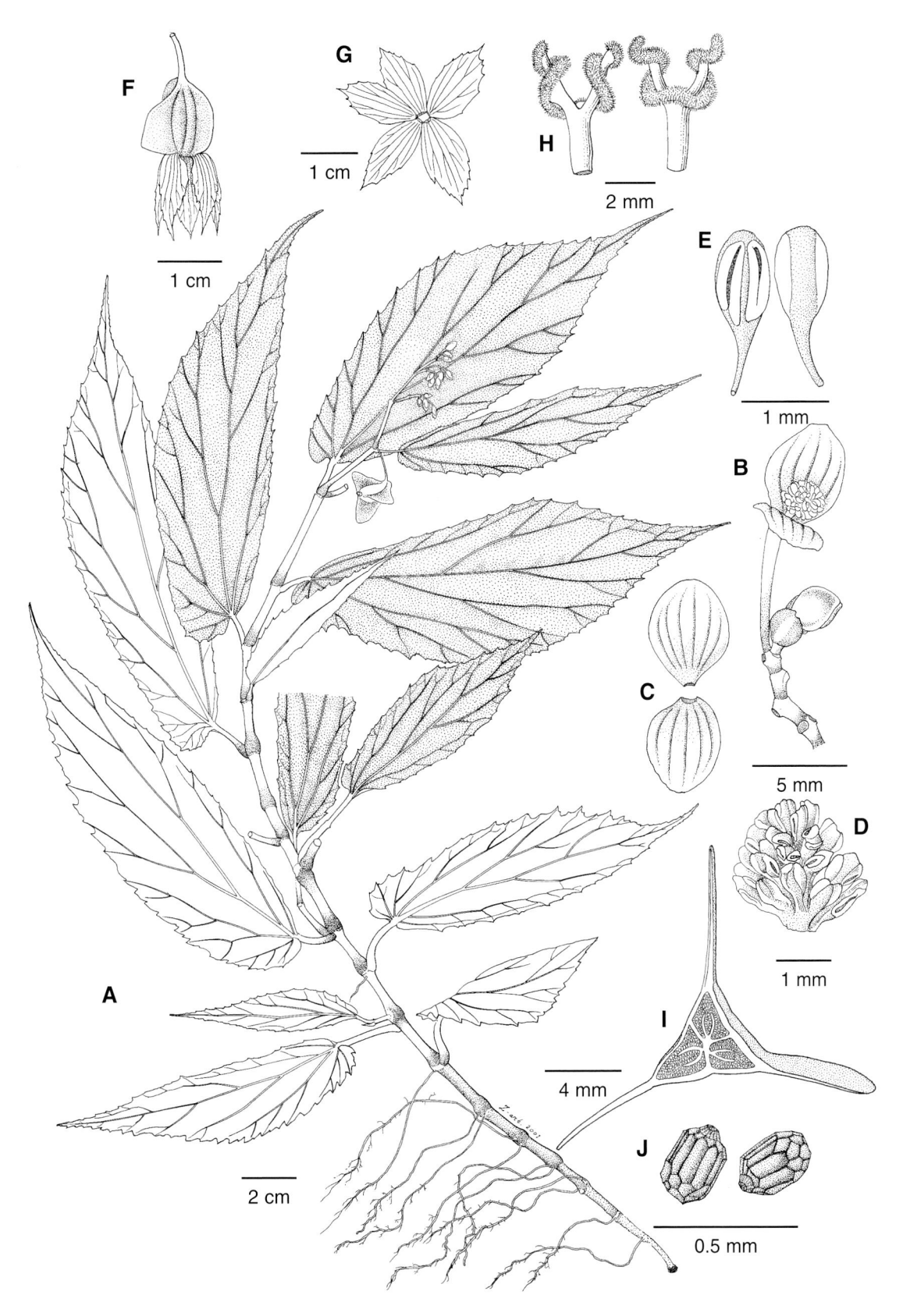

Begonia wrayi Hemsl. **A.** The plant. **B.** Male flower. **C.** Male tepals. **D.** Stamen cluster. **E.** Stamens. **F.** Female flower. **G.** Toothed female tepals. **H.** Styles and stigmas. **I.** T.S. ovary. **J.** Seed. (*RK 5169*)

between locule and wing. **Seeds** barrel-shaped, 0.25–0.3 mm long, collar cells up to ³/₄ of length of seed.

DISTRIBUTION. Peninsular Thailand (in the extreme south east) and Peninsular Malaysia—more common in the north, it reaches its southernmost limit at Pasoh, Negri Sembilan. It has yet to be collected from Penang, Trengganu, Malacca or Johore.

HABITAT. Deep shade in lowland forest up to about 1200 m altitude, frequently by streams.

OTHER SPECIMENS. KEDAH—Baling, *Kiah SFN 35387* (SING), *Nauen SFN 38041* (SING); Gunung Bintang, *Md Haniff 21015* (KLU); Gunung Inas, *Kiew RK 5169* (SING); Gunung Lang, *Kiah SFN 35046* (SING); Sik, *Dolman 21527* (KEP, SING); KELANTAN—Bihai, *Kiew RK 5243* (SING); Gua Gagak, *Kiew RK 3071* (KEP); Kuala Pertang, *Md Haniff 10374* (SING); Merapoh, *Ong 552* (KLU); Sungai Kerteh, *Md Nur & Foxworthy 12116* (KLU); NEGRI SEMBILAN—Bahau, *Carrick 706* (KLU); Jelebu, *Everett 104940* (KEP, SING); Pasoh 4 Felda Scheme, *Kiew & Davison s.n.* 16 Nov 1996 (SING); PAHANG—Batu Kanok, *Kiew RK 1506* (SING); Bukit Sagu, *Henderson 25089* (SING); Fraser's Hill, *Holttum 21646* (KLU), *Md Nur SFN 11103* (KLU); Gunung Senyum, *Henderson SFN 22295* (KLU); Gua Kechapi, *Md Nur & Foxworthy 11911* (SING); Gua Peningat, *Loh FRI 17252* (KEP), *17253* (SING)); Gunung Tahan, *Kiew RK 2485* (KEP), *Ridley 2240* (SING), *Seimund 242* (SING), *Seimund 891* (KLU); Kuantan, *Walker 14155* (KEP, SING); Sungai Lipis, *Brown 29363* (KEP), *Strugnell 20372* (KEP); Sungai Tekai, *Henderson SFN 24850* (SING); Tembeling, *Holttum s.n.* (SING); PERAK—*Wray 3367* (KLU); Grik, *Burkill & Haniff SFN 13636* (SING), *Corner SFN 3162* (SING), *Kiew s.n.* 16 Feb 1992 (KEP), *Mat Asri FRI 26870* (KEP), *Rahim Ismail KEP 95025* (KEP, SING), *Ridley 14801* (SING), Klian Intan, *Spare SFN 36344* (SING); Piah FR, *Ja'amat 39302* (KEP); Slim Hills, *Whitmore FRI 0827* (KEP, SING); Sungkai, *Sow FRI 46171* (KEP); Temenggor, *Latiff 3979* (UKMB), *Ridley 14607a* (SING), *Saw FRI 39940* (KEP); SELANGOR—Chadangan, *Chelliah KEP 98214* (KEP, SING); Genting Bidai, *Ridley 7206* (SING), *15579* (SING); Genting Peras, *Ridley s.n.* (SING); Genting Simpah, *Henderson SFN 8758* (KLU); *Stone 6475* (KLU); *Symington 29817* (KEP); Klang Gates, *Ridley 13430* (SING), *Ridley 13431* (SING); Kuala Kubu Baru-Gap, *Addison SFN 37155* (SING), *Kiew RK 1101* (SING), *RK 1250* (KEP), *RK 1251* (SING), *RK 3799* (KEP), *Ridley 8589* (SING); Semangok, *Ridley 12997* (SING); Ulu Gombak, *Kiew RK 4697* (SING), *Shimuzu et al. 14052* (SING); Ulu Langat, *Phytochemistry Survey KL 580* (KEP); Ulu Selangor, *Goodenough s.n.* (SING).

NOTES. The Common Malayan Begonia is a handsome plant with dark jade green, velvety leaves with a crimson margin that can be coppery red beneath and, when well-grown, has a spreading crown. It is unique among the Peninsular begonias in the tepals in the female flower being toothed.

Apart from the Sparkling Begonia (*Begonia sinuata*), it is the commonest and most widespread begonia in the Peninsula, although it tends to grow as scattered individual plants rather than in populations. It is one of the two cane-like begonias of lowland rain forest. The other is *Begonia holttumii* that has a broader leaf with a long leaf stalk.

It was once reported to be applied externally for enlarged spleen (*Burkill & Haniff SFN 13636*).

Foliage of *Begonia wrayi*.

18. JIEW-HOE'S BEGONIA
Begonia jiewhoei Kiew, sp. nov.
(Tan Jiew-Hoe, patron of plant taxonomy and botanical exploration)

A *Begonia baturongensi* Kiew inflorescentia feminea pedunculata et fructibus brevioribus alias inaequales ferentibus differt. **Typus:** *R. Kiew RK 4916*, 14 March 2000, Gua Musang, Kelantan (holo SING, iso KEP, K)

Stem cane-like, at first erect and up to *c.* 25 cm tall then becoming pendent and up to *c.* 80 cm long and 4–5 mm thick, glossy, deep red becoming brown and woody, little branched, nodes swollen, hairless or sometimes with a fringe of hairs below the stipule; without a tuber. Stipules pale green, ovate, up to 2 × 1 cm, margin not toothed, studded with glandular hairs, tip pointed ending in a long hair, persistent. **Leaves** distant, up to 3–6.5 cm apart, on erect stems held upwards, on pendent stems hanging downwards; stalk glossy, deep purple-red, on erect stems 6–7.5 cm long, on pendent stems 1.5–3 cm long, young leaves with hairs *c.* 2 mm long, grooved above; blade very oblique, velvety, deep malachite green with large grey-silver spots between the veins, beneath usually reddish, sometimes greenish red, succulent in life, thinly leathery when dried, broadly ovate, strongly asymmetric, base unequally heart-shaped, margin minutely toothed, each tooth tipped by

The Jiew-Hoe's Begonia, *Begonia jiewhoei*.

TAN JIEW HOE

BEGONIA JIEWHOEI

a hair up to 2 mm long, tip acute, on erect stems 7–8 × 6–7 cm, broad side 4–4.5 cm wide, basal lobe *c.* 3.5 cm long, on pendent stems decreasing in size toward the tip, 3–6 × 2.5–3.75 cm, broad side 1–2.5 cm wide, basal lobe 1–1.5 cm long, venation palmate-pinnate, 1 pair of veins at the base, 2 pairs along the midrib, branching more than halfway to margin, and 2 veins in the basal lobe, veins slightly impressed above, beneath slightly prominent, reddish and hairy. **Inflorescences** pale red to magenta, hairless, erect, shorter than the leaves; female inflorescences from the lower leaf axils, *c.* 1.8 cm long with a single female flower; mixed inflorescences terminal and in the upper axils, racemose, up to 3 cm long, stalk 1.5–1.7 cm long with one branch, male flowers many, female flowers 1–4, female flowers open first. Bracts pale green, margin with tiny teeth each tipped by a short glandular hair, persistent, broadly ovate, 5–6 × 5–7 mm, upper bracts broadly oval 4.5–5 × 3.5–4 mm. **Male flowers** with a stalk 4–7 mm long; tepals 4, white tinged red, hairless, margin not toothed, tip rounded to slightly acute, outer two broadly ovate, 6–7 × 5–6 mm, inner two narrowly obovate, 4–7 × 1.5–2 mm; stamens many, cluster globose, 2–3 × 3–3.5 mm, stalk 1–1.5 mm; filaments *c.* 0.75 mm long; anthers lemon yellow, narrowly obovate, *c.* 0.5 mm long, tip notched, opening by slits. **Female flowers** with a stalk 7–10 mm long; ovary light green, 6–10 × 10–21 mm, wings 3, equal, locules 3, placentas 2 per locule; tepals (4–)5, rosy red outside, pinkish white inside, margin not toothed with shortly stalked glandular hairs on upper half, tip rounded, outermost broadly oval 5–10 × 4–6 mm, innermost paler pink, oval, 4–7 × 2–2.5 mm; styles 3, free to base, styles and stigmas yellow, 2.5–3 mm long, stigmas spiral. **Fruits** dangling on fine and hair-like stalks 11–14 mm long; capsule 10–15 × 17–30 mm, hairless, locules 3, wings pointed, thinly fibrous to papery, unequal (sometimes equal), larger 8–12 mm wide, smaller two 5–12 mm wide, splitting between locules and wings, styles persistent. **Seeds** barrel-shaped, *c.* 0.3 mm long, collar cells $1/2$–$2/3$ of the seed length.

Leaves of Jiew-Hoe's Begonia: (Left). Upper surface. (Right). Lower surface. (Opposite). Jiew-Hoe's Begonia grows on vertical limestone cliffs.

BEGONIA JIEWHOEI

Begonia jiewhoei. (Above left). Male buds. (Above right). Leaf. (Below). Female flowers.

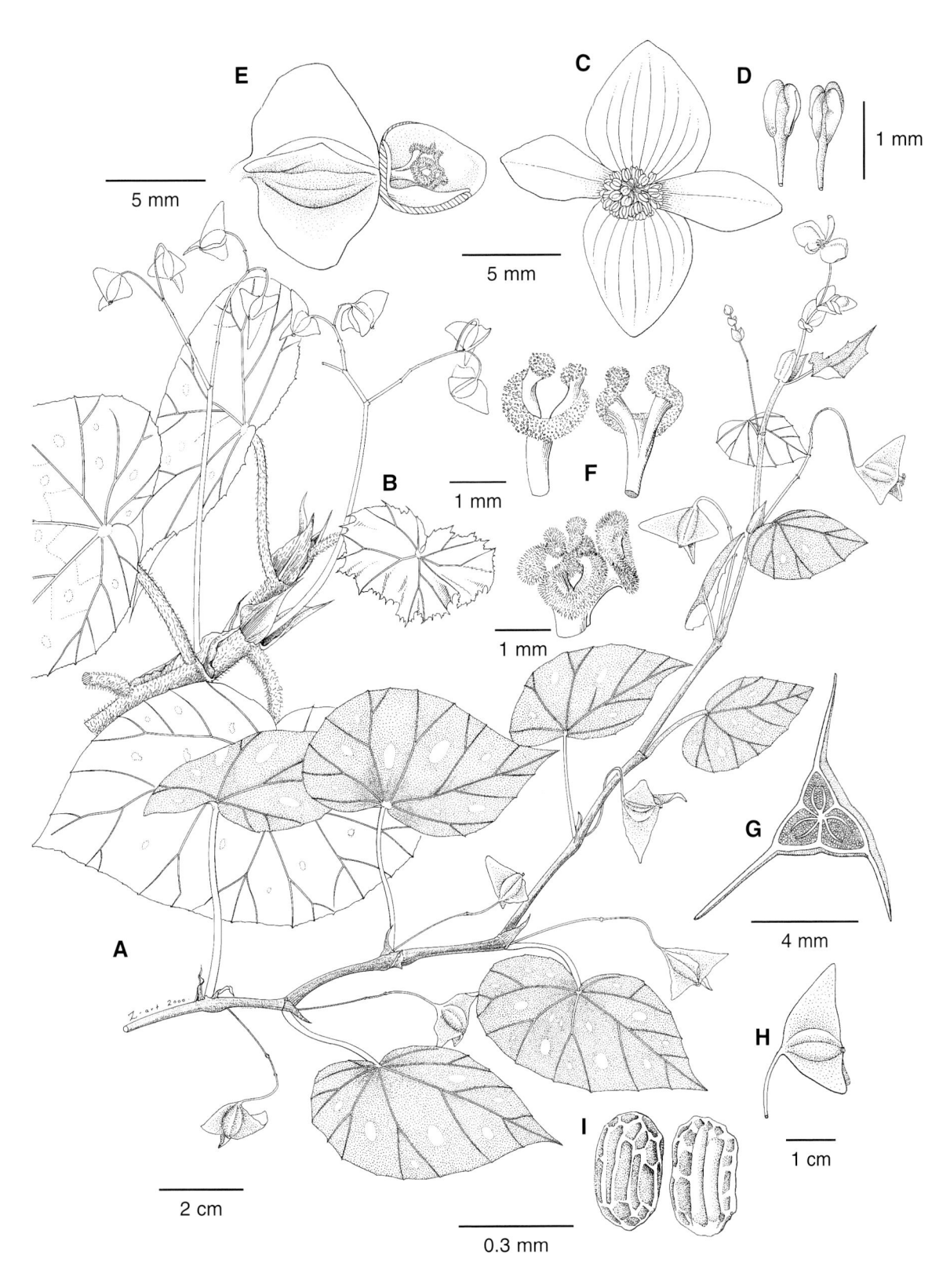

Begonia jiewhoei Kiew **A.** The plant. **B.** Bract. **C.** Male flower. **D.** Stamens. **E.** Female flower. **F.** Styles and stigmas. **G.** T.S. ovary. **H.** Fruit. **I.** Seed. (*RK 4916*)

BEGONIA JIEWHOEI

DISTRIBUTION. Endemic in Peninsular Malaysia—KELANTAN: Bertam Road north of Gua Musang.

HABITAT. In shade, restricted to limestone, in rock fissures or on limestone-derived soil at the base of cliffs.

OTHER SPECIMEN. KELANTAN—Bertam *UNESCO Limestone Expedition* (SING).

NOTES. Jiew-Hoe's Begonia is one of the most beautiful begonias in Peninsular Malaysia with its velvety, malachite-green, silver-spotted leaves and glossy red stems. In the Peninsula, it is the only cane-like begonia that grows on limestone; all other limestone begonias are rhizomatous. Among the cane-like begonias in the Peninsula, it is the only one that becomes pendent with age when its long stems hang down the cliff face. This habit suggests it would grow well in hanging baskets.

It belongs to Sect. *Petermannia* on account of its erect (not rhizomatous) stems, and fruits with three locules, each with two placentas. It is not, however, at all similar to other Peninsular begonias in this section (*B. barbellata*, *B. holttumii* and *B. wrayi*), none of which grows on limestone. Instead, it is closer to begonias that grow on limestone in Sabah, and in particular to *B. baturongensis* Kiew (Kiew, 2001). It resembles *B. baturongensis* in habit, leaf shape and size, in its male flowers having four tepals, and the fruit, which is broader than long and dangles on a fine hair-like stalk. It is distinct from *B. baturongensis* in the female flowers, which are borne on an inflorescence stalk *c.* 18 mm long (female flowers of *B. baturongensis* develop directly from the leaf axil), and the fruits, which are only 10–15 mm long and have unequal wings (as compared with 16–24 mm long with equal wings in *B. baturongensis*).

Like the Bristly Begonia, *B. barbellata*, this species belongs to the Bornean element of the Peninsula's flora and like the Bristly Begonia it is not found west of the Main Range. It is apparently very rare being known from only two localities in the same area.

Foliage of Jiew-Hoe's Begonia, *Begonia jiewhoei*.

19. THE BRISTLY BEGONIA
Begonia barbellata Ridl.
(Latin, *barbellatus*=with short stiff hairs)

Ridley, JFMSM 10 (1920) 135, FMP 1 (1922) 856; Irmscher, MIABH 8 (1929) 116. **Type:** *Ridley s.n.*, February 1917, Chaning Woods, Kelantan (holo K).

Stem reddish brown when young, woody, with densely bristly, hairs reddish brown, nodes thickened, erect, unbranched or once branched, forming clumps by falling and rooting then sending up several shoots, 13–30 cm tall, 4–7 mm thick; without a tuber. Stipules dark reddish brown, densely hairy, narrowly triangular, 5–10 × 1.5–2 mm, margin not toothed, tip ending in a hair, soon falling. **Leaves** distant, 2–6 cm apart, held almost horizontally; stalk dark brown, densely bristly with hooked hairs, 0.75–1.75 cm long, round in cross-section; blade not oblique, plain dull pale green or dark green above and ruby red beneath, above with scattered long hooked red bristly hairs between veins, thin, papery when dried, narrowly obovate, slightly asymmetric and curved, 12–15 × 4.25–6.75 cm, broad side 2.5–4 cm wide, base narrowed, basal lobes hardly developed, 2.5–5 mm long, margin toothed and hairy, tip elongate; venation pinnate, 8 pairs of veins branching close to margin, veins deeply impressed above, beneath prominent, deep ruby red

The Bristly Begonia, *Begonia barbellata*, grows on mud or sandy banks beside slow flowing streams.

BEGONIA BARBELLATA

or covered with brown hairs. **Inflorescences** axillary, flowers clustered with 2 female flowers on stalks 10–20 mm long and 2–3 male flowers on stalks 9–13 mm long, cluster shorter than the leaves, densely hairy, hairs glandular, female flowers open first. Bract pair whitish becoming green, narrowly linear, 12–15 × 1.5–3 mm, densely hairy, hairs *c.* 0.75 mm long, margin not toothed, persistent. **Male flowers** on stalks 5–13 mm long; tepals 4, pure white or white tinged with red, margin not toothed, outer two with dense red hairs outside, rotund, 9–15 × 9–12 mm, tip rounded, inner two narrowly oval, 4–8 × 1.5–2 mm, without hairs, tip slightly pointed; stamens many, cluster globose, 3 mm across, not stalked; filaments 0.5–1 mm; anthers yellow, narrowly oblong, *c.* 1 mm long, tip notched, opening by slits. **Female flowers** with reddish-green stalk 2–8 mm long; ovary white becoming green with hairs on the locules and along the wing margin, oblong, 8–16 × 6–9 mm, wings 3, equal, locules 3, placentas 2 per locule; tepals 5, whitish pale green with a reddish margin, outermost with red hairs outside, broadly ovate, margin entire, tip rounded, 7–10 × 6–8 mm, inner similar but almost without hairs and much smaller 3.5–6 × *c.* 3 mm; styles 3, styles and stigmas yellow, 2–3 mm long, stigmas U-shaped. **Fruits** with stiff, thin stalks 4–9 mm long; capsule held above the leaf, 12–18 × 7–12 mm, sparsely hairy, locules 3, wings 3 equal, thinly fibrous, 2–4 mm wide, splitting between the locules and wings. **Seeds** barrel-shaped, *c.* 0.3 mm long, collar cells *c.* ²/₃ length of the seed length.

Begonia barbellata with a young female flower with an oblong ovary (above) and a male flower (below).

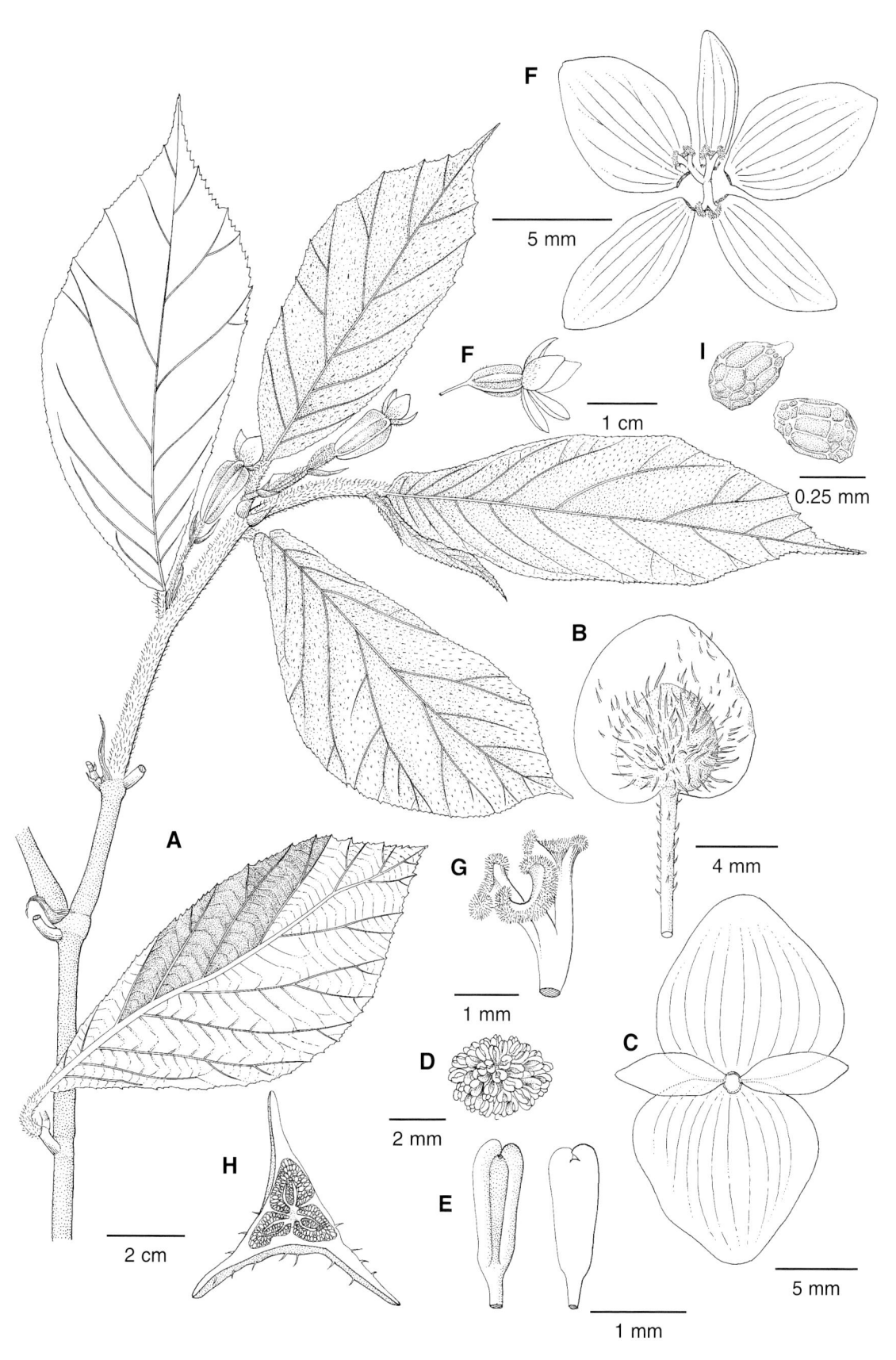

Begonia barbellata Ridl. **A.** The plant. **B.** Male bud. **C.** Male tepals. **D.** Stamen cluster. **E.** Stamens. **F.** Female flower. **G.** Styles and stigmas. **H.** T.S. ovary. **I.** Seed. (*RK 4920*)

BEGONIA BARBELLATA

DISTRIBUTION. Peninsular Thailand (Narathiwat) and Peninsular Malaysia—Kelantan, Trengganu, central Pahang and Johore.

HABITAT. A lowland species up to altitudes of about 100 m, it forms dense clumps on waterlogged muddy or sandy margins of slow flowing streams, as well as growing on stream banks.

OTHER SPECIMENS. JOHORE—Kota Tinggi-Mersing Road, *Sinclair SFN 10155* (SING), Labis, *Jumali & Heaslett s.n.* (SING), *Mohd Shah & Ahmad Shukor MS 2258* (SING), Sungai Kepok, *BH Kiew s.n.* (KEP), Sungai Selai, *BH Kiew s.n.* (SING), Ulu Sedili, *Jumali & Heaslett K 4475* (SINU). KELANTAN—Aring FR, *Anthonysamy SA 1096* (SING), *Kiew KBH 23* (KEP), Sungai Lebir, *Henderson SFN 29566* (SING). PAHANG—Bukit Koh, *Kiew RK 4920* (SING), Bukit Ringit *Ahmad Zainuddin B4* (UKBM), Ulu Wak *Moysey SFN 31060* (SING). TRENGGANU—Batu Biwa, *Kiew RK 2300* (SING), Kuala Sungai Bok, *Kiew RK 5180* (SING), *Mohd Shah et al. MS 3508* (KEP, SING), Sekayu, *Anthonsamy SA 635* (KEP), Ulu Brang, *Moysey & Kiah SFN 33816* (SING), Ulu Setui, *Kiew RK 2258* (KEP, SING), *RK 2698* (KEP).

NOTES. The ruby-red undersides of the leaves make this is an attractive plant. It is unusual for Peninsular Malaysian begonias in growing in swampy places.

The Bristly Begonia is very distinctive and quite unlike any other begonia in the Peninsula in its bristly stems, very short leaf stalks, flowers in clusters, and oblong fruits. It belongs to a group of begonias that is very diverse and common in Borneo, for example, *B. berhamanii* Kiew (Kiew, 2001). Many other non-begonia species that are closely related to Bornean species or also occur in Borneo—the Borneo element of the Peninsula's flora—share the same distribution pattern, i.e., they are found in Johore and on the east coast (Corner, 1960) but do not penetrate further into the Peninsula.

Begonia barbellata.
(Top left). Male flower.
(Bottom left). Young female flower. (Right). Purple underside of the leaf.

20. THE DECORATIVE BEGONIA
Begonia decora Stapf
(Latin, *decora*=nice-looking)

Stapf, Gard. Chron. III, 12 (1892, 19 Nov) 621; W. Watson, Garden & Forest 5 (1892, 23 Nov) 561; Stapf, Kew. Bull. (1893) App. 2, 29; Ridley, FMP 1 (1922) 683; Irmscher, MIABH 8 (1929) 121; Henderson, MWF Dicot (1959) 164 & Fig. 159; Teo & Kiew, Gards' Bull. Singapore 51 (1999) 117. **Type:** *Veitch s.n.*, 10 Sept 1891, cultivated, London (holo K). **Synonym:** *B. praeclara* King, JRASSB 71 (1902) 66; Ridl., JFMSM 4 (1909) 21. **Type:** *Wray 349*, 1890, Gunung Batu Puteh, Perak (lecto K, here designated; iso BM).

Stem rhizomatous, rooting at the nodes, reddish, succulent, unbranched, slender, up to 12 cm long, 6 mm thick, densely covered in long brown hairs *c.* 2 mm long; without a tuber. Stipules reddish, slightly hairy, narrowly triangular, 12–13 × 4–6 mm, margin not toothed, tip pointed, persistent. **Leaves** tufted, up to 5–7 mm apart; stalk reddish at base, greenish in upper two thirds, slightly or densely hairy, 7–17 cm long, 5–6 mm thick, round in cross-section; blade oblique, dark green or

The Decorative Begonia, *Begonia decora*.

BEGONIA DECORA

The Decorative Begonia, *Begonia decora*. (Painting by Wendy Gibbs)

green-bronze, variegated being yellowish green along veins, beneath rosy-purple or deep ruby red and lighter red along veins, densely hairy above, hairs *c.* 3 mm long, curved and raised on conical projections of the blade, undersurface pitted, coarsely velvety in life, papery when dried, dull, obliquely ovate, strongly asymmetric, (6.5–)8.5(–12) × 5.5–7(–10.5) cm, broad side 3.5–5.75 cm wide, base heart-shaped and very unequal, broadly rounded and overlapping, basal lobe 2–4 cm long, margin scalloped and minutely toothed, fringed by hairs, tip elongated; venation palmate-pinnate with *c.* 2 pairs of veins at the base and 1–3 pairs along the midrib, with another 2 in the basal lobe, branching slightly more than halfway to margin, slightly impressed above, beneath prominent and densely hairy. **Inflorescences** axillary, reddish green, hairy, erect, longer than the leaves, 9–21 cm long with two main branches *c.* 1.25 cm long, stalk 7–20 cm long, elongating to 24 cm in fruit, male flowers *c.* 4, female flowers 2, male flowers open first. Bract pair enclosing the developing inflorescence, pale reddish and translucent, broadly ovate, *c.* 19 × 14 mm, margin not toothed, tip rounded, soon falling, upper bracts narrowly oval, *c.* 16 × 4 mm, tip acute. **Male flowers** with pale pink stalk 12–30 mm long; tepals 4, white tinged pink or pale pink, hairless, margin not toothed, tip acute, outer two oval, 27–34 × 18–29 mm, inner two narrowly oval, 20–30

× 9–12 mm; stamens many, cluster hemispherical, 7–9 × 10–12 mm, stalk 1.5–2 mm long; filaments 1.5–2 mm; anthers golden yellow, narrowly oblong, *c.* 2 mm long, tip notched, opening by slits. **Female flowers** with a pale pink stalk 13–16 mm long; ovary pale reddish green, minutely hairy, oblong to obovate, *c.* 11 × 15 mm, wings 3, unequal, locules 2, placentas 2 per locule; tepals 5, pinkish white to rosy pink, without hairs, broadly rounded, margin not toothed, tip rounded, outermost 17–24 × 16–17 mm, innermost similar but smaller, 12–23 × 14 mm; styles 2, styles and stigmas pale yellow or greenish yellow, 3–5 mm long, stigmas spiral. **Fruit** a splash cup pendent on a fine hair-like stalk 15–25 mm long; capsule 13–16 × 30–35 mm long, hairless, locules 2, wings unequal, larger wing thick and fibrous, oblong, tip rounded, 17–26 mm wide, smaller two thinner, rounded, 8–10 mm wide, splitting between the locules and wings. **Seeds** barrel-shaped, light brown, *c.* 0.3 mm long, collar cells *c.* half seed length.

DISTRIBUTION. Endemic in Peninsular Malaysia: Main Range, Perak (Gunung Batu Putih and Tapah Hills) and Pahang (Cameron Highlands).

HABITAT. Montane forest, above 1100 m altitude, in deep shade on earth slopes often close to streams, sometimes creeping up the base of mossy tree trunks.

OTHER SPECIMENS. PERAK—Tapah Hills, *Ng FRI 1360* (KEP, SING); PAHANG—Cameron Highlands, *Anthonysamy SA 410* (KEP), *Chew CWL 848* (SING), *Holttum SFN 31210* (SING), *Kiew RK 1276* (SING), *RK 3117* (KEP), *RK 3118* (KEP), *RK 3129* (KEP), *RK 4133* (SING) *Symington 25984* (KEP), *Tam TSM 10* (KEP), *TSM 15* (KEP); Telom *Ja'amat 27292* (KEP), *Ridley 14005* (SING), *Ridley s.n.* 1901 (SING); Botanic Gardens Singapore 1890, without collector or number, locality wrongly given as Langkawi (SING).

NOTES. The Decorative Begonia first came to the attention of the public when it was exhibited in London at the Begonia Conference in 1892. It was described by Stapf (1892) as 'a pretty dwarf-growing plant with exceedingly ornamental foliage'. It won a First Class Certificate. It was quickly realised that this was a new species, which Stapf named *Begonia decora* and described from the cultivated specimen provided by the Veitch Nursery, who were offering it for sale at 5 shillings per plant.

Begonia decora. (Left). Female flower. (Right). Male flowers.

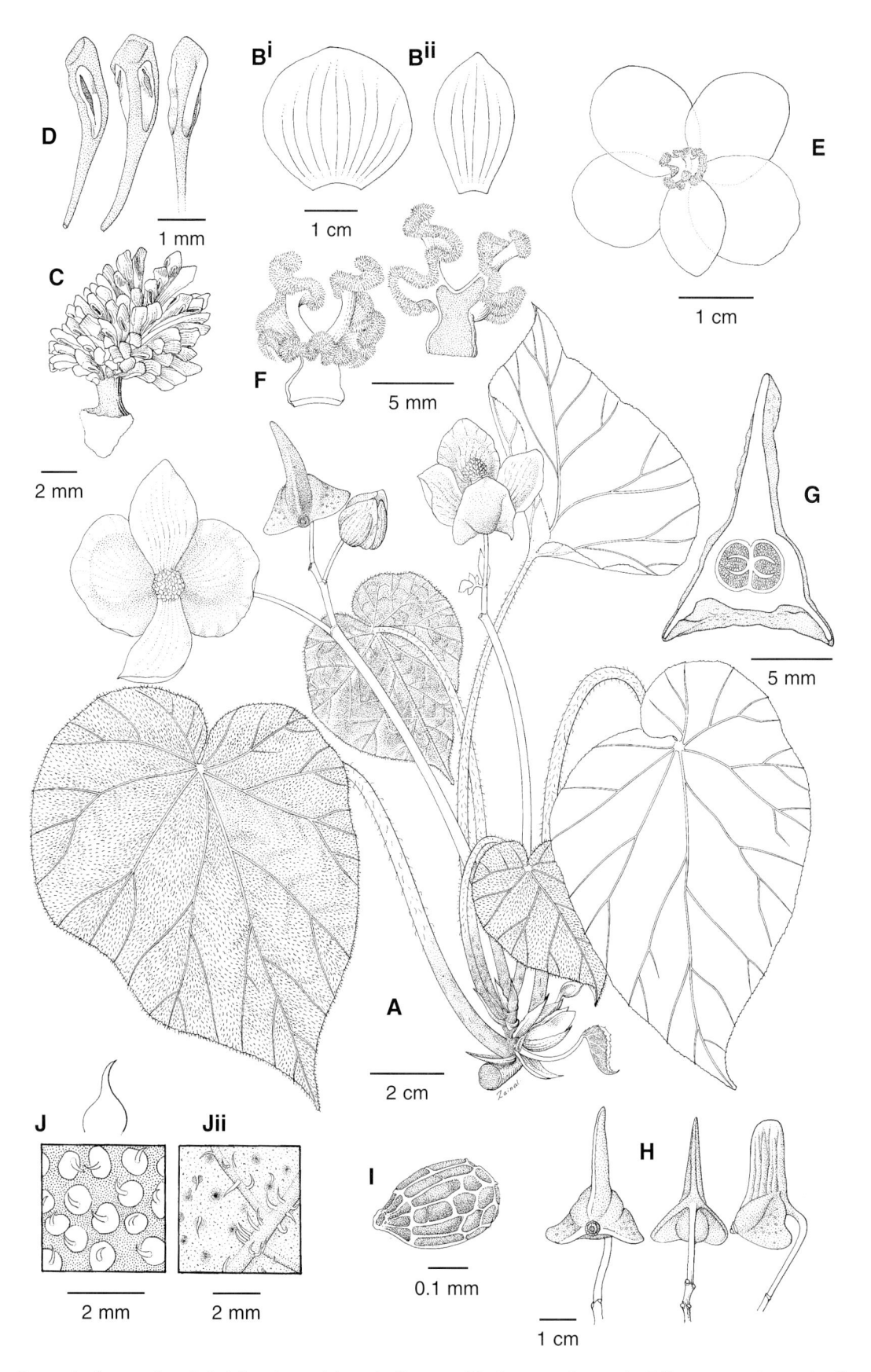

Begonia decora Stapf. **A.** The plant with male flowers. **Bi.** Outer male tepals. **Bii.** Inner male tepals. **C.** Stamen cluster. **D.** Stamens. **E.** Female flower. **F.** Styles and stigmas. **G.** T.S. ovary. **H.** Fruit. **I.** Seed. **Ji.** The upper leaf surface. **Jii.** The lower leaf surface. (*RK 4133*)

BEGONIA DECORA

Male flowers of the Decorative Begonia.

Watson (1892) reported that the plants originated from Perak and were sent by Charles Curtis, Superintendent of the Penang Botanic Garden, who was formerly a professional plant collector for J. Veitch & Co. As Curtis had not collected plants from mountains in Perak at that time, it is likely that L. Wray was the original collector and that the plants originated from Gunung Batu Puteh from where he had collected plants in 1890. (Wray sent plants from this collecting trip to King in Calcutta, who in 1902 described this species as new under the name *B. praeclara*). At that time both Curtis and Wray managed gardens at higher elevations on Penang Hill and Maxwell Hill, respectively, and could grow and exchange mountain plants.

Curtis also sent plants to Ridley in Singapore, where they were erroneously recorded as coming from Langkawi. *Begonia decora* is a mountain plant, confined to the Main Range. By 1891, the Decorative Begonia had died out in the Botanic Gardens, Singapore (Ridley 1909).

The Decorative Begonia was early used in hybridization and towards the end of the nineteenth century was hybridized with *B. rex* to impart a wide range of reds and a metallic sheen to the leaves of Rex cultivars (M.L. Thompson & E.J. Thompson, 1981). It is still in cultivation, both as a species and as hybrids, having been crossed with a variety of species, such as *B. cathayana* Hemsl., *B. deliciosa* Linden *ex* Fotsch, *B. hatacoa* Buch.-Ham. *ex* D. Don and *B. masoniana* Irmsch. All these hybrids have the velvety hairy leaf surface and strong coloration characteristic of *B. decora*. *Begonia decora* also hybridizes in the wild at Cameron Highlands.

Begonia decora. (Left). Fruits. (Right). Male flowers with *Trigona* bee collecting pollen.

Begonia decora × *B. venusta*, a natural hybrid

At Cameron Highlands, where these two species grow together, they interbreed and form hybrid swarms. They share the same pollinator, a stingless bee, *Trigona* sp. The hybrids are fully fertile with 87–100 per cent pollen viability and 98 per cent seed germination (Teo & Kiew, 1999). The hybrids show a complete range of intermediacy between the two parent species: some plants are more like *Begonia decora*, others are exactly intermediate and yet others are more like *B. venusta* indicating that the hybrid plants backcross with plants of the two parent species.

The hybrids can be recognized because, compared with *B. decora*, the leaves are larger, the hairs more widely spaced and less raised, and, compared with *B. venusta*, the leaves are smaller, have at least some scattered hairs and are reddish beneath.

Hybrids are found in all populations where the two species grow together. However, in some populations where *B. venusta* predominates in number, it is difficult to find 'pure' *B. decora* plants in the vicinity. It appears that over time, the species with fewer individuals will gradually be

BEGONIA DECORA × B. VENUSTA

'swallowed up' by the predominant species and the hybrids. Molecular biology shows that this has already happened and genetically pure *B. venusta* plants are no longer found in these hybrid swarms (Kiew *et al.*, 2003). If this continues over a long period, eventually only one variable hybrid will remain. The fact that this has not yet occurred suggests that hybridization is a relatively recent phenomenon. The earliest collection of a hybrid plant was in 1929 (*Symington 20941*).

This hybrid was only recognized in 1983 but even within the last twenty years the trend of *B. decora* being 'swallowed up' has been on the increase. All the hybrid swarms observed are by roadsides or paths in the forest. This opening up of the forest canopy does not appear to have resulted in changes to the begonia populations but rather it may have opened up flyways for the *Trigona* bees to fly over wider areas and effect cross-pollination between contiguous populations of the two species.

One plant was found that had a longer leaf stalk, fewer hairs and a bluish tinge. This appears to be a hybrid between *B. decora* and *B. pavonina*.

HYBRID SPECIMENS: PAHANG—Cameron Highlands, *Kiew RK 1268* (KEP) *RK 1277* (SING), *RK 1278* (SING), *RK 2604* (KEP), *RK 2605* (KEP), *RK 2756* (SING), *RK 2758* (KEP), *RK 2759* (KEP, SING), *RK 3129* (KEP), *RK 3123* (KEP), *RK 3133* (KEP) RK 4128 (SING), RK 4129 (SING); *Perumal et al. FRI 41576* (KEP); *Symington 20941* (KEP); Teo *TLL 10* (KEP).

Leaves of *Begonia venusta* (bottom left); *B. decora* (right) and the natural hybrids between them (top left and centre).

21. THE RED-HAIRED BEGONIA
Begonia wyepingiana Kiew, sp. nov.
(Soong Wye-Ping, naturalist and forest trekker)

A *Begonia decora* Stapf habitu validiore, foliis majoribus non variegatis et tepalis masculis pilosis differt. **Typus:** *Kiew RK 5115*, 30 September 2000, Ulu Selim, Perak (holo SING, iso K, KEP).

Stem rhizomatous, rooting at the nodes, brown, succulent, unbranched, stout, *c.* 7 cm long, (5–)8 mm thick; hairs on stem, stipules and leaf stalk transparent and white, *c.* 5 mm long; without a tuber. Stipules pale magenta, broadly triangular, 12–15 × 5–6 mm, margin not toothed, tip ending in a hair, persistent. **Leaves** tufted, *c.* 5–7 mm apart; stalk green in the upper half, reddish at the base, 8.5–14(–28) cm long, 4–6(–8) mm thick, grooved above; blade oblique, dark malachite green above, deep magenta beneath, not variegated, soft and velvety in life, papery when dried, dull, densely hairy, hairs magenta on raised bases causing indentations on the lower surface, longer and thicker on veins, broadly ovate, strongly asymmetric, 10–15 × 7.5–11(–15.5) cm, broad side 6–9.5 cm wide, base unequally heart-shaped, not overlapping, basal lobe 3–4.5 cm long, margin red, sometimes scalloped on the broader side, toothed, each tooth tipped by a hair, tip elongated; venation palmate-pinnate, 2 pairs of veins from the base, 3–4 pairs along the midrib with another

The Red-haired Begonia, *Begonia wyepingiana*.

BEGONIA WYEPINGIANA

2 in the basal lobe, branching towards the margin, veins impressed above, prominent beneath, the same colour as the blade on both surfaces. **Inflorescences** axillary, greenish with magenta hairs, erect, longer than the leaves, 12–17 cm with two main branches, stalk 11–16 cm long and 4 mm thick, lengthening to 23 cm in fruit, male flowers c. 7, female flowers 2, male flowers open first. **Male flowers** with a pale red stalk c. 13 mm long; tepals 4, margin not toothed, tip rounded, outer two deep pink, outside with deep red hairs c. 2 mm long, broadly ovate, 14–30 × 8–21 mm, inner two paler pink, oval, 13–27 × 6–13 mm, without hairs; stamens many, cluster globose, c. 4 mm across, stalk c. 1 mm long; filaments c. 1 mm long; anthers narrowly oblong, c. 1.5 mm long, tip rounded, opening by slits. **Female flowers** with a deep red stalk; ovary reddish green with red hairs, wings 3, unequal; tepals 4, equal-sized, narrowly oval, pale pink with scattered hairs outside, margin not toothed, tip rounded; styles 2, styles and stigmas pale yellow, stigmas spiral. **Fruit** a splash cup pendent on a stiff stalk 23–27 mm long; capsule 15–18 × 20–30 mm, hairless, locules 2, wings 3, unequal, larger 13–23 mm wide, fibrous, smaller two papery when dry, 6–7 mm wide, splitting between locules and wing. **Seeds** not known.

DISTRIBUTION. Endemic in Peninsular Malaysia, known only from Ulu Selim, Perak.

HABITAT. In a narrow valley on steep slopes above a stream in deep shade at about 1000 m altitude. *Strugnell 20396* records its habitat as 'bamboo swamp'.

Begonia wyepingiana. (Top left). Female flowers. (Top right). Male flowers. (Bottom right). Male flower buds. (Bottom right). Fruit.

BEGONIA WYEPINGIANA

The Red-haired Begonia grows in the steep valley of this rocky stream.

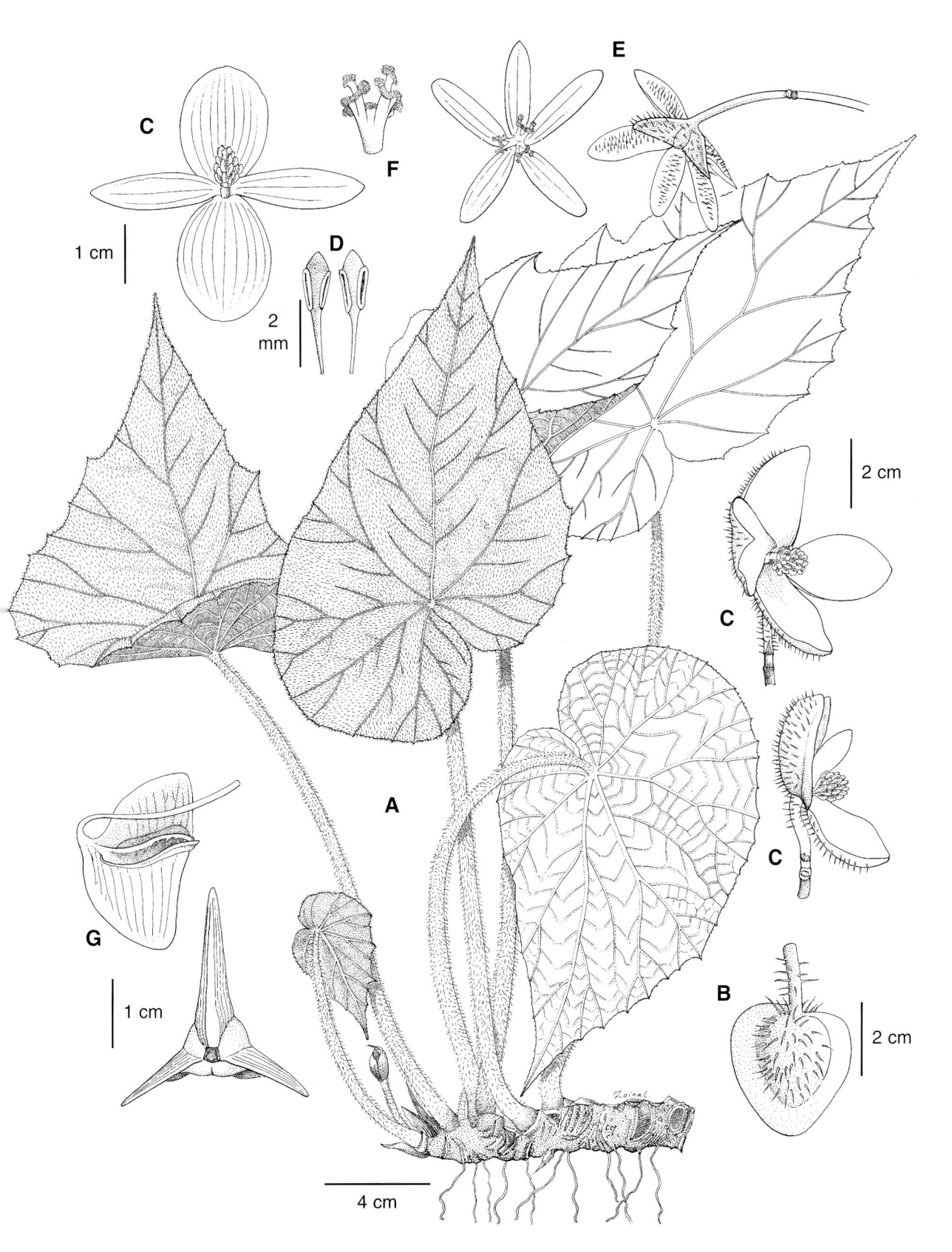

Begonia wyepingiana Kiew **A.** Plant. **B.** Male bud. **C.** Male flower. **D.** Stamens. **E.** Female flower. **F.** Styles and stigmas. **G.** Fruit. (*RK 5115*)

BEGONIA WYEPINGIANA

OTHER SPECIMEN. PERAK—Lipas-Slim Pass at 2800 feet, *Strugnell 20396* (KEP).

NOTES. This new species is a very handsome begonia with a robust habit and dark malachite-green leaves densely covered in magenta hairs, and large deep pink flowers. In its natural habitat it forms striking hemispherical clumps about 40 cm across.

It is very similar to *Begonia decora* in leaf shape and in being densely hairy with the hair bases raised on cones that cause indentations on the lower surface, which looks pitted. However, it is different in its larger leaves, veins that are the same colour as the blade, the hairy tepals of the male flowers (the buds look positively furry) and the larger fruits. The scalloped leaf margin is also a feature not seen in *B. decora*.

In the region, it has more potential than *B. decora* as an ornamental plant because, growing at a lower altitude, it is better suited to lowland conditions and grows well on a free-draining soil mix in light shade. Under natural conditions, small plantlets are common where the old leaf rots on the leaf litter and the blade separates from the petiole. The new plantlets develop from the severed veins.

The Ulu Selim population was discovered by Ms Soong Wye-Ping on one of her forest treks and this handsome begonia is named for her. An earlier specimen, collected in 1930 by Strugnell, is from the same area. This may indicate that this species is rare and local but, on the other hand, this area is inaccessible so these two collections may not reflect the true extent of its distribution.

It belongs to section *Platycentrum* in possessing a rhizomatous habit and fruits with three unequal wings and two locules.

(Above). The Red-haired Begonia grows on steep earth slopes. (Right). The deep crimson lower leaf surface.

22. THE VALLEY BEGONIA
Begonia vallicola Kiew, sp. nov.
(Latin vallis=valley; icola=dwelling)

A *Begonia decora* Stapf foliis plus quam 13 cm longis vix obliquis basi truncatis vel rotundatis (nec cordatis) et basibus pilorum non in conis elevatis orientibus differt. **Typus:** *R. Kiew RK4899*, 15 February 2000, Bujang Melaka, Perak (holo SING, iso K).

Stem rhizomatous, firmly rooted on rocks, fleshy, unbranched to 12 cm long, 7 mm thick, with long brown hairs *c.* 2 mm long; without a tuber. Stipules narrowly triangular, *c.* 8 × 4 mm, midrib and margins fringed by long hairs, tip ending in a hair, soon falling. **Leaves** tufted, *c.* 5 mm apart; stalk pale green, densely hairy, hairs *c.* 3 mm long, 10–18.5 cm long, round in cross-section; blade scarcely oblique, midrib at 0°–10° to stalk, dark green or brown or purple red above, beneath dark red, not variegated, densely hairy above and beneath, hair bases not indented beneath, soft and velvety in life, papery when dried, dull, ovate, slightly asymmetric, 13–18 × 7.5–13 cm, broad side 4–8 cm wide, base truncate to slightly rounded, basal lobe 3–7 mm long, margin minutely dentate, in some plants distinctly scalloped as well, tip attenuated 2–3 cm long; venation palmate-pinnate, 2 pairs of veins at the base, 3(–4) pairs along the midrib with another 1(–2) in the basal lobe, branching two thirds of the way to margin, veins pale green, densely hairy and slightly impressed above, beneath main veins and fine veins are prominent and densely hairy. **Inflorescences** axillary, green or red, without hairs, erect, longer than the leaves, 12–31.5 cm long with two main branches

The Valley Begonia grows on rocks and rock faces bordering this mountain stream.

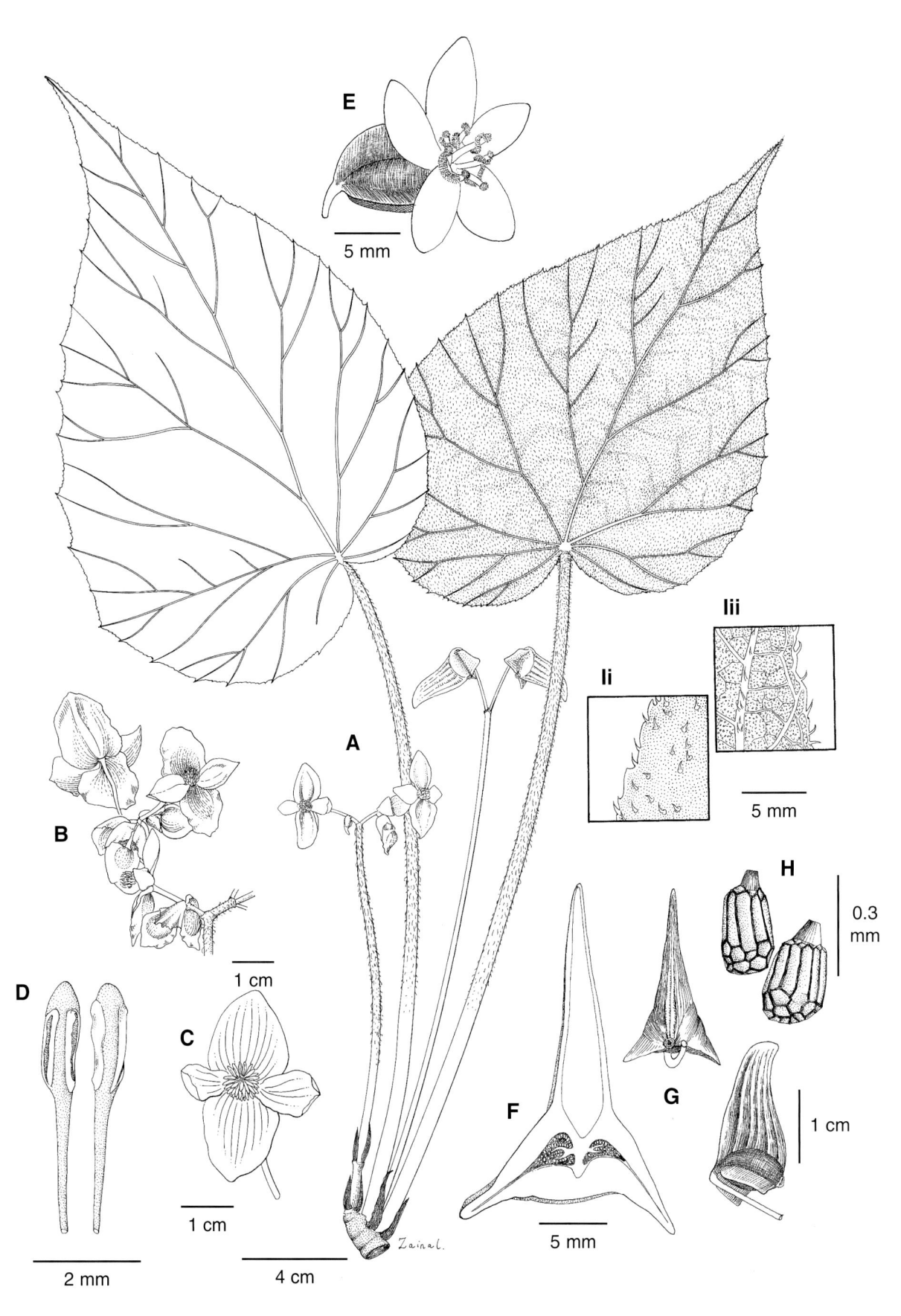

Begonia vallicola Kiew. **A.** The plant. **B.** Inflorescence. **C.** Male flower. **D.** Stamens. **E.** Female flower. **F.** T.S. ovary. **G.** Fruit. **H.** Seed; **Ii.** The upper leaf surface, **Iii.** The lower leaf surface. (*RK 4891*)

BEGONIA VALLICOLA

c. 1–3.5 cm long, stalk 10–28 cm long, male flowers 5–12, female flowers *c.* 2, male flowers open first. Bract pair pale green, broadly ovate, 13–17 × 12–14 mm, margin not toothed, soon falling. **Male flowers** with stalk to 16 mm long; tepals 4, pure white, without hairs, margin smooth, tip rounded, outer two broadly ovate, 17–22 × 16–19 mm, inner two narrowly ovate, 15–20 × 7–9 mm; stamens many, cluster globose, *c.* 4 mm across, stalk *c.* 1 mm long; filaments 1–2 mm long; anthers golden yellow, narrowly oblong, 1.5–2 mm long, tip rounded, opening by slits. **Female flowers** with stalk *c.* 13 mm long; ovary green, *c.* 8 mm long, wings 3, unequal, locules 2, placentas 2 per locule; tepals 5, completely white, without hairs, margin not toothed, tip acute, outer broadly ovate, *c.* 15 × 6 mm, inner oval, *c.* 7 × 5 mm; styles 2, styles and stigmas pale yellow, stigmas spiral. **Fruit** a splash cup pendent on a stiff stalk 17–23 mm long; capsule 10–13 × 18–23 mm, hairless, locules 2, wings 3, unequal, larger oblong, slightly tapered, tip rounded, thickly fibrous, 12–13 × 13 mm, smaller two narrowly oblong, thinly fibrous, 13 × 3–5 mm, splitting between the locules and wings. **Seeds** barrel-shaped, *c.* 0.3 mm long, collar cells more than half the seed length.

DISTRIBUTION. Endemic in Peninsular Malaysia, known only from Gunung Bujang Melaka, Perak.

HABITAT. Growing in lower montane forest along a stream in a narrow valley at *c.* 700 m on the downstream side of damp moss-covered boulders.

OTHER SPECIMENS. PERAK: Gunung Bujang Melaka, *Curtis 3323* (SING), *Curtis 3325* (SING), *Kiew RK 4891* (SING).

The Valley Begonia, *Begonia vallicola*. (Inset). Male flower.

BEGONIA VALLICOLA

NOTES. This new species was first collected by Charles Curtis from Gunung Bujang Melaka in 1898. Like *Begonia decora*, it is a densely hairy plant with dark green leaves. Irmscher (1929) had already noted that it was not typical of *B. decora* in leaf shape and venation but that more specimens were needed to confirm it was indeed different. Recent collections show that it is a new species, which is different from *B. decora* in leaf size, shape of the leaf base, and in the hairs that are not raised on conical projections of the leaf which make indentations on the lower surface. The new species displays variation in leaf margin (some leaves having a distinctly scalloped leaf margin) and in the leaf base, which ranges from rounded to truncate (but is never heart-shaped).

It belongs to section *Platycentrum* in being rhizomatous and in its ovary having two locules and two placentas in each locule. It takes its name from the narrow valley in which it was found.

Characters that distinguish *Begonia vallicola* from *B. decora*

Character	*B. vallicolla*	*B. decora*
Length of leaf blade (cm)	13–18	6.5–12
Angle of leaf stalk to midrib	0°–10° (scarcely oblique)	about 45° (oblique)
Leaf base	truncate or rounded	distinctly heart-shaped
Length of basal lobe (mm)	3–7	20–40
Hair base on lower leaf surface	not indented	indented

The dark-leaved form of the Valley Begonia.

23. THE MOUNTAIN BEGONIA
Begonia alpina L.B. Sm. & Wassh.
(Latin, *alpinus*=growing in the alpine zone on mountains)

L.B. Smith & Wasshausen, Phytologia 54 (1984) 469. **Synonym:** *B. monticola sensu* Ridl., JFMSM 5 (1914) 34 *non* C.DC., FMP 1 (1922) 862; Irmscher, MIABH 8 (1929) 128 **Type:** *Robinson s.n.*, 3 Feb 1913, Gunung Mengkuang Lebah (lecto SING; iso K).

Stem rhizomatous, rooting at the nodes, succulent, without hairs, unbranched, slender, up to 15 cm long, erect for up to 5–6 cm at the tip, 4–6 mm thick; without a tuber. Stipules green or pale reddish brown, lanceolate, 10–13 × 3–4 mm, bristly on the outer surface, margin not toothed, tip elongated into a fine hair, soon falling. **Leaves** tufted, 2–5 mm apart or distant and up to 25 mm apart; stalk pale green, reddish or greenish brown, densely hairy, (7.5–)12(–18) cm, round in cross-section; blade not oblique or sometimes oblique, plain dull or glossy green above or sometimes with a bluish sheen, paler green or reddish brown beneath with microscopic hairs on both surfaces, thin in life, papery when dried, ovate, slightly asymmetric, 9–14.5 × 4.75–8.75 cm, broad side 2.75–4.5 cm wide, base unequal, heart-shaped, sometimes overlapping, basal lobe 1–2 cm long, margin minutely and unequally toothed, teeth sometimes longer at the veins endings, sometimes red, tip elongated *c.* 1.25–2 cm long; venation palmate-pinnate, 2 pairs of veins at the base and 3 pairs along the midrib with 1 minor vein in the basal lobe, veins branching towards the margin, above green or red and deeply impressed, beneath green or reddish brown, prominent and slightly scurfy or bristly. **Inflorescences** axillary, green or dull red, without hairs, erect, longer than the leaf stalks, 5–20 cm long with two main branches 0.5–1.5 cm long, stalk 4–16.5 cm long, elongating up to 23 cm long in fruit, male flowers up to 12, female flowers 2–4, male flowers open first. Bracts green, narrowly linear, *c.* 10 × 2 mm, slightly hairy near the margin, margin finely toothed, soon falling. **Male flowers** with a pink stalk 15–17 mm long, minutely hairy; tepals 4, pale pink, margin not toothed, outer two broadly oval, 15–22 × 10–23 mm, hairy on the outer surface, tip rounded; inner two oval and narrowed to the base, 12–20 × 7–10 mm, hairless, tip slightly pointed; stamens many, cluster globose, 4 × 5 mm, stalk 1–2 mm long; filaments cream *c.* 2 mm long; anthers golden yellow, narrowly oblong, 2–2.7 mm long, tip rounded, opening by slits. **Female flowers** with a pale green stalk, 15–18 mm long; ovary green, 9–13 mm long, minutely hairy, wings 3, rounded, unequal, larger wing 7–10 mm wide, shorter two 6.5–7 mm wide, locules 2, placentas 2 per locule; tepals 5, pure white, glistening, without hairs, margin not toothed, tip rounded, outer

The Mountain Begonia, *Begonia alpina*.

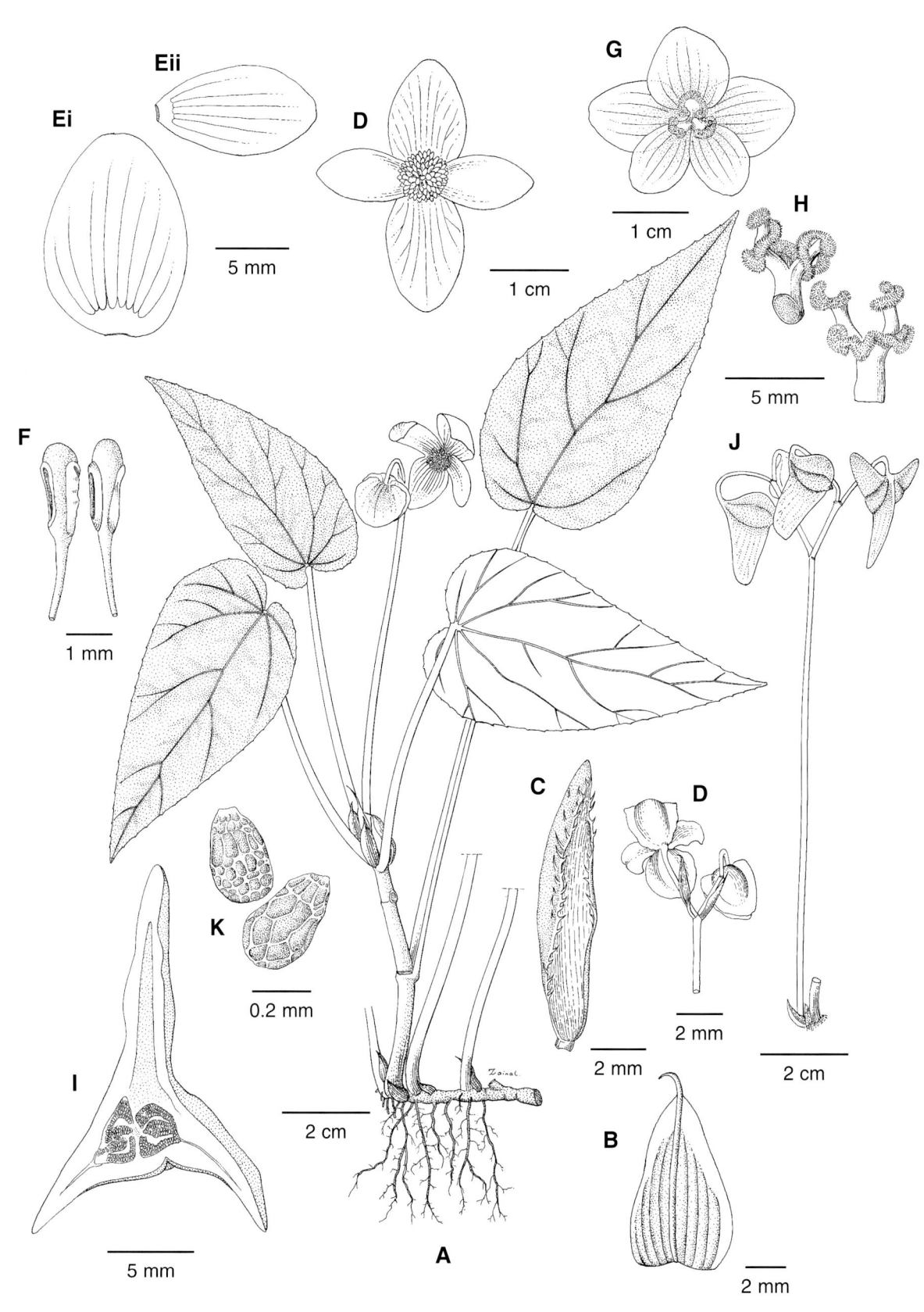

Begonia alpina L.B. Smith. **A.** The plant. **B.** Stipule. **C.** Bract. **D.** Male flower. **Ei.** Outer male tepals. **Eii.** Inner male tepals. **F.** Stamens. **G.** Female flower. **H.** Styles and stigmas. **I.** T.S. ovary. **J.** Fruits. **K.** Seed. (*RK 5151A*)

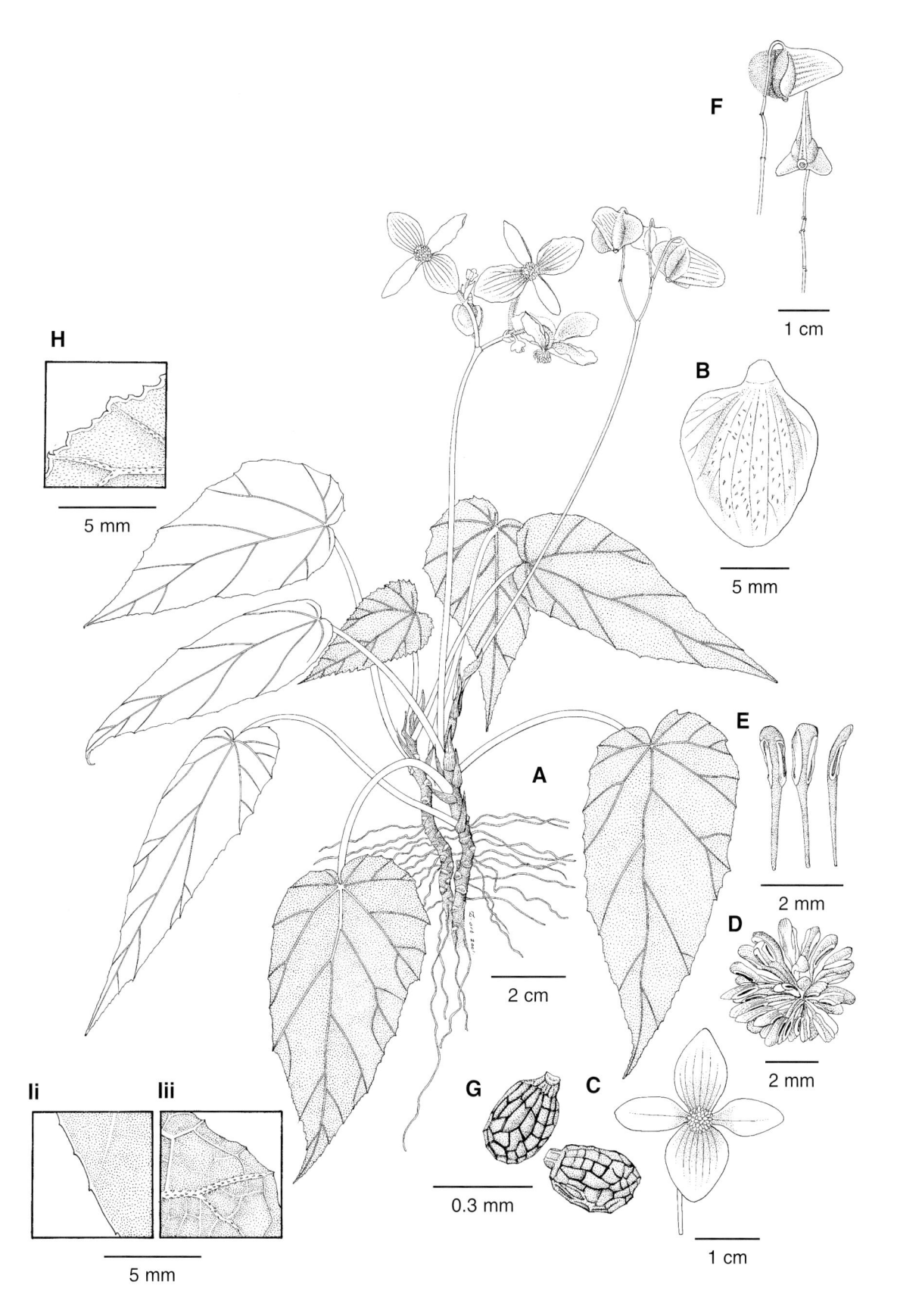

Begonia alpina L.B. Smith. **A.** The plant. **B.** Bract. **C.** Male flower. **D.** Stamen cluster. **E.** Stamens. **F.** Fruit. **G.** Seed. **H.** Leaf margin; **Ii.** Upper leaf surface, **Iii.** Lower leaf surface. (*RK 4885*)

suborbicular, *c.* 13 × 9.5 mm, inner similar but narrower 12–13 × 4–7 mm; styles 2, styles and stigmas dull pale yellow, 3–6 mm long, stigmas spiral. **Fruit** a splash cup pendent on a fine stalk 15–18 mm long; capsule 8–13 × 25–27 mm, hairless, locules 2, wings 3, unequal, larger oblong, 9–16 mm wide, fibrous, apex rounded, smaller two rounded, papery, 5–7 mm wide, splitting between locules and wings. **Seeds** barrel-shaped, *c.* 0.3 mm long, collar cells a third to half seed length.

DISTRIBUTION. Endemic in Peninsular Malaysia, known only from the Genting Highlands area, Pahang.

HABITAT. In valleys, on rocks at the edge of streams at 1000–1700 m altitude.

OTHER SPECIMENS. PAHANG—Genting Highlands, *Bremer & Bremer 1624* (KLU), *Kiew RK 4885* (SING), *RK 5181A* (SING), *Sohmer 9034* (KLU), *Stone 6541* (KLU).

NOTES. The type specimen was collected from Gunung Menuang Lebah, which Burkill (1927) says is not far from Gunung Ulu Kali, the peak at Genting Highlands. It was collected at 5000 ft altitude, that is, in montane forest. Because of this, Ridley called it *Begonia monticola* (dwelling in mountains). However, the name 'monticola' had already been used for another begonia species, which necessitated a name change; 'alpina' was chosen, an unfortunate choice as the mountains in Peninsular Malaysia are too low to support alpine vegetation.

In its habitat, leaf texture and size, it resembles *Begonia perakensis* var. *conjugens*, especially as its leaves are not always markedly oblique. It is distinct, however, in its unequal leaf with a well-developed basal lobe and in its leaves that are less than twice as long as wide. (*B. perakensis* var. *conjugens* has leaves more than twice as long as wide and which do not have a distinct basal lobe). In some plants, the leaves have an attractive bluish sheen.

Begonia alpina. (Above right). Female flower. (Above). Young fruits. (Right). Male flowers.

24. KOK-SUN'S BEGONIA
Begonia koksunii Kiew, sp. nov.
(named in honour of Yap Kok-Sun, nature photographer)

A *Begonia pavonina* Ridl. foliis variegatis, petiolis et laminis brevioribus, floris masculis minoribus differt. **Typus:** *Abdullah Piee & Yap Kok-Sun RK 5212*, 1 July 2001, Peninsular Malaysia, Ulu Perak, Gerik, Sungai Mangga (holo SING).

Stem rhizomatous, rooting at the nodes, brown, succulent, unbranched, slender, *c.* 3 cm long, 2–3 mm thick, without a tuber. Hairs erect, translucent, unbranched, dense and *c.* 2 mm long on stem, stipules, leaf stalk and underside of main veins, around the top of the leaf stalk forming a dense collar of hairs; sparse and *c.* 1 mm long on upper surface of leaf blade and the minor veins. Stipules pale green, narrowly triangular, 4–7 × 2–3 mm, margin not toothed, tip acute ending in a fine long hair, soon falling. **Leaves** tufted, 2–7 mm apart; stalk reddish brown, hairy, 3–8 cm long, round in cross-section; blade very oblique, variegated, dark green with silver band around the margin and extending inwards between the veins, thinly succulent in life, thinly papery when dried, matt, ovate, strongly asymmetric, 4.5–6 × 4.75–5.5 cm, broad side 2.75–3.25 cm, base heart-shaped with lobes overlapping, basal lobe 1.5–2 cm long, margin minutely toothed, each tooth tipped by a hair, tip acute to pointed; venation palmate-pinnate, 1–2 pairs at the base and 2 pairs along the midrib

Kok-Sun's Begonia, *Begonia koksunii*, grows on earth or rocky river banks.

BEGONIA KOKSUNII

Begonia koksunii. (Top left). Male flowers. (Middle bottom left). Female flowers. (Right). The rhizomatous habit.

154

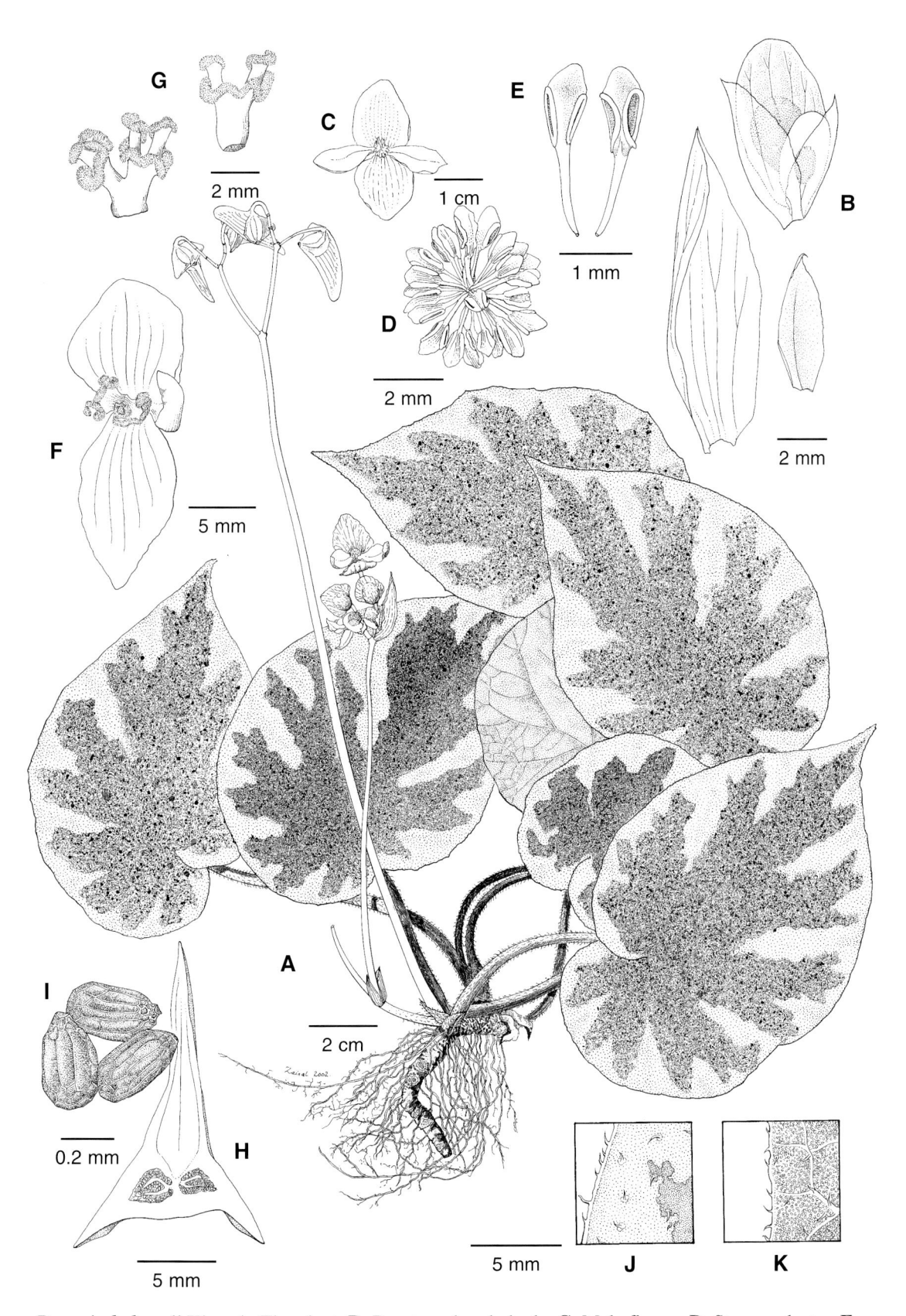

Begonia koksunii Kiew **A.** The plant. **B.** Bracts and male buds. **C.** Male flower. **D.** Stamen cluster. **E.** Stamens. **F.** Female flower. **G.** Styles and stigmas. **H.** T.S. ovary. **I.** Seed. **J.** The upper leaf surface. **K.** The lower leaf surface. (*RK 5212*)

BEGONIA KOKSUNII

Begonia koksunii. (Top). Male flower. (Bottom). Male flower buds.

with 1 vein in the basal lobe, branching towards the margin, veins impressed and dark green above, beneath prominent and reddish. **Inflorescences** axillary, red with scattered 2 mm-long hairs, longer than the leaves, little branched, 7.5–11 cm long with a stalk 6.5–9.5 cm long, lengthening to 23 cm in fruit, branches 2, male flowers 7 to many, female flowers up to 4, male flowers open first. Bracts soon falling. **Male flowers** with a pale red stalk, 7–8(–15) mm long; tepals 4, white with pinkish tinge, deeper pink towards the centre, margin not toothed, tip rounded, outer two minutely and sparsely hairy outside, broadly oval, 12–20 × 10–16 mm, inner two hairless, narrowly oval, 10–17 × 3–7 mm; stamens many, cluster globose, 4–6 mm across, stalk 0.75–1 mm long; filaments 1–2 mm; anthers golden yellow, narrowly obovate, 1–1.5 mm long, tip rounded, opening by slits. **Female flowers** with a stalk 6–8 mm long; ovary 6–8 × 10–18 mm, wings 3, unequal, locules 2, placentas 2 per locule; tepals 3, margin not toothed, tip rounded, outer two broadly oval, 10–16 × 8–14 mm, minutely and sparsely hairy outside, inner one narrowly oval, 7–12 × 4–5 mm; styles 2, styles and stigmas 3–4 mm long, stigmas spiral. **Fruit** a splash cup pendent on a stalk stiff, 12–17 mm long; capsule 12–15 × 21–37 mm, hairless, locules 2, wings unequal, larger oblong, thickly fibrous, tip rounded, 16–25 mm wide, smaller two forming a splash cup, rounded, thinly fibrous, 4–7 mm wide, splitting between the locules and wings. **Seeds** barrel-shaped, *c.* 0.25 mm long, collar cells *c.* 0.75 seed length.

DISTRIBUTION. Endemic in Peninsular Malaysia, known only from Sungai Mangga, Upper Perak.

HABITAT. Locally common on steep earth banks above the river.

OTHER SPECIMEN. *Abdullah Piee & Yap Kok-Sun s.n.*, 23 Aug 2002, Sungai Mangga, Perak (SING).

NOTES. *Begonia koksunii* is one of the most beautiful Malaysian begonias with its silver variegated leaves and rosette habit. It is named in honour of its collector Mr Yap Kok-Sun.

In its medium-sized, ovate leaves, which are almost hairless above, and few-flowered inflorescences with a long stalk and the splash-cup fruit with two locules each with two placentas, it most resembles *B. pavonina* from Cameron Highlands. It is very different from this species in its variegated leaves (those of *B. pavonina* are a uniform yet striking blue-green), its smaller leaf blade and stalk (the blade of *B. pavonina* is more than 8.5 cm by 6 cm wide and the leaf stalk is more than 15 cm long) and the smaller male tepals (the larger pair of tepals of *B. pavonina* are 20–28 mm long).

It falls within sect. *Platycentrum* in possessing a rhizomatous habit, ovaries and fruits with two locules each with two placentas, and fruits with one wing much longer than the other two.

It is a very local species known only from one place. Logging in the area is currently posing a threat to the long-term survival of this species.

25. THE PEACOCK BEGONIA
Begonia pavonina Ridl.
(Latin, *pavoninus*=peacock, referring to the blue-green colour of the leaf)

Ridley, JFMSM 4 (1909) 22, FMP 1 (1922) 863; Irmscher, MIABH 8 (1929) 123; Henderson, MWF Dicot (1959) 165 & Fig 157. **Type:** *Ridley 14125*, Telom (lecto SING, here designated; iso K). **Synonym nova:** *B. robinsonii* Ridl., JFMSM 4 (1909) 22, FMP 1 (1922) 863; Irmscher, MIABH 8 (1929) 125. **Type:** *Ridley 14125* (lecto SING, here designated).

Stem rhizomatous, rooting at the nodes, dark green or reddish, succulent, without hairs, unbranched, slender, up to 10 cm long, 5–8 mm thick; without a tuber. Stipules bright rosy red, hairless, narrowly triangular, 12–15 × 4–5 mm, margin entire, tip ending in a long hair *c.* 1.5–2 mm, soon falling. **Leaves** tufted, 7–12 mm apart; stalk bright rosy red, hairless, 15.5–29 cm long, 5–6 mm thick, round in cross-section; blade oblique, plain mid-green above and beneath or dark blue-green and magenta beneath, bright green shot with metallic blue depending on the angle of the light, thinly succulent in life, thin and papery when dried, slightly glossy above, ovate, strongly asymmetric, 8.5–14 × 6–13.75 cm, broad side 3.5–8.25 cm wide, base heart-shaped, often unequal, basal lobe 2–5 cm long, margin undulating, very minutely toothed, tip pointed *c.* 0.75–1 cm long; veins palmate-pinnate, 1–2 pairs of veins at the base and 2–3 pairs along with midrib with 1–2 in the basal lobe, branching *c.* halfway to margin, veins slightly impressed above, slightly prominent

The striking peacock blue leaves of *Begonia pavonina*.

BEGONIA PAVONINA

The Peacock Begonia, *Begonia pavonina*. (Painting by Wendy Gibbs)

and paler or faintly red beneath, without hairs. **Inflorescences** axillary, peduncle rosy or brownish red, without hairs, erect, as long as or shorter than the leaves, (8.5–)14–25(–32) cm with two main branches 0.5–1 cm long, stalk (8–)13–24(–31) cm long, male flowers 5, female flowers 2, male flowers open first. Bract pair ovate, 1.25 × 1.25 cm, margin entire, soon falling. **Male flowers** with a pink stalk 13–20 mm long; tepals 4, very pale pink outside, almost white inside, hairless, margin not toothed, tip rounded, outer two ovate, 20–28 × 19–20 mm, inner two oval 15 × 6–7 mm; stamens many, cluster globose, 5 × 6 mm, not stalked; filaments *c.* 2 mm long; anther deep yellow, narrowly oblong, *c.* 1 mm long, tip notched, opening by slits. **Female flowers** with a white stalk 16–27 mm long; ovary green, reddish towards the tips of the wings, 5–8 mm long, wings 3, unequal, locules 2, placentas 2 per locule, longer wing 11 mm wide, shorter two 9 mm wide; tepals 5, unequal, white or inner surface slightly pink, outer margin rosy pink, hairless, margin not toothed, tip rounded, outer tepals oval, 18–21 × 11–15 mm, inner tepals narrowly oval, 16–18 × 7 mm; styles 2, styles and stigmas deep yellow, 6 mm long, stigmas spiral. **Fruit** a splash cup pendent on a fine and hair-like stalk, 15–25 mm long; capsule 10–15 × 20–35 mm, hairless, locules 2, wings unequal, larger fibrous, 10–15 mm wide, smaller two 5–9 mm wide, thinly fibrous, splitting between the locules and wings. **Seeds** barrel-shaped, light brown, 0.2–0.3 mm long, collar cells *c.* 0.75 the seed length.

DISTRIBUTION. Endemic in Peninsular Malaysia, known only from Cameron Highlands, Pahang.

HABITAT. Montane forest, 1000–1400 m altitude, locally common on rocks and earth banks near streams, local.

Depending on the angle of the light, the leaves of the Peacock Begonia can either look bright blue (left) or bright green (right). Note the plant on the right is a natural leaf cutting growing from an old fallen leaf.

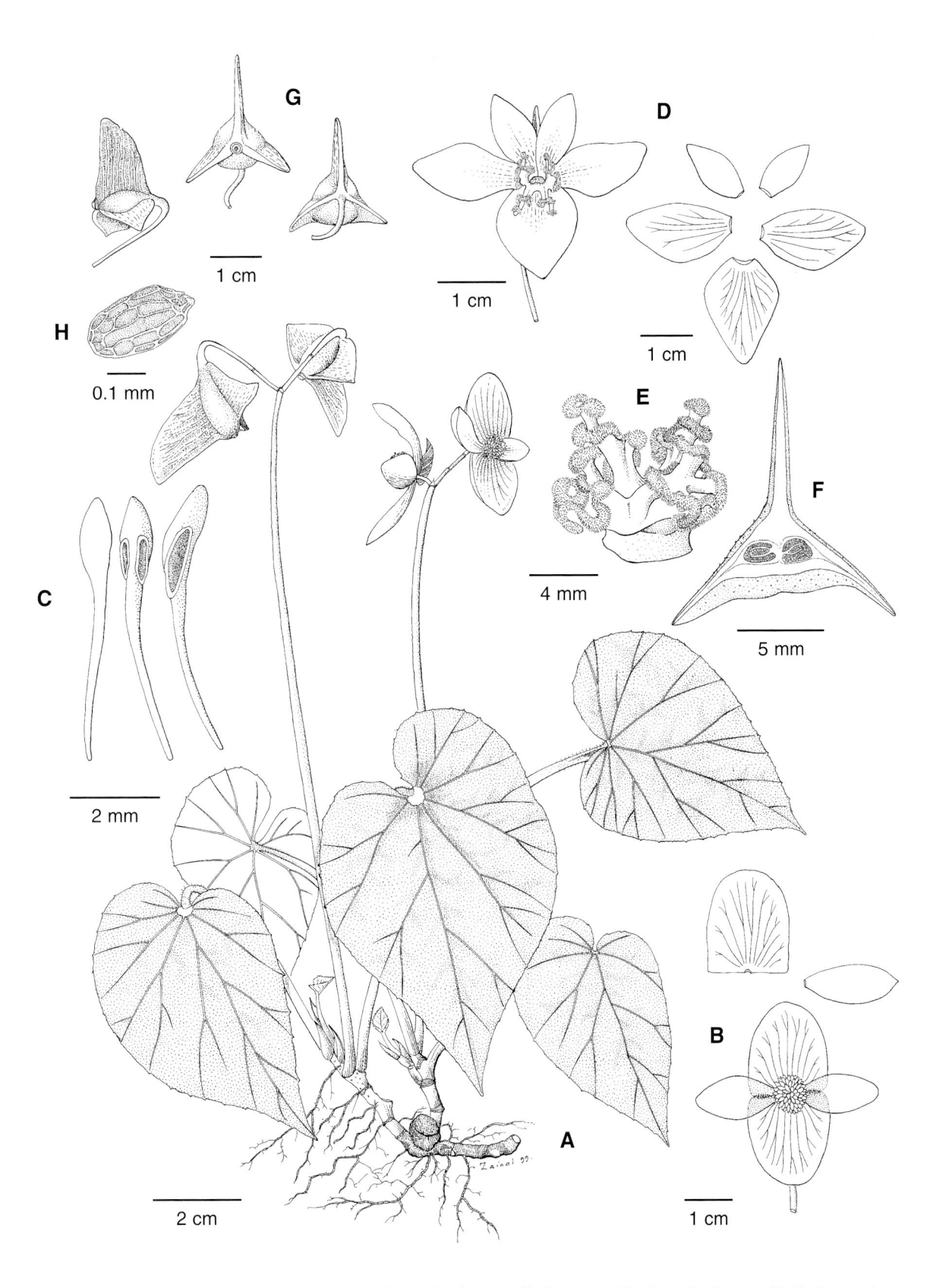

Begonia pavonina Ridl. **A.** The plant. **B.** Male flower. **C.** Stamens. **D.** Female flower. **E.** Styles and stigmas. **F.** T.S. ovary. **G.** Fruit. **H.** Seed. (*RK 4719*)

TAN J H/ALI

BEGONIA PAVONINA

OTHER SPECIMENS. PAHANG—Cameron Highlands, *Batten-Poole s.n.* (SING), *Henderson FMS 10885* (SING), *Iwatsuki et al. M 13716* (SING), *Kiew RK1266* (SING), *RK1267* (KEP), *RK3125* (SING), *RK4719* (SING), *RK4729* (SING), *Kloss 14126* (SING), *Kloss s.n.* (SING), *Md Nur SFN 32874* (SING), *Merton 4156* (KLU), *Ridley s.n. 1908* (SING), *Stone 14407* (KLU), *Student 8147* (KLU).

NOTES. The Peacock Begonia is one of the most remarkable begonias in Malaysia because of its leaves which, just like peacock feathers, change colour from an iridescent blue to bright green depending on the angle of the light. The leaf itself is not coloured blue but instead the iridescence is caused by refraction of light (p. 8).

Ridley (1909) described a very similar species, *Begonia robinsonii*, from the same place, Telom Woods (now called Cameron Highlands), with the same collecting number as his specimen of *B. pavonina*. He reported that these two species were different because *B. robinsonii* had large ovate bracts and *B. pavonina* did not. In *B. pavonina*, the developing inflorescence is enveloped in a pair of large ovate bracts that fall as the flowers open. Whether these bracts are present depends on the age of the inflorescence. *B. robinsonii* cannot therefore be recognized as a distinct species and is here reduced to synonymy.

Doorenbos *et al.* (1998) were uncertain as to which section the Peacock Begonia belonged because they reported that it had a single placenta in each locule. However, examination of the ovary and fruit shows that each locule has two placentas and in this and other characters it is typical of sect. *Platycentrum*.

(Opposite). The Peacock Begonia, *Begonia pavonina,* grows on steep earth banks or rocks in mountain forest. (Above left). Plant with fruits and a female flower. (Top right). Male flower. (Bottom right). Fruit.

163

26. KLOSS'S BEGONIA
Begonia klossii Ridl.
(C. Boden Kloss, zoologist, while subdirector of Federated Malay States Museum, 1908–1923, made collections of mountain plants)

Ridley, J. Linn. Soc. Bot. 41 (1913) 290, FMP 1 (1922) 861; Irmscher, MIABH 8 (1929) 133. **Type:** *Kloss s.n.*, February 1912, Menuang Gasing, Selangor (holo K).

Stem rhizomatous, rooting at the nodes and firmly attached to rocks, reddish brown, succulent, unbranched, stout, *c.* 13 cm long, to 15 mm thick; without a tuber. Stipules reddish, densely hairy, narrowly triangular, 10–12 × 4–5 mm, margin not toothed, tip elongated, soon falling. **Leaves** tufted, *c.* 5 mm apart; stalk reddish with white flecks turning brown with age, up to 31 cm long, 7–12 mm thick, densely hairy with long pale green hairs 2–3 mm long, grooved above; blade oblique or slightly oblique, dull plain light green or yellowish green, sometimes with a bluish hue, sparsely hairy above and beneath, hair bases slightly raised, moderately fleshy in life, papery when dry, broadly ovate, asymmetric, (11–)19–21 × (6–)13–19 cm, broad side 7.5–9.5 cm wide, base either heart-shaped with the basal lobe to 4 cm long or truncate with the basal lobe *c.* 1 cm long, margin minutely toothed, sometimes scalloped, sometimes red, tip elongate *c.* 2 cm long; venation

Begonia klossii. (Right). The rhizomatous habit. (Opposite top left). Female flowers. (Opposite top right). Male flowers. (Above). Male flower.

BEGONIA KLOSSII

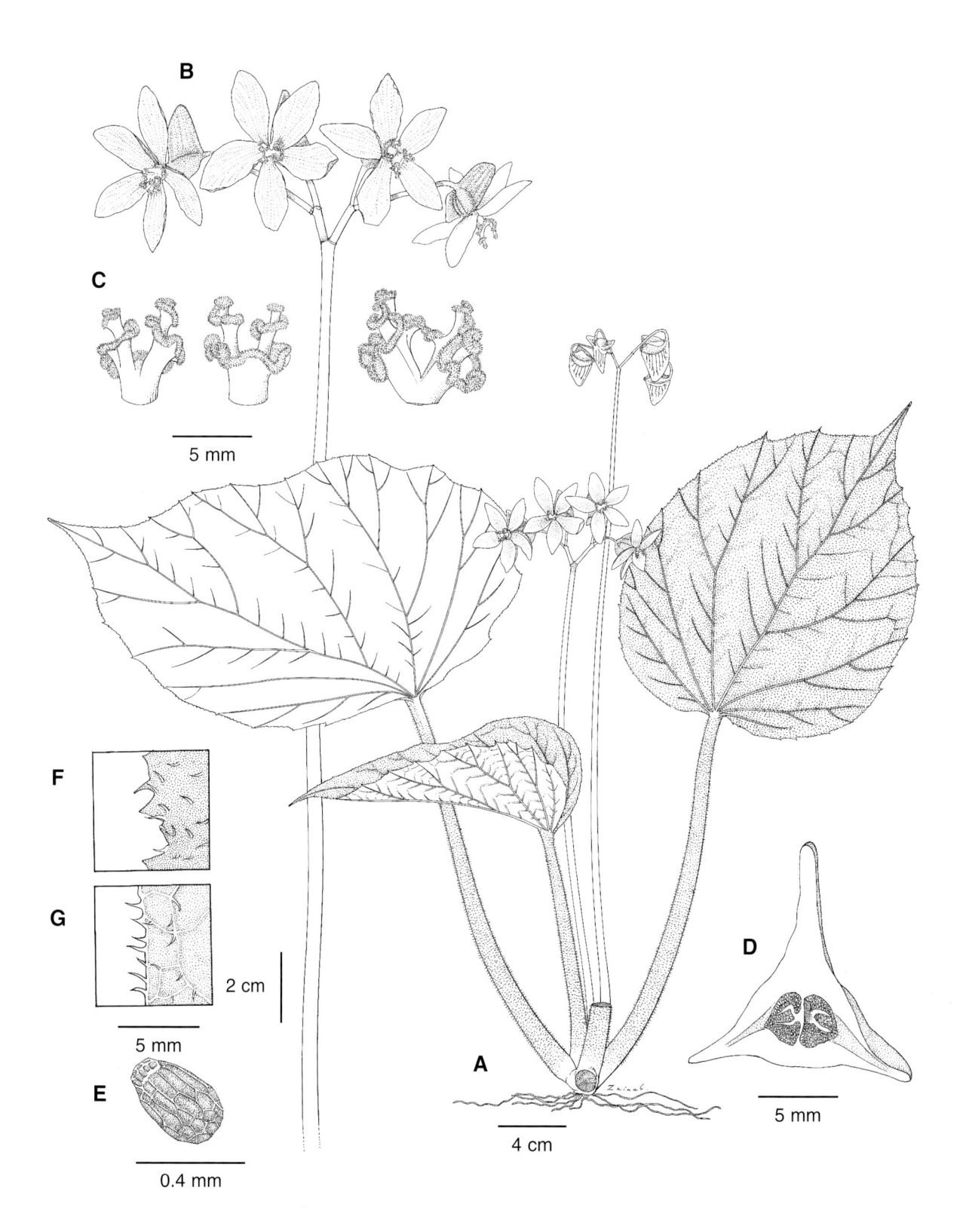

Begonia klossii Ridl. **A.** The plant. **B.** Female flower. **C.** Styles and stigmas. **D.** T.S. Ovary. **E.** Seed. **F.** The upper leaf surface. **G.** The lower leaf surface. (*RK 5200*)

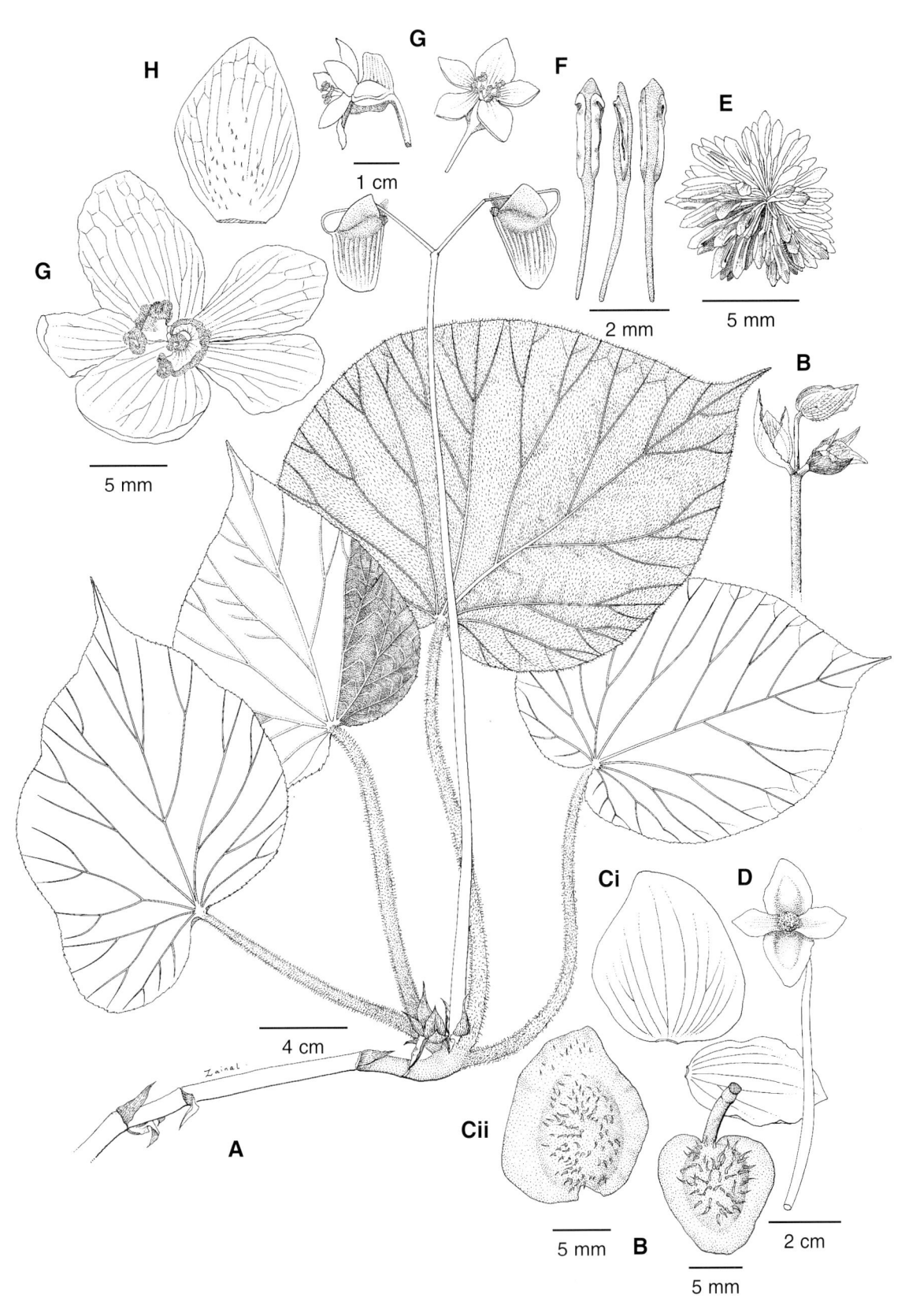

Begonia klossii Ridl. **A.** The plant. **B.** Male buds. **Ci.** Outer surface of male tepal. **Cii.** Inner surface of male tepal. **D.** Male flower. **E.** Stamen cluster. **F.** Stamen. **G.** Female flower. **H.** Outer surface of female tepals.

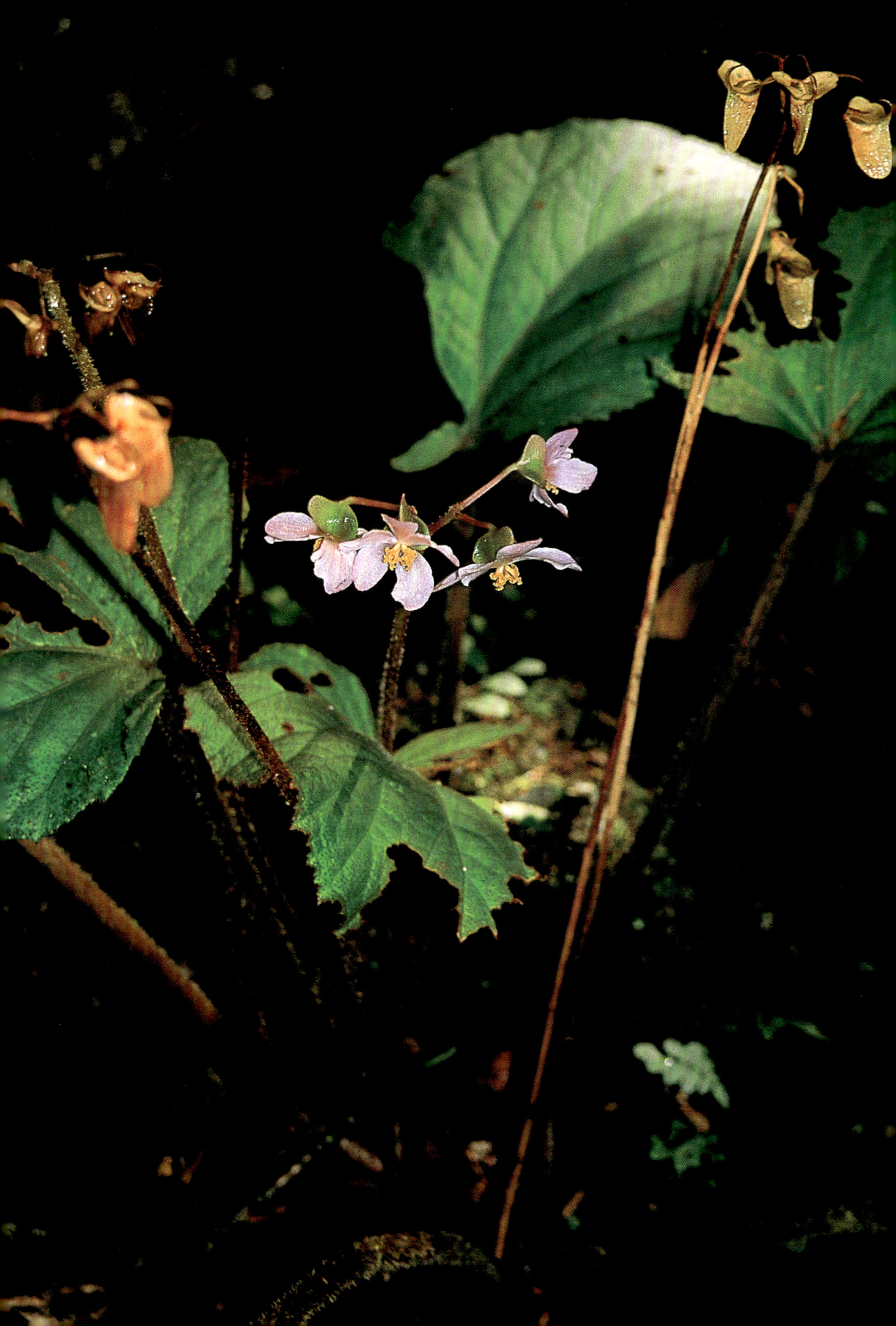

palmate-pinnate, 2–3 pairs of veins at the base and *c.* 3 pairs along the midrib with another 1 in the basal lobe, branching towards the margin, veins impressed above, beneath very prominent with long pale green hairs, same colour as blade or sometimes red. **Inflorescences** axillary, deep red or pale reddish, sparsely hairy, erect, longer than the leaves, 13–28 cm long with two main branches 7–25 mm long, stalk 10–27 cm long, lengthening to 47 cm in fruit, male flowers *c.* 7, female flowers 4, male flowers open first. Bract pair transparent, narrowly triangular, *c.* 10 × 5 mm, margin not toothed, soon falling. **Male flowers** with a greenish-white or pink stalk 8–18 mm long; tepals 4, margin not toothed, tip rounded, outer two pink and sparsely hairy on the outside, hairs translucent and *c.* 0.3 mm long, inner surface white with a pink tip, broadly oblong to slightly ovate, 8.5–14 × 7–10 mm, inner two white, without hairs, narrowly oval, 8.5–10 × 3.5–6 mm; stamens many, cluster globose, *c.* 5 mm across, not stalked; filaments *c.* 1 mm long; anthers golden yellow, narrowly oblong, 1–1.5 mm long, tip rounded, opening by slits. **Female flowers** with a reddish stalk, 28–30 mm long, not hairy; ovary deep green, *c.* 10 mm long, wings 3, unequal, locules 2, placentas 2 per locule; tepals 5, pink, without hairs, narrowly oval, margin not toothed, tip acute, 12–17 × 4.5–8 mm, inner similar but smaller, 10–14 × 4–6 mm; styles 2, styles and stigmas yellow, *c.* 5 mm long, stigmas spiral. **Fruit** a splash cup pendent on a fine stalk 22–30 mm long, capsule 13–17 × 20–21 mm, hairless, locules 2, wings 3, unequal, larger oblong, obliquely rounded, fibrous, 10–19 mm wide; smaller two obliquely triangular, tip rounded, drying papery, 4–8 mm wide, splitting between the locules and the two shorter wings. **Seeds** barrel-shaped, *c.* 0.4 mm long, collar cells *c.* half the seed length

DISTRIBUTION. Endemic in Peninsular Malaysia, the Gombak valley area, Selangor.

HABITAT. Growing on stream banks or well above the water level on the downstream side of large boulders in mountain streams at 1000–1300 m altitude.

OTHER SPECIMENS. SELANGOR —Sungai Pisang, *Kiew RK 3251* (SING), *RK 5200* (SING); Quarzite ridge east of Batu Caves, *B.H. Kiew RK 2099* (SING); Ulu Yam-Batu Dam Road, *Kiew RK 5285* (SING).

NOTES. Kloss's Begonia is closest to *Begonia maxwelliana* but is different in its striking densely hairy petioles and its smaller flowers, nor does it have the brown hairs on the underside of the veins that are characteristic of *B. maxwelliana*.

It was first collected by Kloss (in whose honour it was named) in 1912 from Gunung Menuang [Mengkuang] Gasing at the head of the Langat Valley on a peak now called Gunung Nuang.

(Opposite). Kloss's Begonia grows on boulders in mountain forest. (Left). Fruit of *Begonia klossii*.

27. THE GOPENG BEGONIA
Begonia paupercula King
(Latin, *pauperculus*=poor little thing, perhaps referring to the small size of its flowers)

King, JASB 71 (1902) 64; Ridley, FMP 1 (1922) 862; Irmscher, MIABH 8 (1929) 122. **Type:** *King's Collector 5952*, April 1884, Gopeng, Perak (lecto K, here designated).

Stem rhizomatous, clinging to rocks and rooting at the nodes, red, succulent, without hairs, unbranched, stout, 25–30 cm long, to 1.5 cm thick; without a tuber. Stipules narrowly triangular, 16–19 × 3–4 mm, without hairs, margin not toothed, tip pointed, ending in a fine hair, persistent. **Leaves** tufted, 2–5 mm apart; stalk greenish red towards the base, 11.5–33(–44) cm long, slightly grooved above; blade strongly oblique, sometimes slightly oblique and then more symmetric, plain light green with a red patch at the junction of the blade and stalk, thinly succulent in life, papery when dried, matt, hairless above, sometimes with microscopic sparse hairs beneath, almost rotund to broadly ovate, 14–22 × 12–27 cm, very asymmetric, broad side 10–14.5 cm wide, base unequally heart-shaped, basal lobe 4–5 cm long or sometimes almost symmetric with the base more or less equal with the basal lobes *c.* 1.5 cm long, margin with larger teeth at the vein endings

The Gopeng Begonia grows on boulders above rocky streams in hill forest.

BEGONIA PAUPERCULA

Begonia paupercula. (Left). Male flowers. (Right). Habit.

and with minute teeth in between, tip elongate to 2 cm long; venation palmate-pinnate, 2–3 pairs at the base and 2–3 pairs along the midrib with another 1 in the basal lobe, branching *c.* halfway to margin, veins slightly impressed above, beneath prominent, the same colour as the blade, usually without hairs, sometimes with fine brown hairs on main veins. **Inflorescences** axillary, green, erect, few-flowered, 16–33 cm long, shorter than the leaves, branching up to three times, branches 0.6–1.1 cm long, stalk 11.5–32 cm long, in fruit extending to 40 cm and longer than the leaves, male flowers 5–11, female flowers 3–7, male flowers open first. Bracts oval, *c.* 15 × 6 mm, margin entire, tip pointed, soon falling. **Male flowers** with a rosy pink stalk 12–20 mm long; tepals 4, white or rosy pink, margin not toothed, tip pointed, outer two broadly oval to rotund, (12–)20–21 × 8–17 mm, inner two narrowly oval, 11–17 × 3–5 mm; stamens many, cluster globose, 5–6 mm across, not stalked or with stalk to 0.5 mm long; filaments 1–2.3 mm long; anthers golden yellow, narrowly oblong, 1.5–2 mm long, tip rounded, opening by slits. **Female flowers** with a stalk *c.* 15 mm long; ovary *c.* 8 × 12 mm, wings 3, unequal, locules 2, placentas 2 per locule; tepals 5, oval, unequal, margin not toothed, tip rounded, without hairs, outermost rosy pink, *c.* 10 × 7 mm,

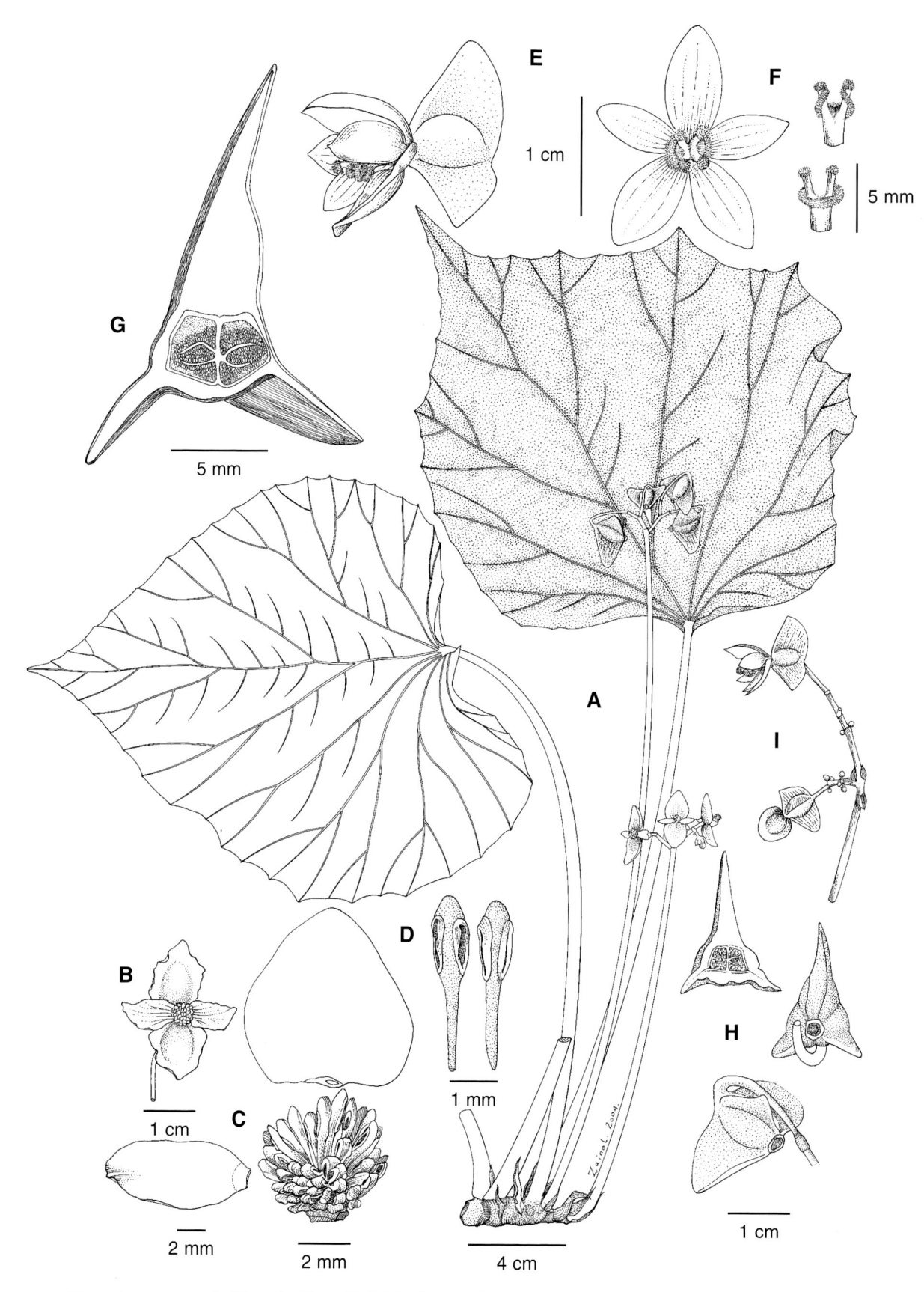

Begonia paupercula King **A.** Plant. **B.** Male flower. **C.** Stamen cluster. **D.** Stamens. **E.** Female flower. **F.** Styles and stigmas. **G.** T.S. ovary. **H.** Fruit. **I.** Inflorescence. (*RK 4892*)

BEGONIA PAUPERCULA

innermost almost white, 6 × 4 mm; styles 2, styles and stigmas yellow, 5 mm long, stigmas spiral. **Fruit** a splash cup pendent on a stalk stiff up to 24 mm long; capsule 11–15 × 29–33 mm, locules 2, wings unequal, larger oblong, tip rounded, 12–19 mm wide, thick and fibrous, smaller two rounded, 4–8 mm wide, thinly fibrous, splitting between the locules and the wings. **Seeds** barrel-shaped, 0.2–0.3 mm long, collar cells less than half the seed length.

DISTRIBUTION. Endemic in Peninsular Malaysia, known only from the Gopeng area in Perak.

HABITAT. It grows on rocks on the banks of torrential streams at 300–700 m altitude.

OTHER SPECIMENS. PERAK—Gunung Bujang Melaka, *Curtis s.n.* 1895 (SING), *Kiew RK 4892* (SING), *Mohd Shah et al. MS 3378* (SING), *Ridley 9687* (SING); Sungai Groh (east of Gopeng), *Ng FRI 1586* (KEP).

NOTES. The reason that King called this begonia 'paupercula' is obscure as it is a robust begonia with large leaves. Perhaps, because he described *Begonia venusta* at the same time and that species has magnificent large flowers, the largest of any begonia in the Peninsula, those of the Gopeng Begonia in comparison appeared poor, small ones.

As King originally noted, some plants have leaves that are very oblique and strongly asymmetric, like those of *B. venusta* and *B. maxwelliana*, while other leaves are scarcely oblique and are almost symmetric. Plants with both leaf shapes are found in the same population so there is no doubt that they belong to the same species. Unlike *B. venusta* and *B. maxwelliana*, the Gopeng Begonia is strictly rhizomatous and does not produce shoots with long internodes.

King's original description recorded the male flower as having only two tepals but the type specimen has four. The type specimen is recorded as from limestone (which the collector, Kunstler, wrote as 'limbs' on the label). This is an error as Gunung Bujang Melaka is a granite mountain and, although there are limestone hills close to Gopeng, this species does not grow on them.

The Gopeng Begonia, *Begonia paupercula*. (Top). Fruits. (Middle). Male flowers. (Bottom). Female flowers.

28. THE MAXWELL HILL BEGONIA
Begonia maxwelliana King
(from Maxwell Hill, Perak, from where it was first collected)

King, JASB 71 (1902) 66; Ridley, FMP 1 (1922) 863; Irmscher, MIABH 8 (1929) 118; Henderson, MWF Dicot (1959) 168 & Fig 162. **Type:** *Wray 119*, August 1885, Perak (lecto K, here designated).

Stem rhizomatous, usually rooted to rocks but sometimes semi-erect and up to 45–60 cm tall and falling and rooting at the nodes, reddish brown, succulent becoming woody, with sparse bristles, unbranched, stout, 1.5–1.75 cm thick, drying to 0.5–1 cm thick; without a tuber. Stipules minutely hairy outside, broadly triangular, 20–26 × *c.* 7 mm, margin fringed by hairs, not toothed, tip elongate to 4 mm long, soon falling. **Leaves** on the rhizome tufted and 7–10 mm apart, on semi-erect stems distant and up to 23 cm apart; stalk pink or reddish brown flecked with white, hairs dense, short (0.5–1 mm long) and appearing like brown scurf, 14–42 cm long, round in cross-section; blade very oblique, pink or light purple-brown or dark fawn when young maturing to dull, plain mid-green above and paler beneath, thinly fleshy in life, thinly leathery when dried, broadly

The Maxwell Hill Begonia grows on boulders beside streams.

ovate to ovate-rotund to suborbicular, strongly asymmetric, 14–22 × 11–24 cm, broad side 7–14 cm wide, base unequally heart-shaped, basal lobe 2–8 cm long, margin scarcely toothed, teeth only conspicuous at the vein endings, tip elongate to 2 cm long; venation palmate-pinnate with 2–3 pairs at the base and another 3–4 pairs along the midrib, (1–)2(–3) veins in the basal lobe, branching *c.* halfway to margin, veins impressed above, beneath prominent and densely covered in short rusty-brown hairs. **Inflorescences** axillary, pink to reddish brown, without hairs or with sparse rusty hairs, erect, shorter than the leaves in the male flowering phase, (19.5–)30–40 cm long with two main branches 1.5–7.5 cm long, stalk 18–37 cm long in female and fruiting phases, male flowers many, female flowers 8, male flowers open first. Bract pair ovate, up to 3 × 1.5 cm, minutely hairy, persistent. **Male flowers** with a stalk 8–10 mm long; tepals 4, white tinged pink, without hairs, margin not toothed, tip rounded, outer two ovate to broadly oval, 8–13(–25) × 6–7(–15) mm, slightly hairy outside; inner two narrowly oval 8–9(–11) × 2.5–3(–4) mm, hairless; stamens many, cluster globose, *c.* 4 mm across, scarcely stalked; filaments 1.5–2.2 mm long; anthers yellow, narrowly oblong, 2–2.5 mm long, tip rounded, opening by slits. **Female flowers** with a brownish red stalk to 25 mm long; ovary brownish red, (4–)5–15 × 2.5–12 mm, hairy, wings 3, unequal, locules 2, placentas 2 per locule; tepals (4–)5, white-pinkish to rosy pink, hairless, margin not toothed, tip acute, outer oval, 8–16(–21) × 4–9(–12) mm, inner similar but smaller; styles 2, styles and stigmas dull orange, *c.* 6 mm long, stigmas spiral. **Fruit** a splash cup pendent on a thin, stiff stalk 17–25 mm long; capsule 12–17 × 25–42 mm, hairless, locules 2, wings unequal, larger oblong, tip rounded, *c.* 25 mm wide, thick and fibrous, smaller two rounded, 8–10 mm wide, thin and fibrous, splitting between the locules and wings. **Seeds** barrel-shaped, *c.* 0.3 mm long, collar cells *c.* half the seed length.

The Maxwell Hill Begonia, *Begonia maxwelliana*. (Top left). Female flowers. (Bottom left). Male flowers. (Right). Young fruits. (Following page). The Maxwell Hill Begonia, *Begonia maxwelliana*.

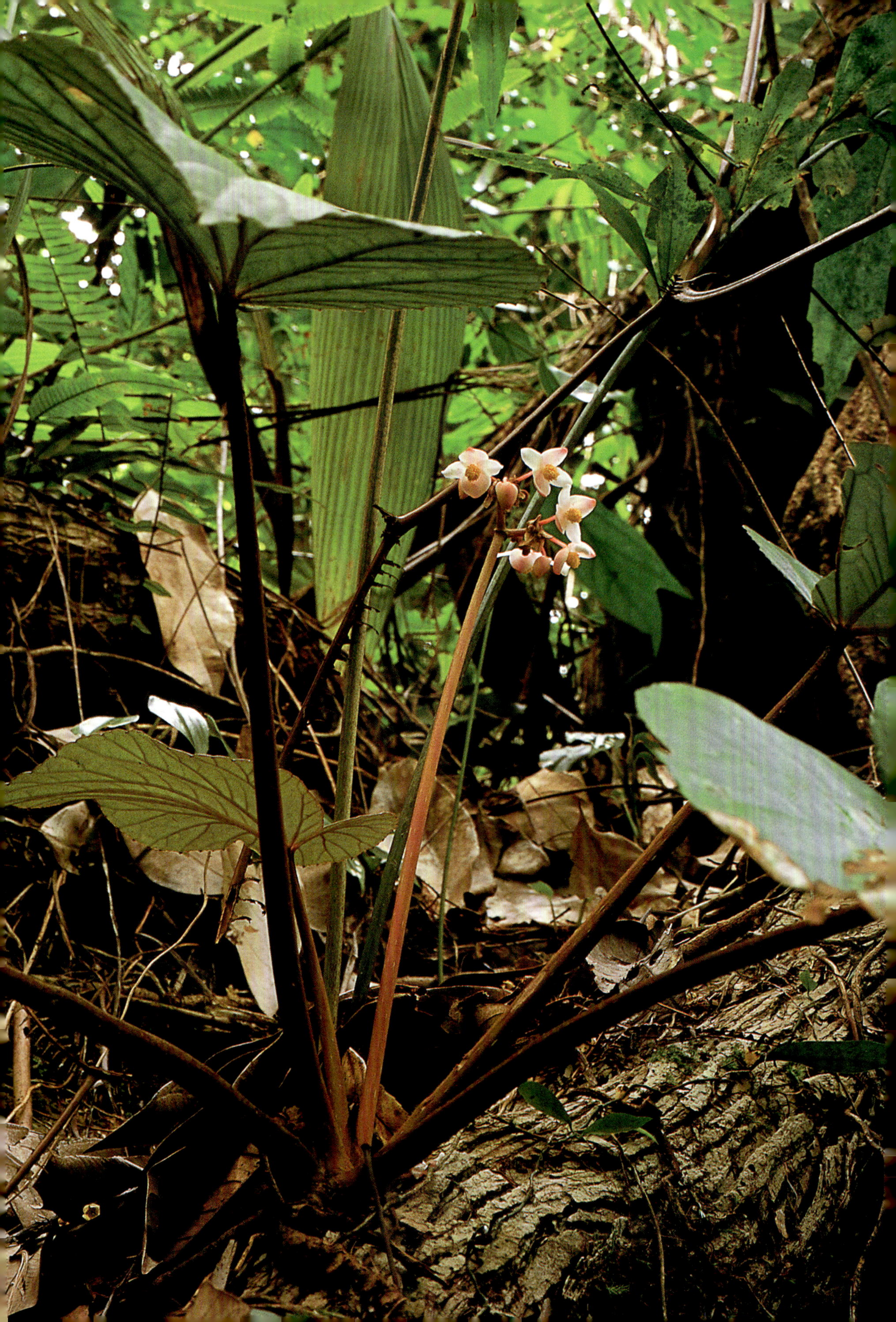

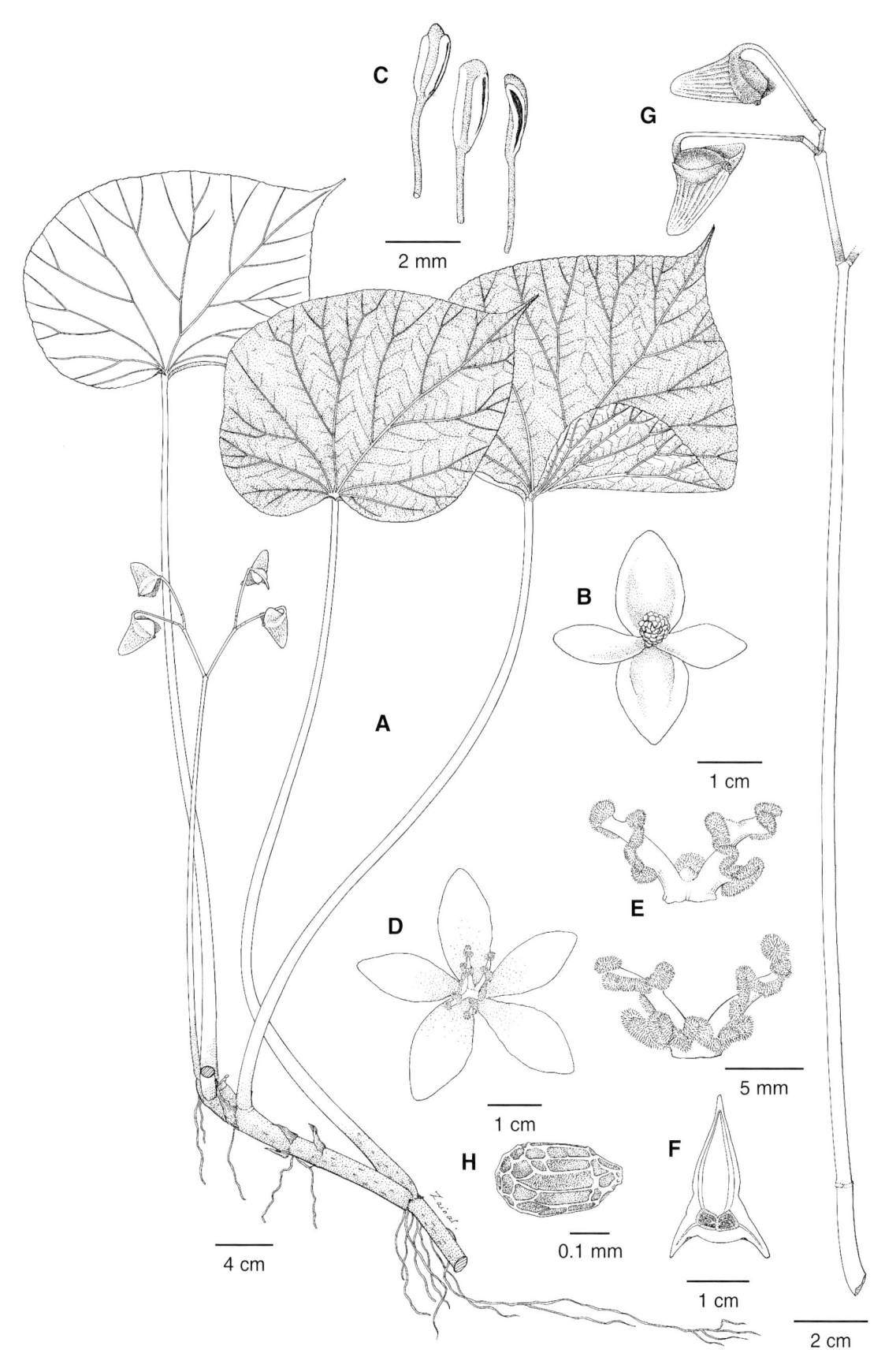

Begonia maxwelliana King. **A.** The plant. **B.** Male flower. **C.** Stamens. **D.** Female flower. **E.** Styles and stigmas. **F.** T.S. ovary. **G.** Fruit. **H.** Seed. (*RK 4906*)

BEGONIA MAXWELLIANA

DISTRIBUTION. Endemic in Peninsular Malaysia: Perak, Pahang and Kelantan.

HABITAT. On the earth banks of streams, on rocks and mossy tree bases from the lowlands at 300 m to lower montane forest at 1300 m altitude on Maxwell Hill.

OTHER SPECIMENS. KELANTAN—Bihai, *Kiew RK 5236* (SING), Sg. Kerchik, *Kiew RK 5236* (SING); PENANG—Government Hill, *Curtis 103* ?cultivated (SING); PERAK—Bk. Kinta, *Kiew RK 2594* (SING); Gopeng, *Kiew RK 2622* (KEP), *Saw FRI 37690* (KEP), G. Bubu, *Hou 643* (SING), *Kiew RK 2573* (SING), G. Inas Sg. Kupang, *Whitmore FRI 4671* (KEP); Ipoh, *Anthonysamy SA 1045* (SING); Kledang Saiong, *Symington 25720* (KEP); Maxwell Hill, *Anderson 122* (SING), *Anthonysamy SA 1163* (KEP), *Curtis s.n. 1888* (SING), *Curtis s.n. 1901* (SING), *Burkill & Haniff SFN 12755* (SING), *Kiew RK 4906* (SING), *Kiew RK 5104* (SING), *Rahimatsah N2* (KEP), *Ridley s.n. 1891* (SING), *Ridley s.n. 1892* (SING), *2931* (SING), *12803* (SING), *Sidek SK 445* (SING), *Sinclair & Kiah SFN 38782* (SING); Papan *Kiew RK 2560* (KEP), *Kiew RK 5171* (SING); Tapah Hills, Sg Who, *Ng FRI 1349* (KEP).

NOTES. The Maxwell Hill Begonia is distinctive in its rusty-brown hairs on the lower surface of the leaf. It belongs to the group of robust begonias with large unequal leaves and semi-erect stems, but unlike the others, *B. fraseri*, *B. longicaulis* and *B. venusta*, it is generally found as a rhizomatous plant clinging to rocks and boulders but on Maxwell Hill on wet stream banks it sometimes develops semi-erect shoots. Maxwell Hill is the highest altitude at which it is found. In the foothills, it is widespread at 300–500 m altitude. Common in Perak, it is one of the few begonias found on both sides of the Main Range. It is not found further south than Jeruntut, Pahang.

Female flowers of *Begonia maxwelliana*.

29. THE CABBAGE-LEAVED BEGONIA
Begonia venusta King
(Latin, venustus=beautiful, referring to the large flowers)

King, JASB 71 (1902) 65; Ridley, FMP 1 (1922) 862; Irmscher, MIABH 8 (1929) 126; Teo & Kiew, Gard. Bull. Singapore. 51 (1999) 117. **Type:** *Wray 1598*, Ulu Batang Padang, Perak (lecto K, here designated). **Synonym:** *B. megapteroidea* King, JASB 71 (1902) 65. **Syntypes:** *Wray 1450, 157*, Perak (not seen).

Robust begonia, all parts without hairs except for a few hairs on the margin of the stipules and bracts. **Stem** reddish, succulent, at first with a slender rhizome 8–16 mm thick, sending up weak semi-erect shoots to 60 cm tall, which fall and root at the nodes; without a tuber. Stipules green, narrowly triangular, 20–30 × 5–7 mm, margin not toothed, tip elongate and ending in a hair, persistent and clustered at nodes on erect stems. **Leaves** at first tufted and 3–15 mm apart, then in semi-erect shoots distant and 10–26 cm apart; stalk red or less frequently green, flecked white, 22–50 cm long, grooved above; blade very oblique, glossy, plain dark green, paler green or reddish beneath, thick and rubbery in life, papery when dried, broadly ovate, strongly asymmetric, 10–18.5 × 9.5–19.5 cm, broad side 5.5–11 cm wide, base unequally heart-shaped, not overlapping, basal lobe 3–6(–9) cm long, margin minutely toothed, tip acute; venation palmate-pinnate, 2–3 pairs at the base and (1–)2(–3) pairs along the midrib with another 2(–3) veins in the basal lobe, branching *c.* halfway to the margin, veins impressed above, beneath prominent and pale green. **Inflorescences** axillary, red, as long as the leaves, 15–25 cm long with two main branches 1.5–2.5 cm long, stalk 13–22.5 cm long, in fruit elongating to 40(–67) cm and taller than the leaves, male flowers *c.* 9, female flowers 4, male flowers open first. Bract pair pale pink, semi-translucent, broadly ovate, 12–25 × 9–10 mm, tip pointed, soon falling.

Begonia venusta. (Above). Male flowers. (Below). Female flowers.

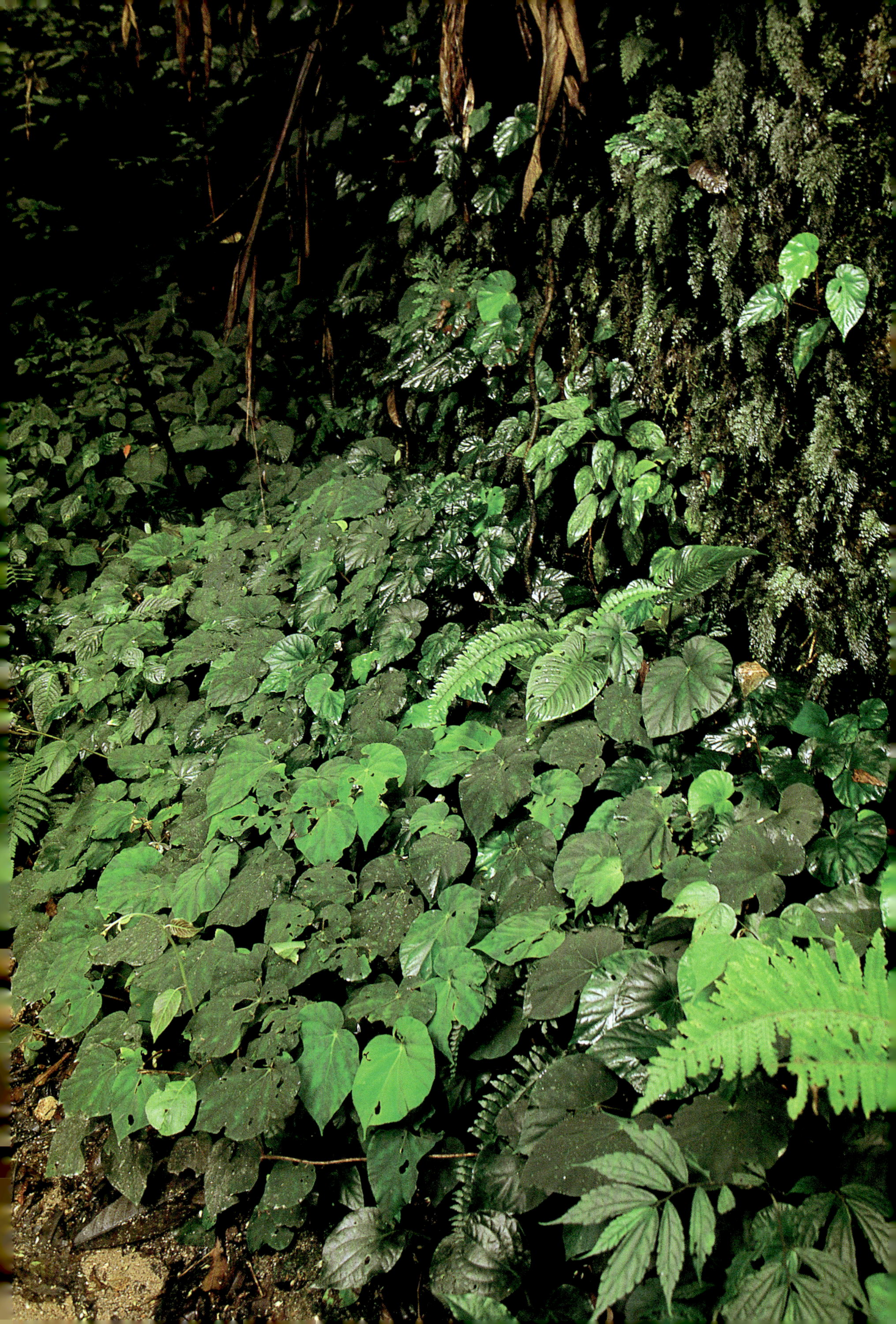

BEGONIA VENUSTA

(Above). The Cabbage-leaved Begonia, *Begonia venusta*. (Painting by Wendy Gibbs). (Opposite). The Cabbage-leaved Begonia forms drifts on valley slopes.

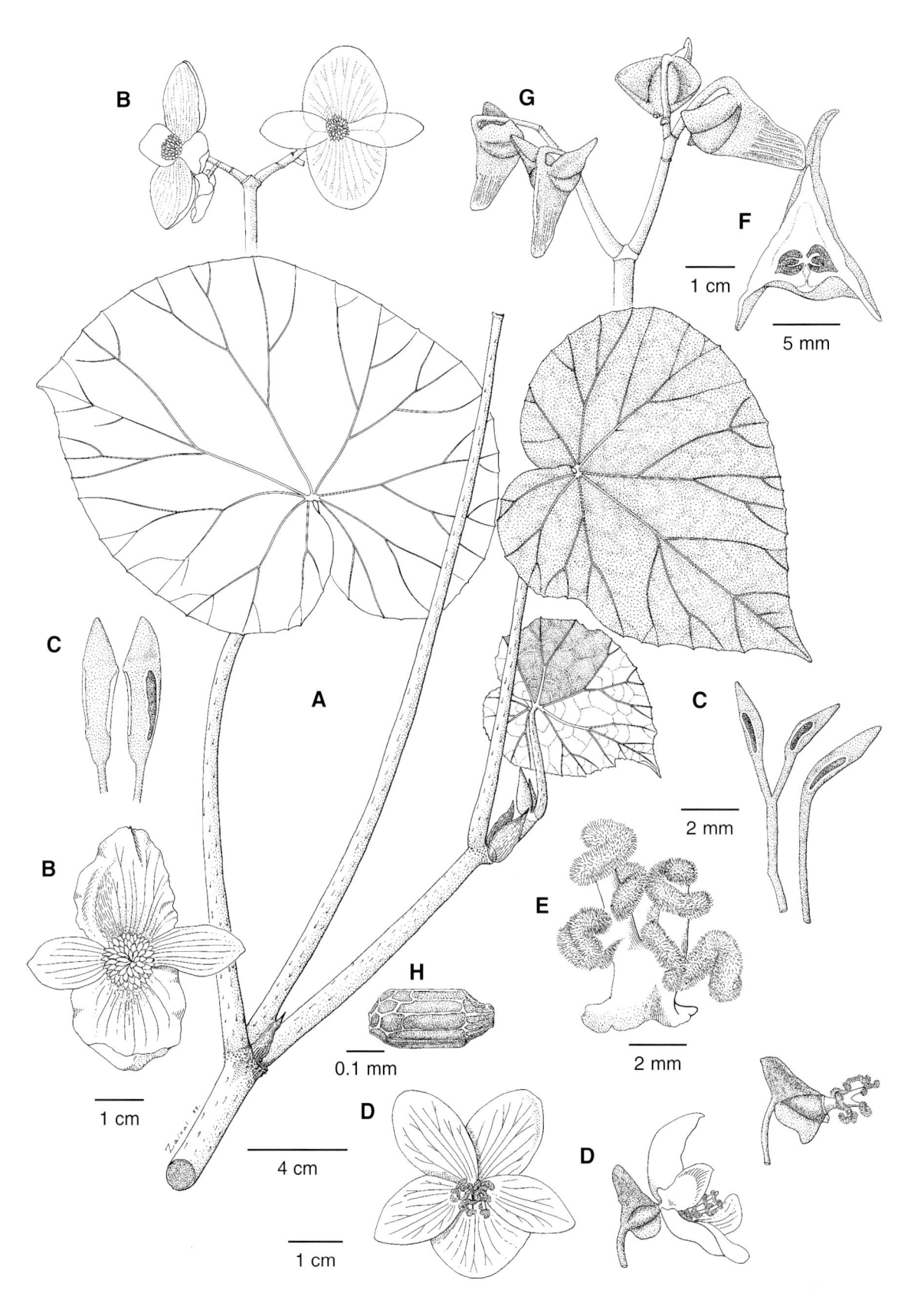

Begonia venusta King **A.** The plant. **B.** Male flower. **C.** Stamens. **D.** Female flower. **E.** Styles and stigmas. **F.** T.S. ovary. **G.** Fruit. **H.** Seed. (*RK 4728*)

BEGONIA VENUSTA

Male flowers with a deep pink stalk 20–36 mm long; tepals 4, white or white tinged pink, without hairs, margin not toothed, tip rounded, outer two broadly oval, 18–36 × 17–38 mm, inner two oval, 16–30 × 7–21 mm; stamens many, cluster hemispherical, 6–8 mm across, almost sessile; filaments 2(–3) mm long; anthers yellow, narrowly oblong, 1–1.5(–2) mm long, tip pointed, opening by slits. **Female flowers** with a deep pink stalk 10–20 mm long; ovary bronzy-green, 8–11 mm long, wings 3, unequal, locules 2, placentas 2 per locule; tepals (4–)5, white or pale pinkish white, broadly oval, margin not toothed, tip rounded, outermost 18–20 × *c.* 14 mm, innermost similar but smaller *c.* 15 × 6 mm; styles 2, styles and stigmas yellow, 8–9 mm long, stigmas spiral. **Fruit** a splash cup pendent on a stiff stalk 28–35 mm long, capsule 15–20 × 30–45 mm long, locules 2, wings 3, unequal, larger wing thickly fibrous, 18–34 mm wide, oblong tapered to a rounded tip, smaller two ovate, 6–12 mm wide, thickly papery, 6–12 mm wide, splitting between the locules and the wings. **Seeds** barrel-shaped, light brown, *c.* 0.3 mm long, collar cells *c.* half the seed length.

DISTRIBUTION. Endemic in Peninsular Malaysia: Perak (Ulu Batang Padang) and Pahang (Cameron Highlands).

HABITAT. It grows in montane forest at 1725 to 1890 m altitude on earth slopes or in wet gullies in damp sandy or peaty areas. It invades banks of ditches edging vegetable plots in Cameron Highlands.

OTHER SPECIMENS. PAHANG: Cameron Highlands G. Batu Gangan, *Jaamat 27035* (KEP); G. Brinchang, *Anthonysamy SA 359* (KEP), *Garcia FRI 32746* (KEP), *Kiew RK 1269A* (KEP), *RK 1269B* (SING), *RK 2751* (SING), *RK 2753* (KEP, SING), *RK 4728* (SING), *RK 5187* (SING), *Saw FRI 34381* (KEP), *Tam TSM 8* (KEP); G. Irau, *Kloss s.n.* 1908 (SING); Path 4 *Anthonysamy SA 300* (SING); Sg. Bertam, *Anthonysamy SA 411* (KEP), *Kiew RK 2603* (SING), *RK 2757* (KEP), *RK 3116* (SING); Sg Burong, *Holttum SFN 31365* (SING); Robinson's Falls, *Kiew RK 3124* (SING); Tanah Rata Sg. Terolak, *Jaamat 27575* (KEP).

NOTES. This is the most rampant of all begonias in Peninsular Malaysia. In suitable conditions where there is wet soil in forest, it forms knee-deep drifts to the exclusion of all other herbs. It is well named as it has the largest flowers of any in the Peninsula with male flowers 4 to 7 cm long and 2 to 3 cm wide and the female flowers measure about 4 cm across. Its common name derives from its large rubbery leaves that have the texture and glossiness of cabbage leaves. Unless supported by trees or rocks, its semi-erect stem falls to the ground and roots. Supported, the stem can reach about half a metre in height.

Wherever it grows, it readily hybridizes with *Begonia decora* (see under that species). The hybrids are easily recognized because they have smaller, darker green leaves with scattered hairs on the upper surface.

Irmscher (1929) recorded the ovary as having 3–5-branched placentas in each locule. This is in error as there are only two per locule.

Young female flowers of *Begonia venusta*.

30. THE FRASER'S HILL BEGONIA
Begonia fraseri Kiew
(L.J. Fraser, a trader operating in the Fraser's Hill area in 1890s, after whom the hill is named)

Kiew, Sandakania 6 (1995) 64. **Type:** *R. Kiew RK 3831*, 17 July 1995, Fraser's Hill, Pahang (holo KEP).

Stem semi-erect, falling and rooting at the nodes, green or reddish brown, succulent, without hairs, little branched, nodes swollen, 14–60 cm tall and *c.* 6 mm thick; without a tuber. Stipules in pairs, without hairs, narrowly triangular, 10–20 × 3.5–5 mm, margin not toothed, tip narrowed, ending in a hair, persistent. **Leaves** well-spaced, 12–27.5 cm apart; stalk reddish brown, without hairs, 7.5–24.5 cm long, grooved above; blade oblique, glossy, plain dark green, sometimes with a bluish tinge above, thinly leathery in life, papery when dried, ovate, strongly asymmetric, 11.5–16 × 7–12.5 cm, broad side 5–7 cm wide, base unequally heart-shaped, basal lobe large and rounded, 2.75–4.5 cm long, margin with minute teeth at the vein endings, tip elongate *c.* 1.5 cm long; venation palmate-pinnate, 2(–3) pairs at the base and 2–3 pairs along the midrib with 1–2 in the basal lobe, branching *c.* halfway to margin, slightly impressed above, beneath prominent with matted brown hairs. **Inflorescences** axillary, reddish green, without hairs, erect, longer than the adjacent leaf, (2–)6–10.5 cm long with two main branches *c.* 1.25 cm long, stalk 4–9 cm long, lengthening to 11 cm in fruit, male flowers up to 7, female flowers 2–4, male flowers open first. Bracts narrowly triangular, *c.* 5 × 1.5 mm, tip acute, soon falling. **Male flowers** with a stalk 7–10(–22) mm long; tepals 4 (rarely 3), white tinged pink, margin not toothed, tip rounded, without hairs, outer two rotund, 11–17 × 14–17 mm, inner two (or rarely one) narrowly oval, (6–)10–13 × 5–8 mm; stamen cluster *c.* 5 × 4 mm, not stalked; filaments 2–3 mm long, anthers pale yellow, narrowly oblong, *c.* 2 mm long, tip rounded, opening by slits. **Female flowers** with a reddish stalk 2–3 mm long; ovary green tinged red, wings 3, unequal, locules 2, placentas 2 per locule; tepals 4, margin not toothed, tip rounded, without hairs, white or pink, deep pink at the tip, broadly oval, *c.* 10 × 7 mm, innermost narrowly obovate, white, *c.* 5 × 3 mm; styles 2, styles and stigmas golden yellow, *c.* 4 mm long, stigmas spiral. **Fruit** a splash cup pendent on a thick and flexible stalk *c.* 25 mm long, capsule 15–17 × 28–36 mm, without hairs, locules 2, wings unequal, the larger oblong with rounded tip, thick and fibrous, 12–19 mm

(Right). Male flowers of *Begonia fraseri*. (Opposite). The semi-erect stem of the Fraser's Hill Begonia, *Begonia fraseri*, clings to a tree trunk for support.

BEGONIA FRASERI

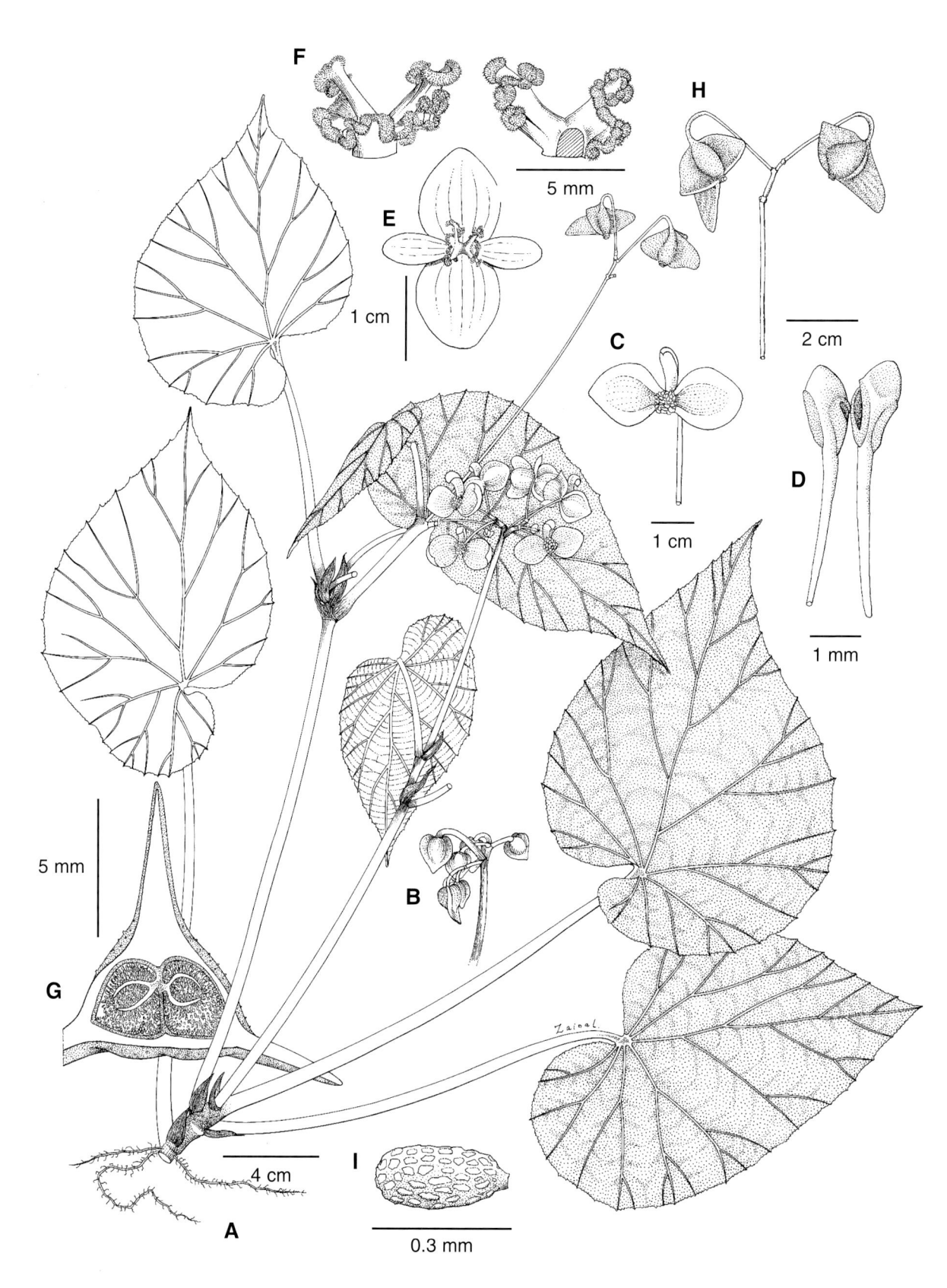

Begonia fraseri Kiew. **A.** The plant. **B.** Male bud. **C.** Male flower. **D.** Stamens. **E.** Female flower. **F.** Styles and stigmas. **G.** T.S. ovary. **H.** Fruit. **I.** Seed. (*RK 3829*)

wide, the smaller two bluntly triangular, thinly fibrous, 8–13 mm wide, splitting between the locules and wings. **Seeds** barrel-shaped, *c.* 0.3 mm long, collar cells *c.* half the seed length.

DISTRIBUTION. Endemic in Peninsular Malaysia. Pahang: Fraser's Hill and Gunung Benom.

HABITAT. Lower montane forest at about 1000–1700 m altitude in deep shade on well-drained earth slopes, sometimes scrambling over rocks or up the base of trees, often close to streams.

OTHER SPECIMENS. PAHANG—Fraser's Hill, *Addison SFN 37198* (SING), *SFN 37381* (SING); *Allen s.n.* 1953 (SING); *Burkill & Holttum SFN 8428* (SING); *Corner SFN 33176* (SING), *Kiew RK 1246* (SING), *RK 1247* (KEP, SING), *RK 3268* (SING), *RK 3562* (KEP), *RK 3829* (SING); *Purseglove P4315* (SING), *Siswa/i Tn-73* (UKMB); G. Benom *Native Collector FMS Mus s.n.* (SING).

NOTES. The Fraser's Hill Begonia belongs to the group of semi-erect begonias, *B. longicaulis*, *B. venusta* and to a lesser extent *B. maxwelliana*, that fall and root at the nodes unless supported by rocks or tree trunks. It is not common, being known from a few, small scattered populations.

Occasionally, as illustrated here, a plant may produce all of its male flowers with three tepals because one of the laterals does not develop. Typically the male flowers have four tepals.

Begonia fraseri. (Top left). Male flowers. (Top right and bottom). Female flowers. Note the abnormal male flowers have only three tepals.

31. THE GUNUNG TAHAN BEGONIA
Begonia longicaulis Ridl.
(Latin, *longicaulis*=long-stem)

Ridley, JSBRAS 75 (1917) 35, FMP 1 (1922) 120; Irmscher, MIABH 8 (1929) 120. **Type:** *Ridley s.n.*, July 1911, Gunung Tahan, Pahang (lecto SING, here designated; iso K).

Stem at first rhizomatous, stout *c.* 7–8 mm thick, then semi-erect and sparingly branched, falling and rooting at the nodes, red, succulent, *c.* 30–40 cm tall and 4–5 mm thick, all parts of the plant without hairs; without a tuber. Stipules narrowly triangular, 13–23 × 3–5 mm, margin not toothed, tip tapering to a point, persistent. **Leaves** distant and 6.5–16 cm apart; stalk (12–)16–25 cm long, grooved above; blade very oblique, plain, glossy bright green, thinly succulent, papery when dried, ovate, strongly asymmetric, 10.5–13.5 × 10–13.5 cm, broad side 6–8 cm wide, base deeply heart-shaped, slightly unequal, basal lobe 3(–5) cm long, margin minutely toothed, teeth longer at the vein endings, tip abruptly narrowed *c.* 1(–2) cm long; venation palmate-pinnate, (1–)2 pairs at base and another (2–)3 pairs along midrib, 1–2 veins in the basal lobe, branching *c.* halfway to margin, slightly impressed above, prominent beneath. **Inflorescences** axillary, erect, longer than the leaves, 23–37 cm long with two main branches, stalk 21–32 cm long, male flowers 5–10, female flowers 2–4, male flowers open first. Bract pair ovate, *c.* 22 × 10 mm, margin not toothed, tip pointed, soon falling. **Male flowers** with a stalk 10–25 mm long; tepals 4, pink, without hairs, margin not toothed, tip rounded, outer two broadly ovate, *c.* 20 × 15–20 mm, inner two oblong *c.* 17 × 10 mm; stamens many, cluster globose, 5–8 mm across, stalk *c.* 1 mm long; filaments *c.* 2 mm; anthers yellow, narrowly oblong, *c.* 1.75 mm long, tip rounded, opening by slits. **Female flowers** with a

(Above). The Gunung Tahan Begonia is known only from the gully near the summit of Gunung Tahan. (Opposite). The semi-erect stem of the Gunung Tahan Begonia scrambles over rocks. Male flower (inset).

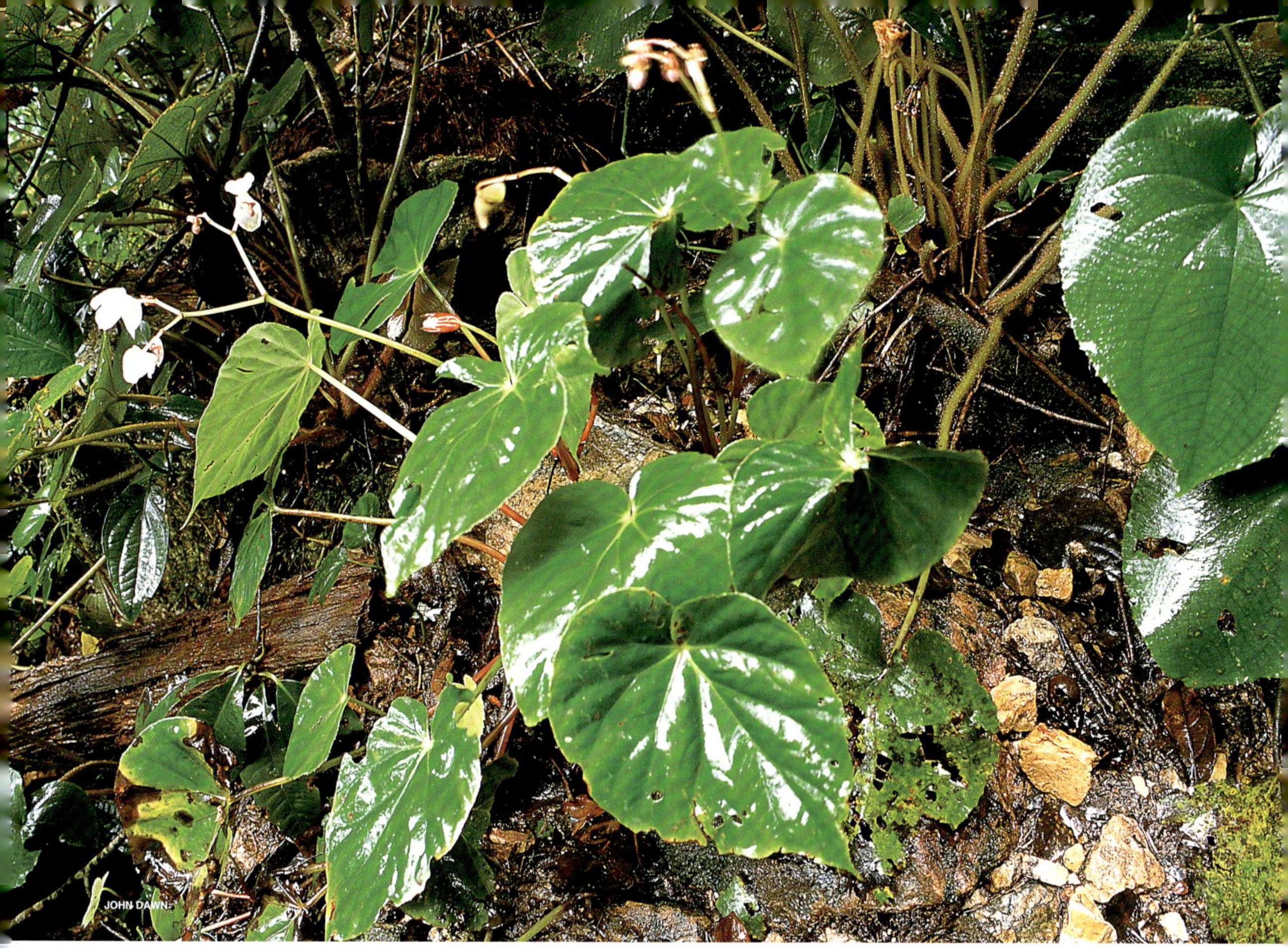
JOHN DAWN

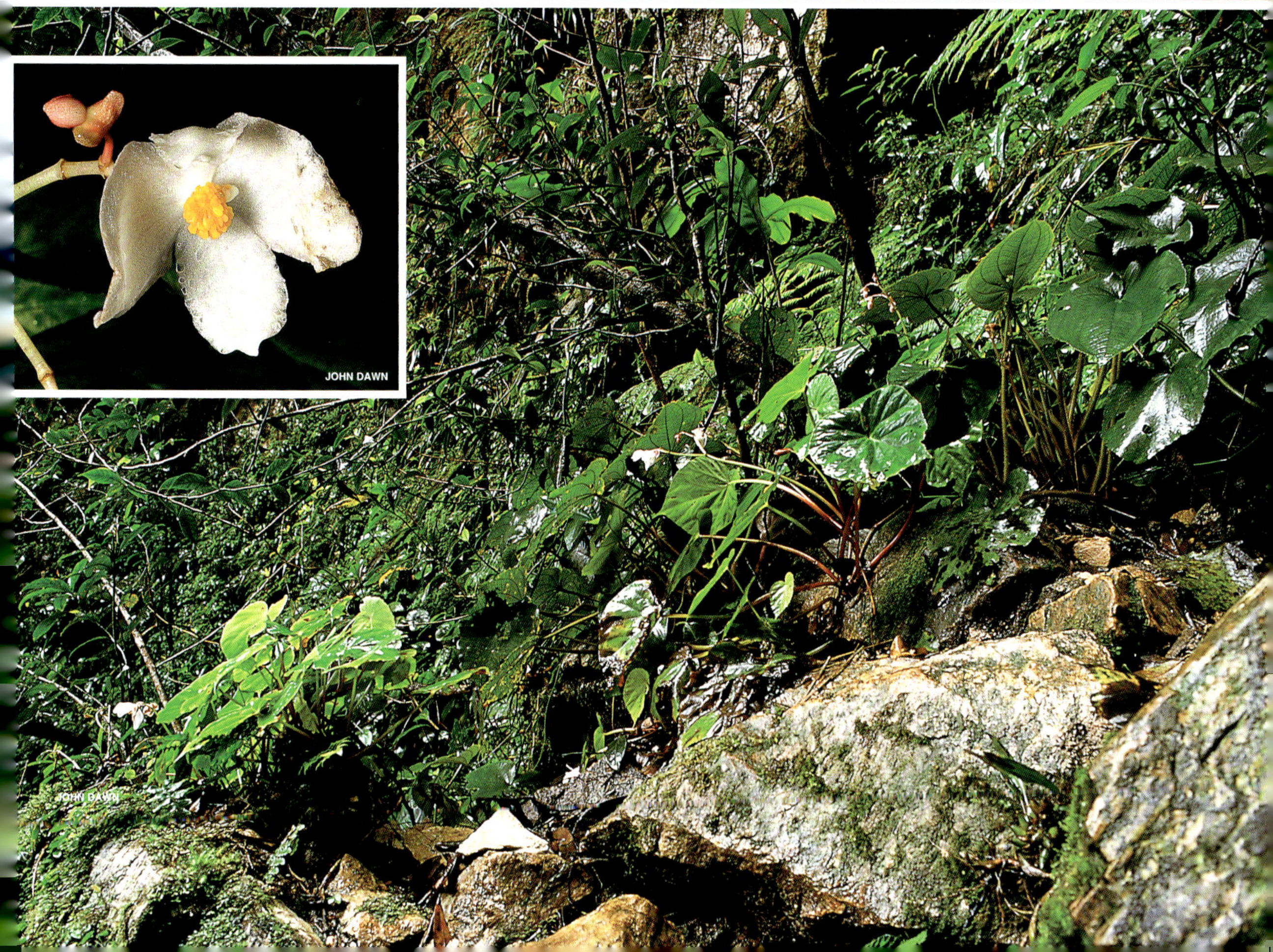
JOHN DAWN

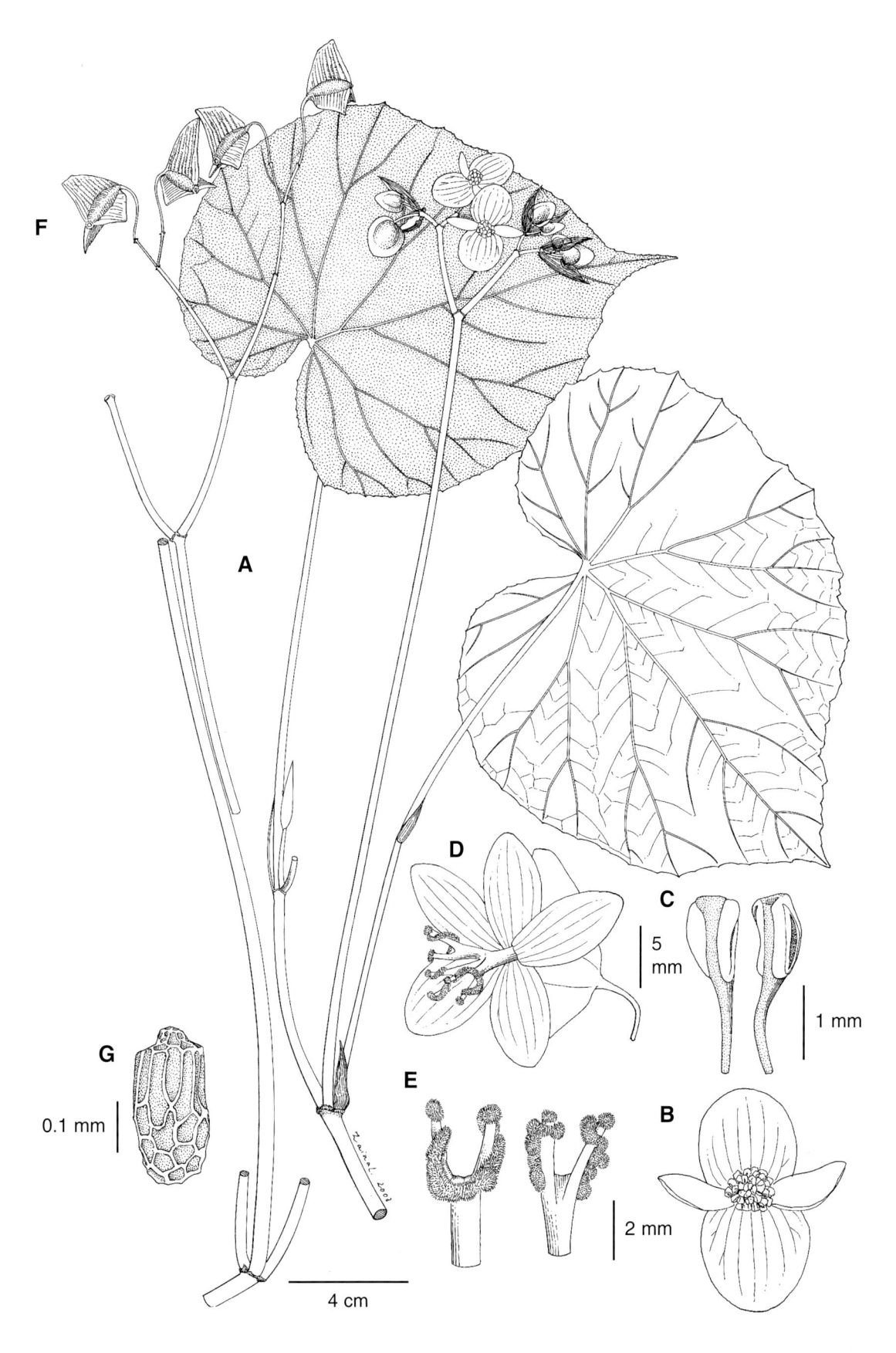

Begonia longicaulis Ridl. **A.** The plant. **B.** Male flower. **C.** Stamens. **D.** Female flower. **E.** Styles and stigmas. **F.** Fruit. **G.** Seed. (*RK 2454*)

stalk to 15 mm long; ovary *c.* 7 mm long, wings 3, unequal, locules 2, placentas 2 per locule; tepals 5, pink, margin not toothed, tip rounded, outer broadly oval, 12–14 × *c.* 6 mm, inner similar but smaller; styles 2, styles and stigmas yellow, 6–7 mm long, stigmas spiral. **Fruit** a splash cup pendent on a stiff stalk *c.* 20 mm long, capsule 15 × 28–30 mm, locules 2, wings unequal, larger fibrous, 15–18 mm wide narrowing to 2 mm long at the tip, smaller two rounded, 8–10 mm wide, splitting between the locules and wings. **Seeds** barrel-shaped, *c.* 0.3 mm long, collar cells *c.* half the seed length.

DISTRIBUTION. Endemic in Peninsular Malaysia, known only from the gully on Gunung Gedong, Gunung Tahan, Pahang.

HABITAT. It grows on boulders or fallen logs in the damp gully on sandstone rocks at about 2000 m altitude.

OTHER SPECIMENS. PAHANG—Gunung Tahan, *Kiew RK 2454* (KEP, SING); *FRI 1505* (KEP, SING); *Ng FRI 1505* (KEP, SING), *FRI 20967* (KEP); *Wong & Wyatt-Smith WYK 13* (KEP).

NOTES. The Gunung Tahan Begonia has a confusing history. Ridley included plants of *Begonia venusta* from Gunung Korbu, Perak, in his original description of this species. Irmscher identified Ridley's Gunung Tahan specimen as *B. robinsonii* (now a synonym of *B. pavonina*), which is a small rhizomatous begonia. It is, however, distinct among the semi-erect begonias in its relatively smaller leaves on long petioles and in its long internodes with conspicuous large stipules. Like *B. fraseri*, it is basically a weak climber leaning against rocks and tree trunks for support.

(Right). Leaf and (below) fruits of *Begonia longicaulis*.

32. LOW'S BEGONIA
Begonia lowiana King
(Sir Hugh Low, Resident of Perak, 1877–1889)

King, JASB 71 (1902) 67; Ridley, FMP 1 (1922) 864; Irmscher, MIABH 8 (1929) 137. **Type:** *Wray 1567*, Gunung Berumban, Pahang (lecto K, here designated).

Stem pale green, erect and cane-like, succulent, arising from a basal woody rhizome, little branched, up to 130 cm tall, flowering at 30 cm, 1–1.5 cm thick at base, softly and densely hairy, hairs glandular, nodes reddish brown, swollen and brittle; without a tuber. Stipules pale green, broadly ovate, 15–20 × 8–10 mm, margin very hairy but not toothed, tip rounded, soon falling. **Leaves** distant, up to 9–18 cm apart; stalk green, reddish towards base, hairy, 7–26 cm long, 5–10 mm thick, round in cross-section; blade very oblique, softly hairy above, hairs slightly sticky and usually translucent or sometimes magenta, blade plain dull pale green, soft in life, thinly papery when dried, broadly ovate, asymmetric, l6–19.5 × 14–18.5 cm, broad side 9–11 cm wide, base deeply heart-shaped, lobes sometimes overlapping at base, basal lobe rounded, 7–9 cm long, margin scalloped in upper half and minutely toothed, tip elongated; venation palmate-pinnate, 2–3

The soft, densely hairy leaves of Low's Begonia, *Begonia lowiana*.

Low's Begonia, *Begonia lowiana*. (Painting by Wendy Gibbs)

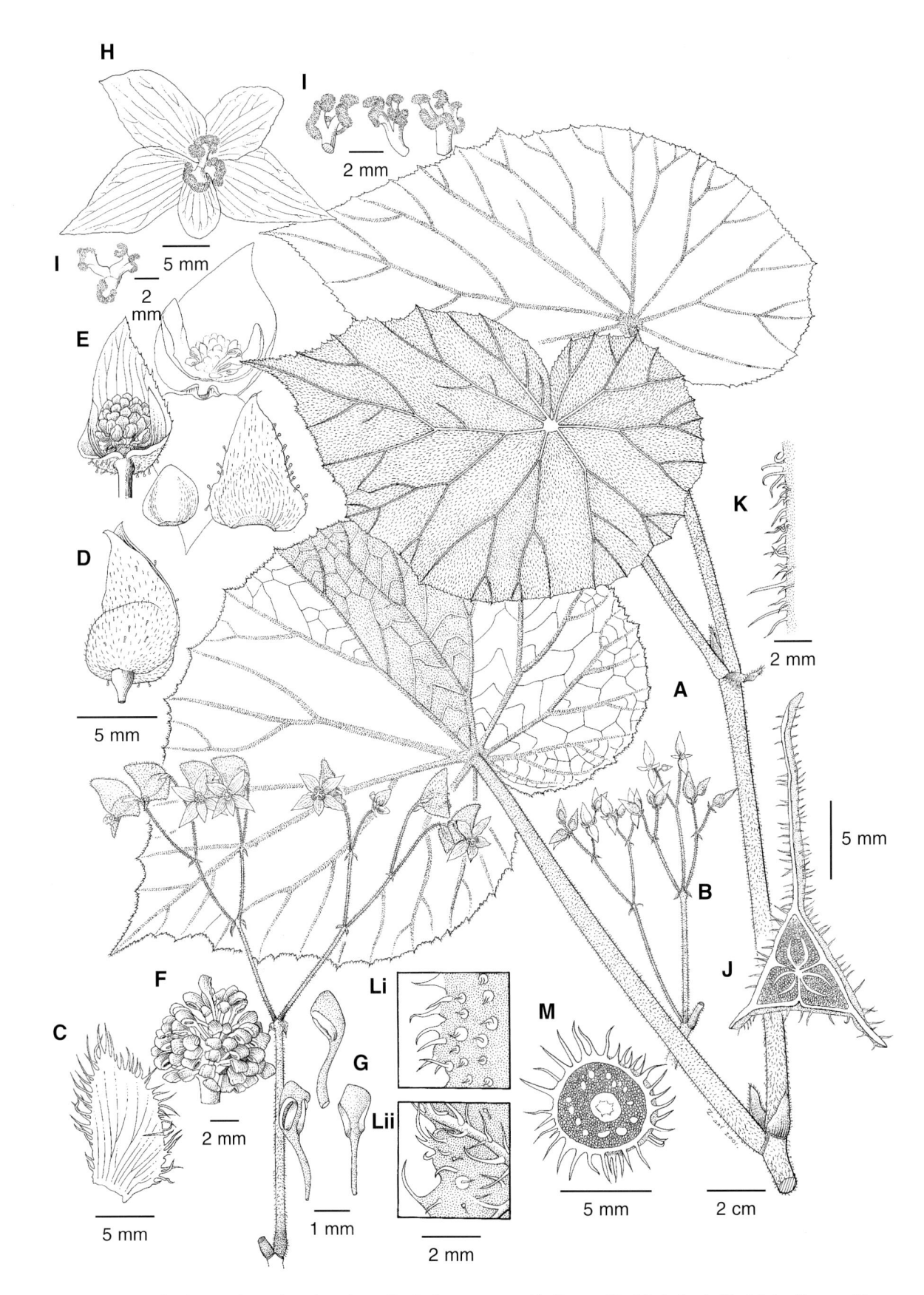

Begonia lowiana King. **A.** The plant. **B.** Inflorescence. **C.** Bract. **D.** Male bud. **E.** Male flower. **F.** Stamen cluster. **G.** Stamens. **H.** Female flower. **I.** Styles and stigmas. **J.** T.S. ovary. **K.** The leaf margin. **Li.** The upper leaf surface. **Lii.** The lower leaf surface. **M.** T.S. of leaf stalk. (*RK 4726*)

pairs at the base and 2–3 pairs along the midrib with 2–3 veins in the basal lobes, branching *c.* halfway to the margin, veins impressed above, beneath prominent, hairy and the same colour as the blade. **Inflorescences** usually axillary, sometimes terminal, greenish at the base, rosy pink in the upper branches, entire inflorescence (branches, bracts and pedicels) densely hairy with glandular hairs, erect, shorter than the leaves, up to 8 cm with two main branches *c.* 1.5 cm long with a stalk *c.* 6.5 cm long, male flowers many, female flowers 4, male flowers open first. Bract pair pale green with rosy tinge to deep rosy pink, ovate, up to 10 × 6 mm, hairy, margin not toothed, soon falling. **Male flowers** with a rosy pink, hairy stalk 15–50 mm long; tepals 4, margin not toothed, tip pointed, outer two pale rosy pink, rarely almost white, ovate, very hairy outside, 15–20 × 10–13 mm, inner two white tinged pale pink, narrowly ovate, hairless, *c.* 14 × 8 mm; stamens many, cluster hemispherical, 4 × 6–7 mm, stalk *c.* 1 mm; filaments 1–1.5 mm, anthers pale yellow, narrowly obovate, *c.* 1.25 mm long, tip rounded, opening by slits. **Female flowers** with a rosy pink stalk 10–33 mm long; ovary whitish becoming deep pink, densely hairy, 12–15 mm long, wings 3, unequal, locules 3, placentas 2 per locule; tepals 5, white at the base, rosy pink towards tip, ovate, margin not toothed, outer two *c.* 20 × 10–11 mm, hairy outside, tip pointed, inner two without hairs, tip rounded, *c.* 12 × 5–9 mm, the innermost much smaller *c.* 5 × 3 mm; styles 3, styles and stigmas pale yellow and slightly greenish, 5–6 mm long, stigmas spiral. **Fruit** dangling on a fine and hair-like stalk 30–37 mm long, capsule 12–22 × 23–35 mm, becoming hairless, locules 3, wings unequal, tip rounded, reddish green, thinly fibrous, larger wing up to 23 mm wide, smaller

Begonia lowiana. (Left). Male flower. (Top right). Female flower. (Bottom right). Fruit. Note the pointed tepals, which are unique among the Peninsula's begonias.

two 5–7 mm wide, splitting between the locules and wings, styles persistent. **Seeds** barrel-shaped, *c.* 0.3 mm long, collar cells *c.* half the seed length.

Distribution. Endemic in Peninsular Malaysia. Pahang: Cameron Highlands and Gunung Benom.

Habitat. In upper montane forest at *c.* 2000 m altitude. On Gunung Brinchang, Cameron Highlands, it is common in bamboo forest.

Other specimens. PAHANG—Gunung Benom, *Native Coll. FMS Mus. s.n.* (SING); Cameron Highlands: Gunung Beremban, *Chew 777* (SING), Gunung Brinchang, *Anthonysamy SA 326* (KEP), *SA 412* (KEP), *Kiew RK 1270* (KEP), *RK 1271* (KEP), *RK 4726* (SING), *RK 5227* (SING), *Saw FRI 34380* (KEP), *Spare SFN 3610* (SING), Gunung Irau, *Symington 38583* (KEP).

Notes. Low's Begonia is one of the two softly hairy begonias with pale green leaves and sticky hairs. The other is *Begonia jayaensis*, a much shorter plant that grows in the lowlands on limestone. *Begonia lowiana* has a soft stem, which is supported by growing against steep earth banks or tree trunks. It is the only begonia in the Peninsula that has pointed tepals.

Begonia lowiana is one of the cane-like begonias in the Peninsula.

33. THE CAVE BEGONIA
Begonia jayaensis Kiew, sp. nov.
(Latin, *-ensis*=originating from)

A *Begonia lowiana* King foliis brevioribus (latitudine longitudinem excedenti) et fructibus trilocularibus alias aequales ferentibus dignoscenda. **Typus:** *R. Kiew & S. Anthonysamy RK 2906*, 11 May 1990, Sungai Bring, Kelantan (holo SING; iso K, KEP, L)

Stem pale green, weak and succulent, erect, up to 22 cm tall, falling and rooting, unbranched, slender *c.* 3 mm thick, densely and softly hairy, hairs translucent, multicellular, glandular and sticky, *c.* 2 mm long; without a tuber. Stipules pale green, densely hairy, narrowly triangular, 7–10 × 2–4 mm, margin not toothed, densely hairy, tip acute ending in a long hair, persistent. **Leaves** distant, up to 1.5–4.5 cm apart; stalk pale green, densely hairy, 5–12 cm long, round in cross-section; blade oblique, plain light green, the dense covering of hairs giving a grey-green appearance, sometimes purplish brown beneath, soft in life, extremely thin and papery when dried, broadly ovate sometimes reniform, strongly asymmetric, 6–8.5(–10) × 6.5–11 cm, broad side 3.5–7 cm wide, base deeply heart-shaped, lobes sometimes overlapping, basal lobe 1–2.5 cm long, margin toothed and fringed by hairs, tip sometimes shortly acute; venation palmate, 2–3 pairs of veins at the base and 1–2 pairs along the midrib, with 1 or 2 veins in the basal lobe, branching *c.* one third of the way to the margin, veins densely hairy on both surfaces, plane above, beneath

(Above). The Cave Begonia, *Begonia jayaensis*, grows in and around caves and is known only from limestone on the Nenggiri River (following page).

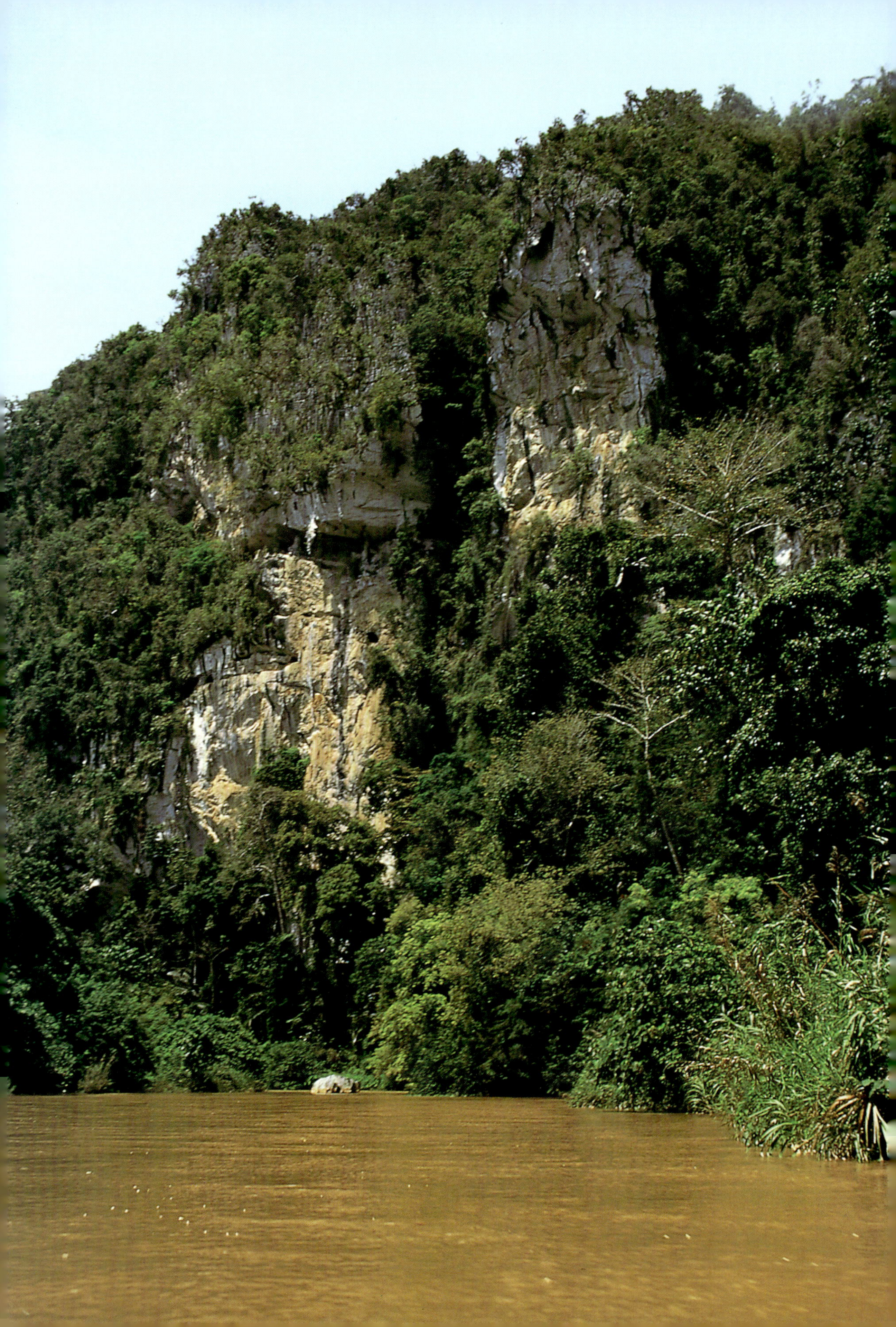

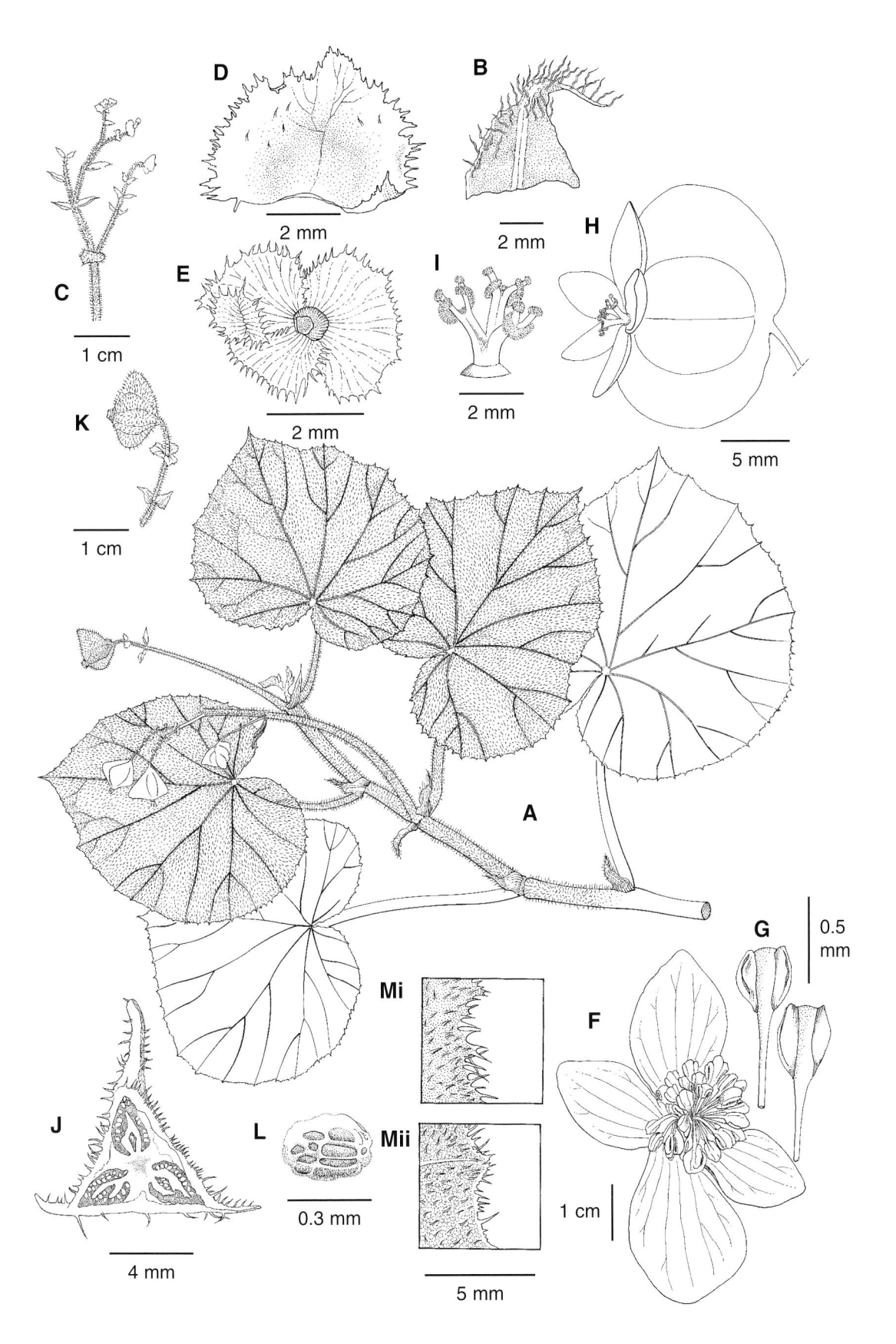

Begonia jayaensis Kiew. **A.** The plant. **B.** Stipules. **C.** Inflorescence branch. **D.** Bract. **E.** Bracteole. **F.** Male flower. **G.** Stamens. **H.** Female flower. **I.** Styles and stigmas. **J.** T.S. ovary. **K.** Fruit. **L.** Seed. **Mi.** The upper leaf surface. **Mii.** The lower leaf surface. (*RK 4912*)

prominent and the same colour as the blade. **Inflorescences** terminal, sometimes axillary, pale green, reddish at the nodes, densely hairy, erect, longer than the leaves, 9–19(–29) cm long with up to 4 branches 0.5–3 cm long, stalk 3–6(–10) cm long, male flowers many, female flowers 1(–2), male flowers open first. Bracts pale green, broadly ovate, *c.* 4 × 5 mm, margin not toothed, soon falling; bracteoles in pairs and clasping the inflorescence branches, greenish white, broadly ovate, *c.* 1.5 × 2 mm, persistent. **Male flowers** with a pale pink stalk *c.* 8–13 mm long; tepals 4, greenish white in bud, white when open, hairless, margin not toothed, tip rounded, outer two rotund, 5–8 × 5–8 mm, inner two narrowly oval, 5–7 × 2–3.5 mm; stamens *c.* 17, cluster lax, globose, *c.* 2 mm across, not stalked; filaments *c.* 0.75 mm long; anthers golden yellow, broadly obovate, *c.* 0.5 mm long, tip deeply notched, opening by slits. **Female flowers** with a pale green stalk *c.* 6 mm long; ovary green, hairy, *c.* 11 × 10 mm, wings 3, almost equal, 2–3 mm wide, locules 3, placentas 2 per locule; tepals 4 or 5, greenish white or white, without hairs, margin not toothed, tip slightly pointed, outer ovate-rotund, 6–7 × 6–7 mm, innermost similar but smaller; styles 3, styles and stigmas yellow, *c.* 2 mm long, stigmas spiral. **Fruits** pendent on a fine, hair-like stalk 5–10 mm long, capsule 7–11 × 4–19 mm, hairy, locules 3, wings equal, rounded, thinly fibrous, *c.* 2–6 mm

(Above). The Cave Begonia, *Begonia jayaensis*. (Opposite). *Begonia jayaensis*. (Left). Male flowers and fruit. (Right). Female flowers.

wide, splitting between the locules and wings, tepals persisting in the fruit. **Seeds** barrel-shaped, *c.* 0.3 mm long, collar cells *c.* half the seed length.

DISTRIBUTION. Endemic in Peninsular Malaysia. Kelantan: Sungai Nenggiri (Gua Jaya and Gua Chawan) and Sungai Jenera (Sungai Bring limestone).

HABITAT. Restricted to limestone, it grows on guano in caves, on the dry base of cliff faces and on stalactites above the cave mouth in light shade.

OTHER SPECIMENS. KELANTAN—Sungai Nenggiri: Gua Chawan, *Kiew & Anthonysamy RK 2892* (SING), Gua Jaya, *Kiew & Anthonysamy RK 2906* (SING), *RK 2890* (SING), Kiew *RK 4912* (KEP, L, SING).

NOTES. Among the Peninsular species, only the Cave Begonia and *Begonia lowiana* have soft, thin leaves with sticky hairs. However, they differ in many features as *B. lowiana* is a large plant that grows over a metre tall and has thick stems and large leaves, its flowers are larger, rosy red and the tepals are always pointed and hairy outside, and the fruit is larger and has three locules but one wing is much larger than the other two.

Begonia jayaensis belongs to sect. *Diploclinium* on account of its fruit having three locules and two placentas per locule. It is the second Peninsular begonia in this section. It is also unusual in the low number of stamens (less than 20 per flower) and persistent bracteoles.

It appears to be an ephemeral plant, dying down and growing up again from seed. Many seedlings can always be found. Within the cave mouth on guano in Gua Jaya, it forms dense drifts with all the plants having their leaves orientated toward the light.

It is extremely local, being known from just three limestone hills and is considered a rare and endangered species (Kiew, 1991).

34. CORNER'S BEGONIA
Begonia corneri Kiew
(E.J.H. Corner, Assistant Director, Botanic Gardens Singapore, 1929–1941).

Kiew, Bot. Jahrb. Syst. 113 (1991) 272. **Type:** *Corner SFN 30748*, 20 Nov 1935, Sungai Nipa, Trengganu (holo SING, iso SING).

Stem creeping above the ground and rooting at the nodes, nodes not swollen, reddish purple or brown, succulent, hairs dense, brown and 2–2.5 mm long, *c.* 28 cm long, unbranched, slender *c.* 4 mm thick; without a tuber. Stipules narrowly triangular, 10–18 × 3–4 mm, margin not toothed, hairy, tip acute, persistent. **Leaves** distant, 2–5 cm apart; stalk reddish purple with dense brown hairs, 3.5–6 cm long, *c.* 2 mm thick, grooved above; blade slightly oblique, glossy, plain green or yellowish green, thin in life, papery when dried, oval to ovate, slightly asymmetric, 7.5–11 × 6.5–9 cm, broad side 4.5–5.5 cm wide, base slightly heart-shaped, basal lobes more or less equal, *c.* 2–3 mm long, margin reddish, not toothed, fringed by brown hairs, tip acute; venation palmate-pinnate, 2 pairs of veins at the base and 2–3 pairs along the midrib, branching *c.* halfway to the margin, with another pair in the basal lobes, veins impressed above, beneath prominent and reddish purple with dense brown hairs, colour showing through to the upper surface. **Inflorescences** axillary from the axils of the older leaves or on the bare stem after the leaves have fallen, without hairs, shorter

Foliage of Corner's Begonia, *Begonia corneri*.

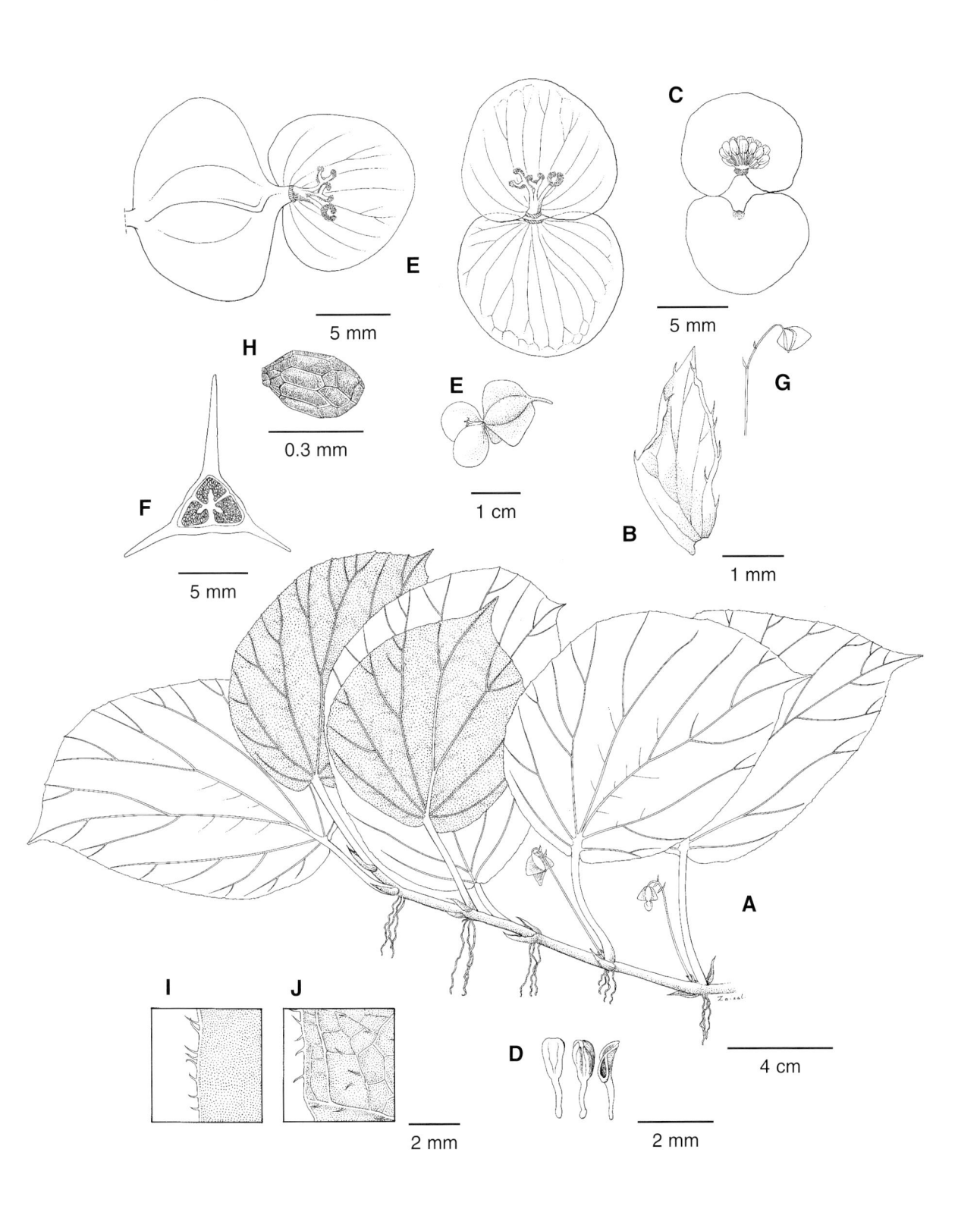

Begonia corneri Kiew **A.** The plant. **B.** Stipule. **C.** Male flower. **D.** Stamens. **E.** Female flower. **F.** T.S. ovary. **G.** Fruit. **H.** Seed. **I.** The upper leaf surface. **J.** The lower leaf surface. (a, b, i *SFN 30748*; c–h *Piee & Yap s.n.*)

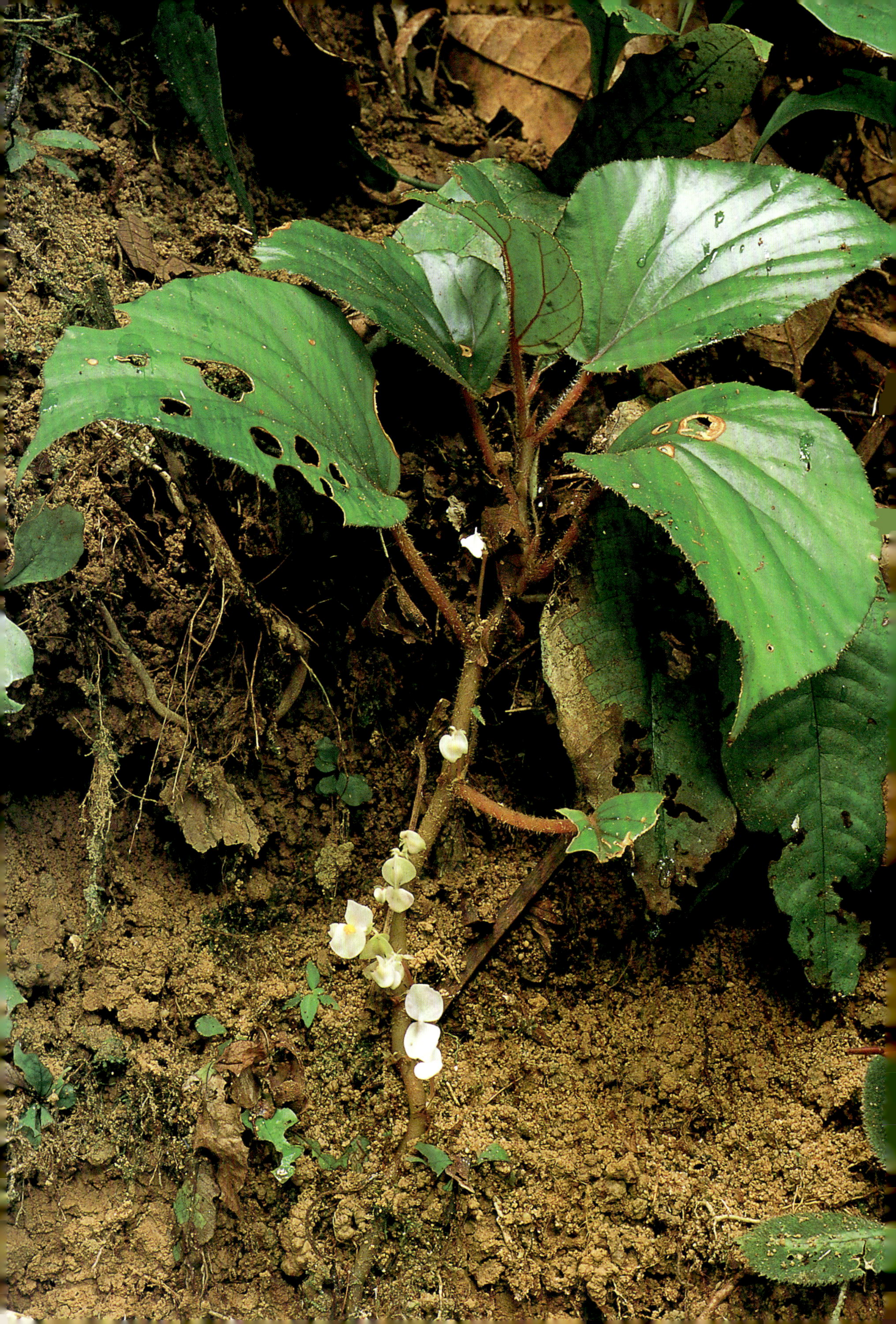

than the leaves, 2–3 cm long, unbranched, few-flowered with *c.* 5 male flowers and 1 female flower, male flowers open first. Bract pair reflexed, narrowly triangular, *c.* 2 × 1 mm, margin with brown hairs, soon falling. **Male flowers** with a white, minutely hairy stalk 5–8 mm long; tepals 2, pure white, broadly ovate, hairless, 8–10.5 × 11–14 mm, margin not toothed, tip rounded; stamens many, cluster globose *c.* 3–4 mm across, without a stalk; filaments *c.* 1–1.5 mm long; anthers obovate, yellow, *c.* 1–1.5 mm long, tip slightly notched, opening by slits. **Female flowers** with a minutely hairy stalk 4–5 mm long, ovary whitish green, 8–10 mm long, wings 3, equal, locules 3, placenta 1 per locule, tepals 2, hairless, broadly rotund, 6–9 × 8.5–10 mm, margin not toothed, tip rounded; styles 3, styles and stigmas *c.* 2.5 mm long, stigmas U-shaped. **Fruits** pendent on a thin recurved stalk 4–5 mm long with a tuft of hairs at base of the capsule, capsule 9–12 × 12–15 mm, hairless, locules 3, wings 3, equal, rounded, 4–5 mm wide, thin and papery, splitting between the locules and wings, styles persistent. **Seeds** barrel-shaped, 0.25–0.3 mm long, collar cells more than half the seed length.

DISTRIBUTION. Endemic in Peninsular Malaysia—Trengganu, known only from Sungai Nipa, Cukai.

HABITAT. On shaded earth banks near rivers, below 85 m altitude.

OTHER SPECIMEN. TRENGGANU—Sungai Nipa, *Piee & Yap s.n.* 12 July 2002 (SING).

NOTES. Corner's Begonia is extremely rare and known only from two collections from the same area. Most of the forest in the type locality, Sungai Nipa, has been logged and the riverbanks are now exposed to sunlight making them an unsuitable habitat for this begonia. Currently Corner's Begonia is known from a single population. It is a critically endangered species, being both rare and extremely local.

It is a very distinctive species in its creeping habit with widely spaced leaves, a habit in Peninsular begonias shared only by *Begonia thaipingensis*, which is, however, quite different as it has more rounded leaves and much longer inflorescences compared with those of *B. corneri*. Among the Peninsula's begonias, Corner's Begonia is unique in its short inflorescences with a single fruit, which are produced from the lower leaf axils or from the bare stem.

(Left). *Begonia corneri*. Male and female flowers. (Opposite). *Begonia corneri* with short inflorescences arising from the stem.

35. THE LARUT BEGONIA
Begonia forbesii King
(H.G. Forbes, botanist and ethnologist, collected in Sumatra 1880–1881)

King, JASB 71 (1902) 58; Ridley, FMP 1 (1922) 855; Irmscher, MIABH 8 (1929) 159. **Type:** *Wray 2476*, August 1888, Sungai Larut, Perak (lecto K, here chosen).

Stem rhizomatous, rooting at the nodes, wiry, without hairs, unbranched, slender, 4–7 cm long, 3–4 mm thick; without a tuber. Stipules narrowly triangular, in bud densely hairy on the margin, becoming hairless, 12–13 × *c.* 2 mm, margin not toothed, tip ending in a hair, persistent. **Leaves** tufted, up to 5 mm apart; stalk dark brownish red, minutely hairy, 8–13.5 cm long, round in cross-section; blade oblique, plain, dull mid-green above, whitish green or sometimes slightly reddish beneath, thinly fleshy in life, papery when dried, broadly ovate, slightly asymmetric, 7.5–10.5 × 6–11.5 cm, broad side 3.5–7 cm wide, base heart-shaped, basal lobe 0.5–2 cm long, margin with minute teeth, in larger leaves scalloped between the veins in the upper half of the leaf, tip elongated; venation palmate, 2–3 pairs of veins, branching towards margin with another 1–2 veins in the basal lobe, veins slightly impressed above, beneath slightly prominent, slightly darker than blade. **Inflorescences** axillary, pale reddish brown, minutely hairy when young, longer than the leaves, 6.5–10.5 cm long with two main branches 0.7–2 cm long, stalk 5.75–9 cm long, lengthening to 15.5 cm in fruit, male flowers many but with only *c.* 4 open at any one time, female

(Above). The Larut Begonia, *Begonia forbesii*. (Opposite). The Larut Begonia is known from a single waterfall.

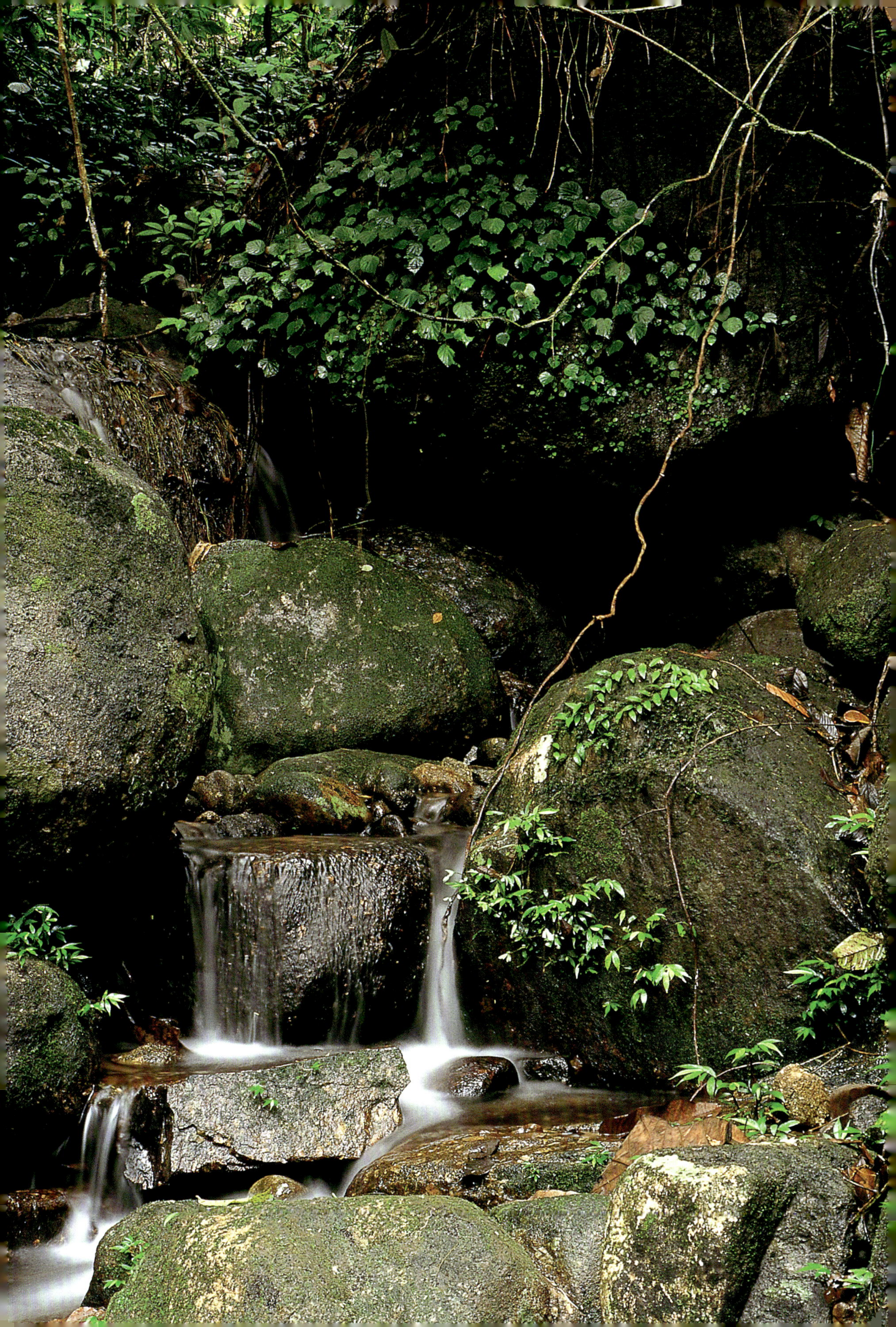

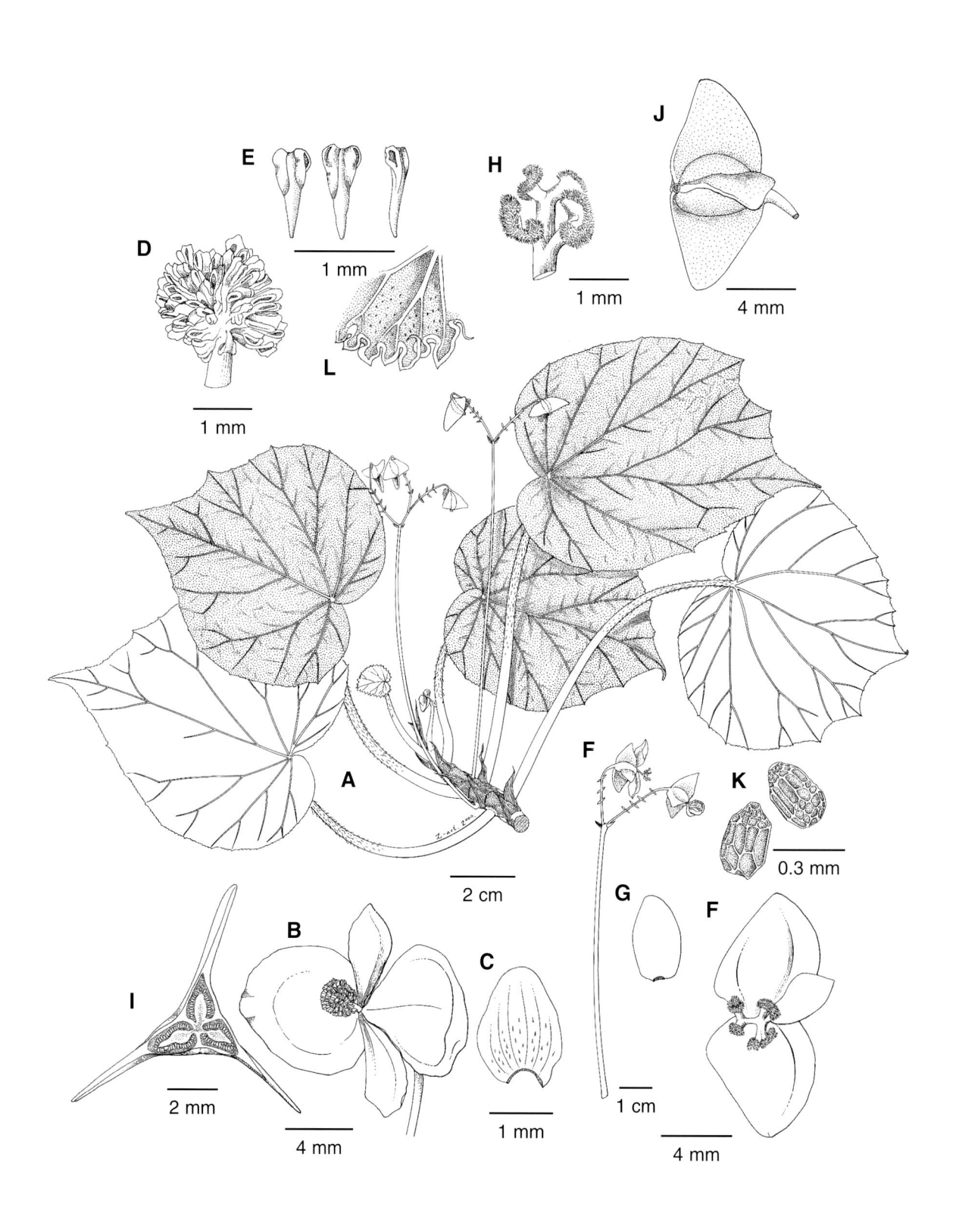

Begonia forbesii King. **A.** The plant. **B.** Male flower. **C.** Male tepals. **D.** Stamen cluster. **E.** Stamens. **F.** Female flower. **G.** Female tepal. **H.** Styles and stigmas. **I.** T.S. ovary. **J.** Fruit. **K.** Seed. **L.** The lower leaf surface. (*RK 4909*)

BEGONIA FORBESII

flowers 4, male flowers open first. Bract pair reddish, narrowly ovate, *c.* 4 × 1.5–2 mm, margin not toothed, persistent. **Male flowers** with a pink stalk *c.* 10 mm long; tepals 4, margin not toothed, tip rounded, outer two white with a pinkish tip, minutely hairy outside, rotund, 8–10 × 8–9 mm, inner two white, hairless, obovate, *c.* 7 × 4 mm; stamens many, cluster globose, *c.* 2.5–3 mm across, stalk *c.* 0.5 mm; filaments *c.* 0.5 mm long; anthers yellow, narrowly obovate, *c.* 0.75 mm long, tip notched, opening by slits. **Female flowers** with a stalk 7–8 mm long; ovary pale rosy pink, 4–6 mm long, wings 3, equal, locules 3, placenta 1 per locule; tepals 3–4, rosy pink, hairless, margin not toothed, tip rounded, outer two rotund, 7–7.5 × 7–8 mm, innermost narrowly oval, *c.* 6 × 2.5 mm; styles 3, styles and stigmas yellow, *c.* 2 mm long, stigmas U-shaped. **Fruits** dangling

Begonia forbesii. (Top). Rhizomatous habit. (Bottom left). Male flowers. (Bottom middle). Young female flower. (Bottom right). Fruits.

on a fine, hair-like stalk 7–9 mm long, capsule 5–7 × 13–18 mm, hairless, locules 3, wings 3, equal, tip rounded, thin and papery, 4–7 mm wide, splitting between the locules and wings. **Seeds** barrel-shaped, 0.3 mm long, collar cells *c.* half the seed length.

DISTRIBUTION. Sumatra (doubtful) and Peninsular Malaysia—in the Peninsula known only from Sungai Larut, Perak.

HABITAT. Rocky banks beside a waterfall in a small stream in light shade.

OTHER SPECIMEN. PERAK—Sungai Larut, *Kiew RK 4909* (KEP, SING).

NOTES. The Larut Begonia was originally described from two specimens, one from Sungai Larut, Perak, and the other from Sumatra (*Forbes 2666*). Unfortunately, I have not been able to find the Forbes's specimen (it is not in the BM, K or SING herbaria), so it cannot be ascertained whether the two specimens really belong to the same species, but in view of the high levels of endemism in Asian begonias, it is highly unlikely. In addition, the original description reported the species as having dense rusty hairs, which the Perak plants do not have. Since Wray's specimen was originally cited, it is here chosen as the lectotype for *B. forbesii* and the description refers only to the Perak population. King recorded the female flowers with four tepals but flowers in the population we observed had only three. The specimens we collected have fruits with a single placenta in each of the locules, firmly placing it in Section *Reichenheimia*.

In Peninsular Malaysia, the Larut Begonia was collected in 1888 and was only recently rediscovered, again at Sungai Larut, by Lim Swee-Yian and fellow birdwatchers. It is extremely rare, currently known from a single population, which is in danger of its habitat being destroyed by encroachment by farming. It is a very vulnerable species as any opening of the canopy would wipe out the entire population.

A female flower of *Begonia forbesii*.

36. LENG-GUAN'S BEGONIA
Begonia lengguanii Kiew, sp. nov.
(Saw Leng-Guan, Curator of the Herbarium, Forest Research Institute Malaysia, 1985–present)

Begoniae yappii Ridl. affinis, sed foliis ovatis usque rotundatis margine minute denticulatis (nec ovalibus et integris), et floribus roseis (nec albis) differt. **Type:** *Saw Leng Guan FRI 36295*, 1st July 1988, Bukit Rengit, Pahang (holo KEP).

Stems rhizomatous, brown, succulent, unbranched, slender, up to 10 cm long, 3–4 mm thick, hairs sparse, white and translucent; without a tuber. Stipules green, densely hairy, narrowly triangular, 8–9 × 2–2.5 mm, margin not toothed, densely fringed with hairs, tip ending in a hair, persistent. **Leaves** tufted, 4–8 mm apart; stalk reddish brown, sparsely hairy, (2.5–)5(–11.5) cm long, shallowly grooved above, hairs long and white; blade oblique, plain greenish-grey with a red patch at the base of the blade at the junction with the stalk, scintillating above, slightly succulent, drying extremely thin and papery, broadly ovate to rotund, slightly asymmetric, 4–5.5 × 5–8.5 cm, broad side 2–3.25 cm wide, base slightly heart-shaped, basal lobes almost equal, 2–5(–7) mm long, margin minutely toothed, fringed by hairs, tip acute, *c.* 5 mm long; venation palmate, veins 2–3 pairs branching *c.* halfway to the margin, impressed above, beneath plane, sparsely hairy, in young leaves reddish brown, in mature leaves the same colour as blade. **Inflorescences** axillary, reddish brown, sparsely hairy, longer than the leaves, 8–13.25 cm long with two main branches 0.5–1.5 cm

(Above). Leng-Guan's Begonia, *Begonia lengguanii*. (Following page). *Begonia lengguanii* grows on rock faces beside waterfalls.

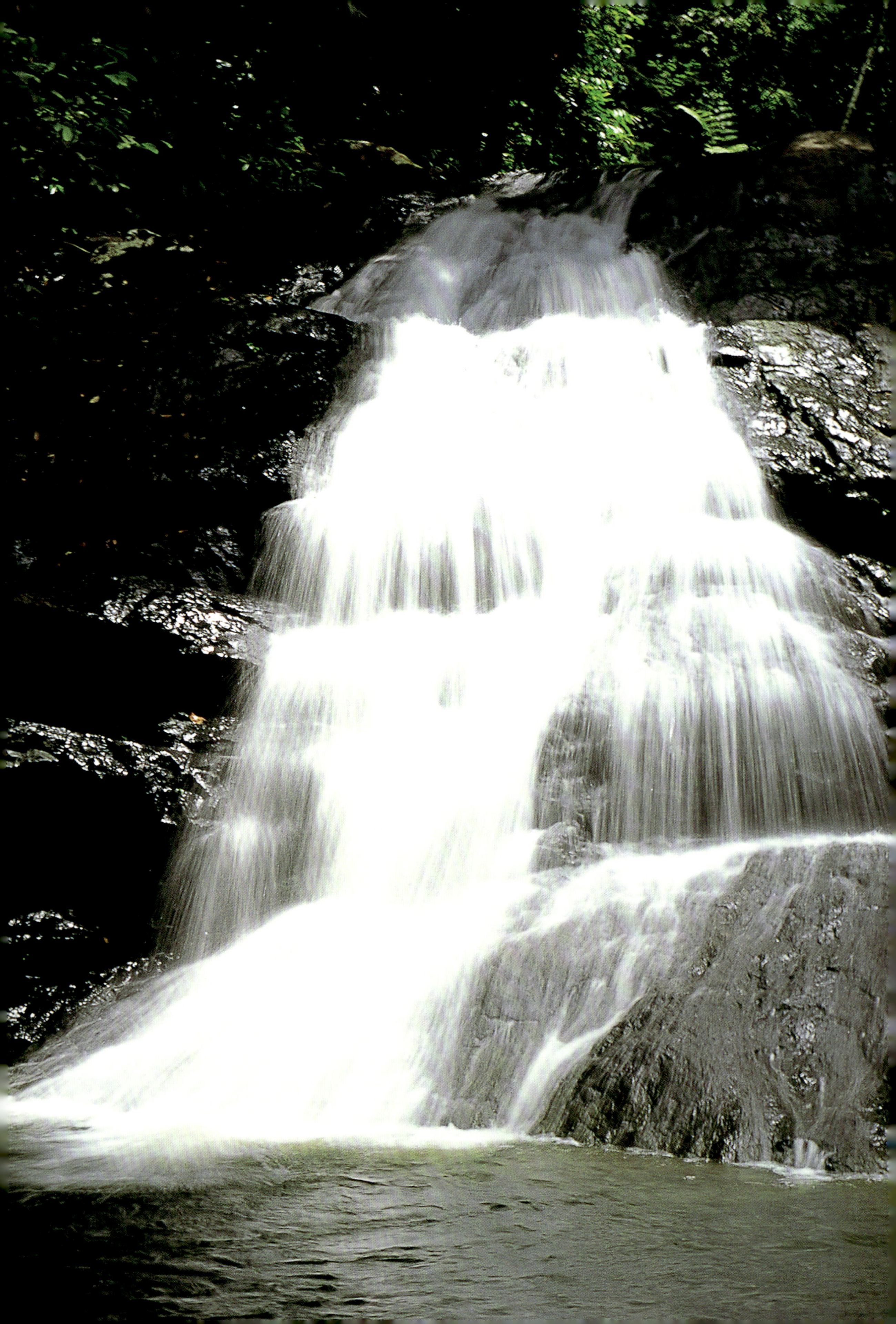

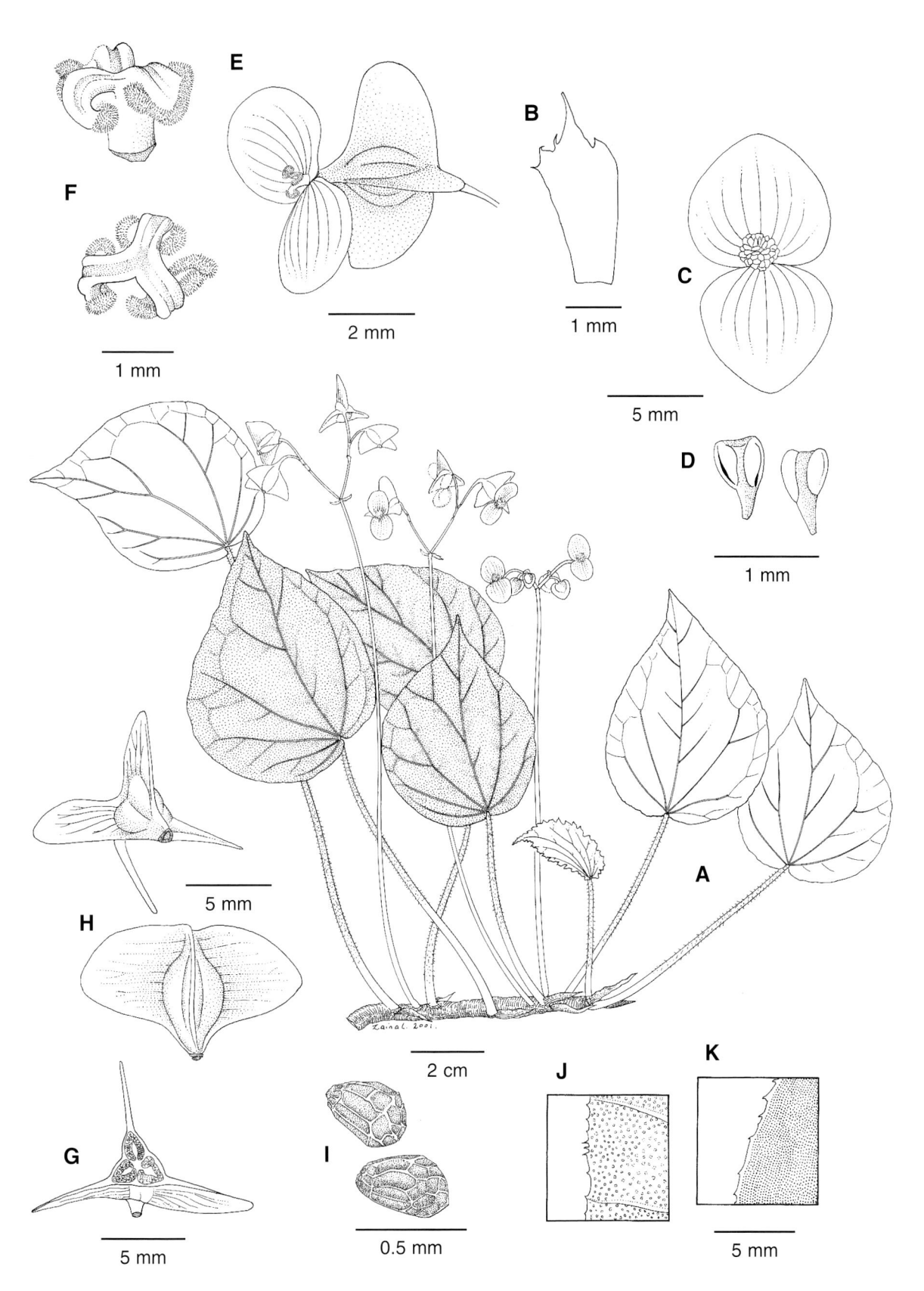

Begonia lengguanii Kiew **A.** The plant. **B.** Stipule. **C.** Male flower. **D.** Stamens. **E.** Female flower. **F.** Styles and stigmas. **G.** T.S. ovary. **H.** Fruit. **I.** Seed. **J.** The upper leaf surface. **K.** The lower leaf surface. (*RK 5201*)

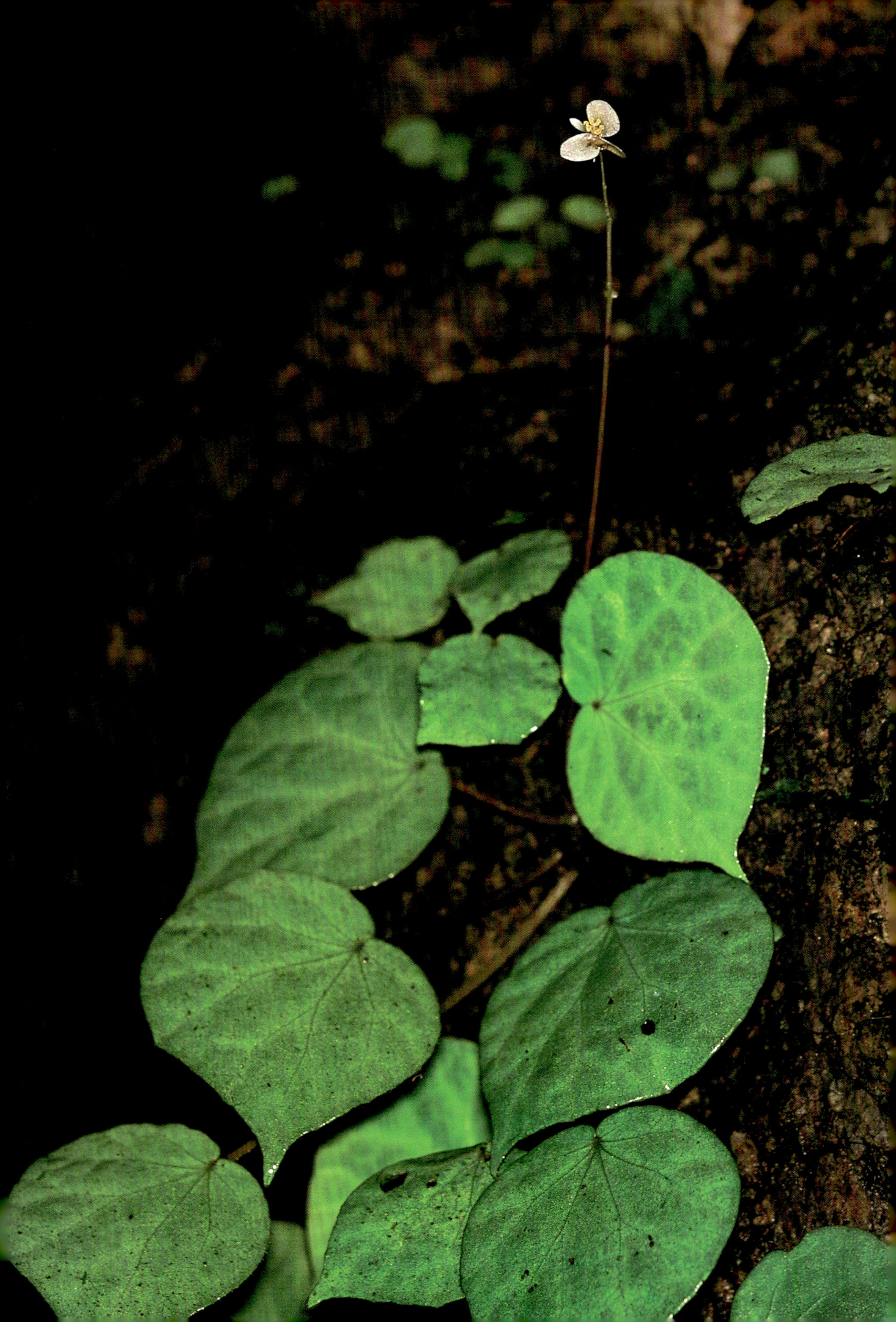

long, stalk 7–12 cm long, male flowers *c.* 5, female flowers 2, male flowers open first. Bract pair green, obovate, 2–3(–4) × 1–2 mm, margin fringed by hairs, persistent. **Male flowers** with a stalk (4–)7(–12) mm long; tepals 2, outside deep pink, inside pale pink, darker toward the centre, outside with microscopic glandular hairs, rotund, 6–9 × 7–10 mm, margin not toothed, tip slightly acute; stamens many, cluster globose, 1.5–2 mm across, stalk 1–2 mm long; filaments 0.25–0.75 mm long; anthers yellow, broadly obovate, *c.* 0.5–0.75 mm long, tip slightly notched, opening by slits. **Female flowers** with a green or reddish-green stalk 7–9 mm long; ovary pinkish, 6–11 mm long, wings 3, equal, locules 3, placenta 1 per locule; tepals 2, deep pink, hairless, broadly ovate, 4.5–6 × 5–7 mm, margin not toothed, tip slightly acute; styles 3, styles and stigmas golden yellow, 1.5–2.75 mm long, stigmas U-shaped. **Fruits** pendent on a fine, hair-like stalk 7–10(–15) mm long, capsule (5–)8(–10) × 11–15 mm, hairless, locules 3, wings 3, equal, triangular tapering to a rounded tip, thinly fibrous, 4–7 mm wide, splitting between the locules and wings. **Seeds** barrel-shaped, *c.* 0.4 mm long, collar cells almost as long as the seed.

DISTRIBUTION. Endemic in Peninsular Malaysia, known only from Bukit Rengit and the Lanjang Forest Reserve in south Pahang.

HABITAT. On wet rocks in shade by waterfalls in lowland forest at *c.* 70 m altitude.

OTHER SPECIMENS. PAHANG—Bukit Rengit in the Krau Game Reserve, *Kiew RK 5201* (SING); Lancang F.R., *B.H. Kiew s.n.* Sept 1986 (SING); Lancang, Bukit Tapah, *Ahmad Zainuddin C24* (UKMB).

NOTES. The species is named for Dr Saw, who first drew my attention to this pretty little begonia.

Begonia lengguanii belongs to sect. *Reichenheimia* in possessing fruits with three equal wings and each of the three locules having a single unbranched placenta. Within this section, it most resembles *B. yappii* in its leaves that are longer than wide, are almost equal sided and are not variegated. It is distinct from this species in its leaves that are ovate or rotund with a minutely toothed margin (*B. yappii* has broadly oval leaves that are not toothed), its pink male and female flowers (not white) and the tepals of the male flower that are as wide as long (*B. yappii* has tepals wider than long).

(Opposite). Leng-Guan's Begonia, *Begonia lengguanii*. (Above left). Fruits. (Centre). Female flowers. (Right). Male flower.

37. THE RAJAH BEGONIA
Begonia rajah Ridl.
(Sanskrit, raja=king)

Anon., Gard. Mag. (18 Aug 1894) 485 *nomen*; Ridley, Gard. Chron. 3, Ser 16 (25 August 1894) 213 & Fig 31, Kew Bull. (1895) App. 2, p. 34; Rolfe, Kew Bull. Misc. Info. (1914) 327; Ridley, FMP 1 (1922) 855; Irmscher, MIABH 8 (1929) 96; Kiew, Nature Malaysiana 14 (1989) 66; Kiew, Begonian 56 (1989) 53. **Type:** *Native Collector s.n.*, 1892, Trengganu [Tringganu] (holo K *ex* SING). **Synonym:** *Begonia peninsulae* Irmsch. ssp. *peninsulae* MIABH 8 (1929) 98. **Type:** 'Tringganu' without collector, number or date (holo K).

Stems rhizomatous, 6–8 mm thick; without a tuber. Stipules broadly triangular, *c.* 11 × 6 mm, persistent. **Leaves** tufted; stalk reddish pink, hairy, 2.5–25 cm long, round in cross-section; blade slightly oblique, glossy green and variegated bronze-green to purple-brown between the veins, veins sometimes yellowish green, beneath paler, blade prominently raised between the veins (bullate), stiffly succulent in life, rounded or kidney-shaped, asymmetric, 3–5.5(–15) × 4.5–9.5(–16) cm, base heart-shaped, basal lobes 2–4 cm long, margin scalloped and fringed by hairs, tip short and acute; venation palmate with 2–3 pairs of veins branching towards margin and another pair in the basal lobes, deeply impressed above, beneath prominent and sparsely hairy. **Inflorescences** axillary, rosy red, shortly hairy, longer than the leaves, 10–25 cm long with two main branches 3–8 cm long, stalk 7–12 cm long, male flowers many, female flowers up to 4, male flowers open first. Bract pair pale green, obovate, 2–8 × 1–4 mm, persistent. **Male flowers** with a stalk 10–27 mm long; tepals 4, margin not toothed, tip rounded, outer two white, pale rosy red outside, rotund, 4–12 × 5–12 mm, inner two white, obovate, 5–11 × 1.2–5 mm; stamens many, cluster globose; filaments 0.5–0.7 mm long; anthers broadly obovate, *c.* 0.5 mm long, tip slightly notched, opening by slits. **Female flowers** with a whitish green ovary, 4–8 mm long, wings 3, equal, locules 3, placenta 1 per locule; tepals 3 (rarely 4), pale pink, hairless, ovate, margin not toothed, 6–11 × 5–8.5 mm; styles 3, styles and stigmas, 2–3 mm long, stigmas spiral. **Fruits** dangling on a fine, hair-like stalk 9–10 mm long, capsule 6–7 × 5–6 mm, hairless, locules 3, wings equal, 4–5 mm wide, splitting between the locules and wings. **Seeds** not known.

DISTRIBUTION. Endemic in Peninsular Malaysia: Trengganu (without exact locality) and Johore.

HABITAT. Clinging to rocks along streams.

OTHER SPECIMEN. JOHORE—Sungai Selai, *Sam et al. FRI 47082* (KEP).

NOTES. *Begonia rajah* was collected in 1892 from Trengganu and grown in the Botanic Gardens, Singapore. From there, it was taken to England where, on 14 August 1894, it received a First Class Certificate from the Floral Committee of the Royal Horticultural Society. Illustrations of it appeared almost at once in the Gardeners' Chronicle and Gardeners' Magazine and it became a prized ornamental species. The Gardeners' Magazine of 25 August 1894 described it as 'one of the most distinct and handsome species introduced of late; its attractiveness is due to the shade of pale buff-green, bronze-green and reddish-brown, which contrast so effectively in the different lights and shades that play upon the curiously undulated surface of the leaf.'

It is not known exactly from where the original specimen was collected as the label only recorded it as from 'Tringganu' (=Trengganu) collected by a Native Collector. In 1986, I found a begonia in Trengganu with leaves with similar variegation and I thought I had rediscovered *Begonia rajah* (Kiew, 1989). However, the number of tepals in the male and female flowers is

BEGONIA RAJAH

different and the leaf was smaller and not rounded. It proved to be a distinct species, *Begonia reginula*.

In August 2002, Sam Yen-Yen made a surprising discovery when she encountered *Begonia rajah* growing in Johore, very distant from its original locality in Trengganu. This time there is no doubt it is the real *B. rajah* as the Johore plants have the characteristic bullate leaf with bronzy variegation and the male flowers have four tepals and the females three.

The Rajah Begonia has persisted for over a hundred years in cultivation, although it has a reputation of being difficult to grow, so it is not common. It has been successfully hybridized with begonias, such as the hardier *B. goegoensis* N.E. Brown, imparting bronzy highlights to the leaf of the hybrid.

Begonia rajah Ridl. (Top). Inflorescence with male flowers. (Centre). Male Flowers. (Below). Female flowers. (Right). Watercolour painting by Charles de Alvis. (Reproduced with permission of the Singapore Botanic Gardens)

38. THE QUEEN BEGONIA
Begonia reginula Kiew, sp. nov.
(Latin: reginula=little queen)

Begoniae rajah Ridl. affinis sed floribus masculis tepalis 2 (nec 4) et floribus feminis tepalis 2 (nec 3 vel 4) provisis differt. **Typus:** *Kiew RK 2278*, 29 April 1986, Ulu Setui, Trengganu (holo SING).

Begonia rajah sensu Kiew, Nature Malaysiana 14 (1989) 67, front cover; Kiew, Begonian 56 (1989) 54 *non B. rajah* Ridl.

Stems rhizomatous, green or reddish, wiry, without hairs, unbranched, slender, up to 15 cm long, 3–5 mm thick; without a tuber. Stipules pale green or reddish, narrowly triangular, 5–7 × 2–3 mm, margin fringed by hairs, not toothed, tip ending in a long hair, persistent. **Leaves** tufted, *c.* 2–3 mm apart; stalk rosy red to reddish brown, sparsely hairy, 2.5–6.5(–12) cm long, grooved above; blade

The Queen Begonia, *Begonia reginula*.

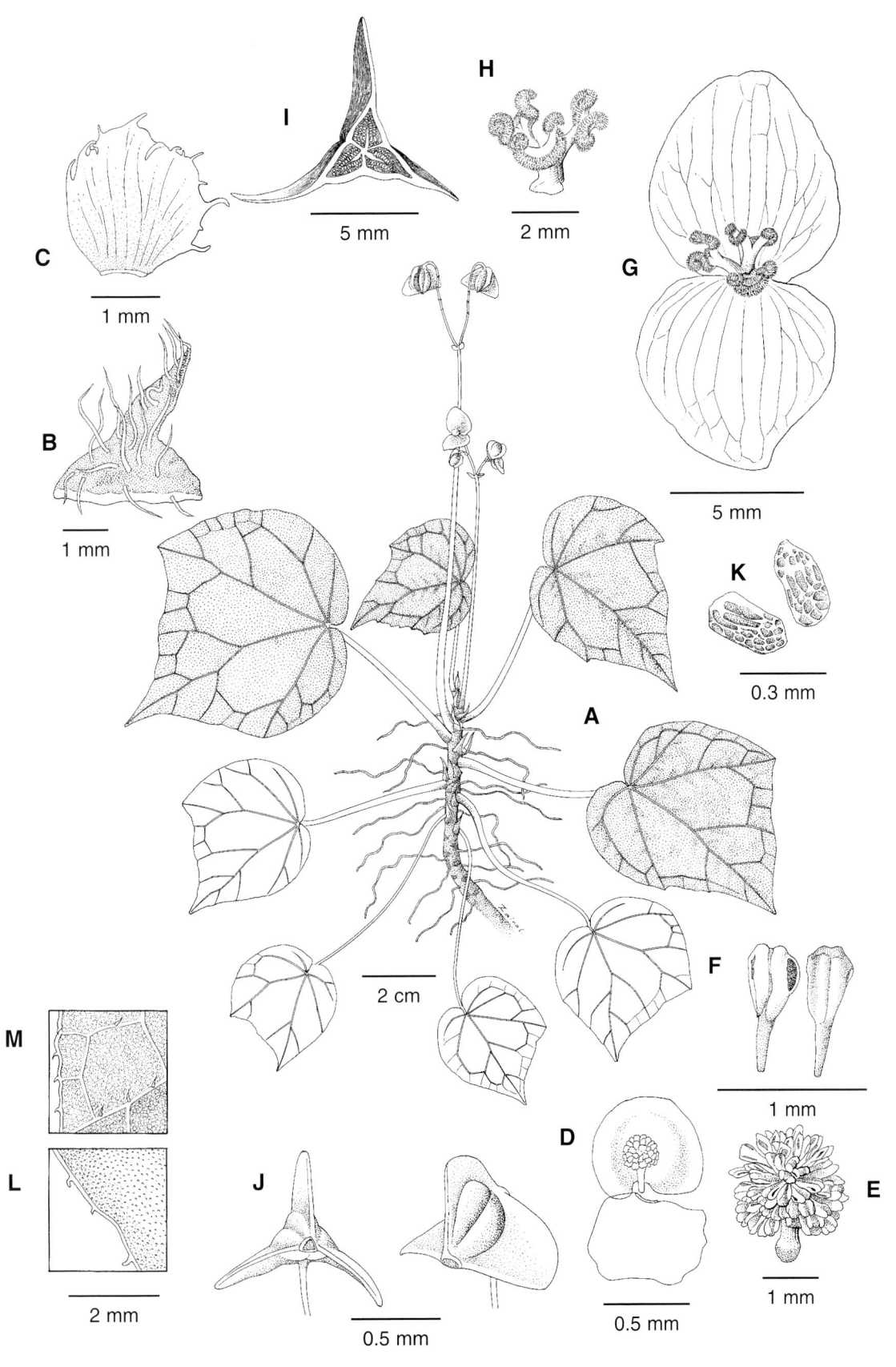

Begonia reginula Kiew. **A.** The plant. **B.** Stipule. **C.** Bract. **D.** Male flower. **E.** Stamen cluster. **F.** Stamens. **G.** Female flower. **H.** Styles and stigmas. **I.** T.S. ovary. **J.** Fruit. **K.** Seed. **L.** The upper leaf surface. **M.** The lower leaf surface. (*RK 5175*)

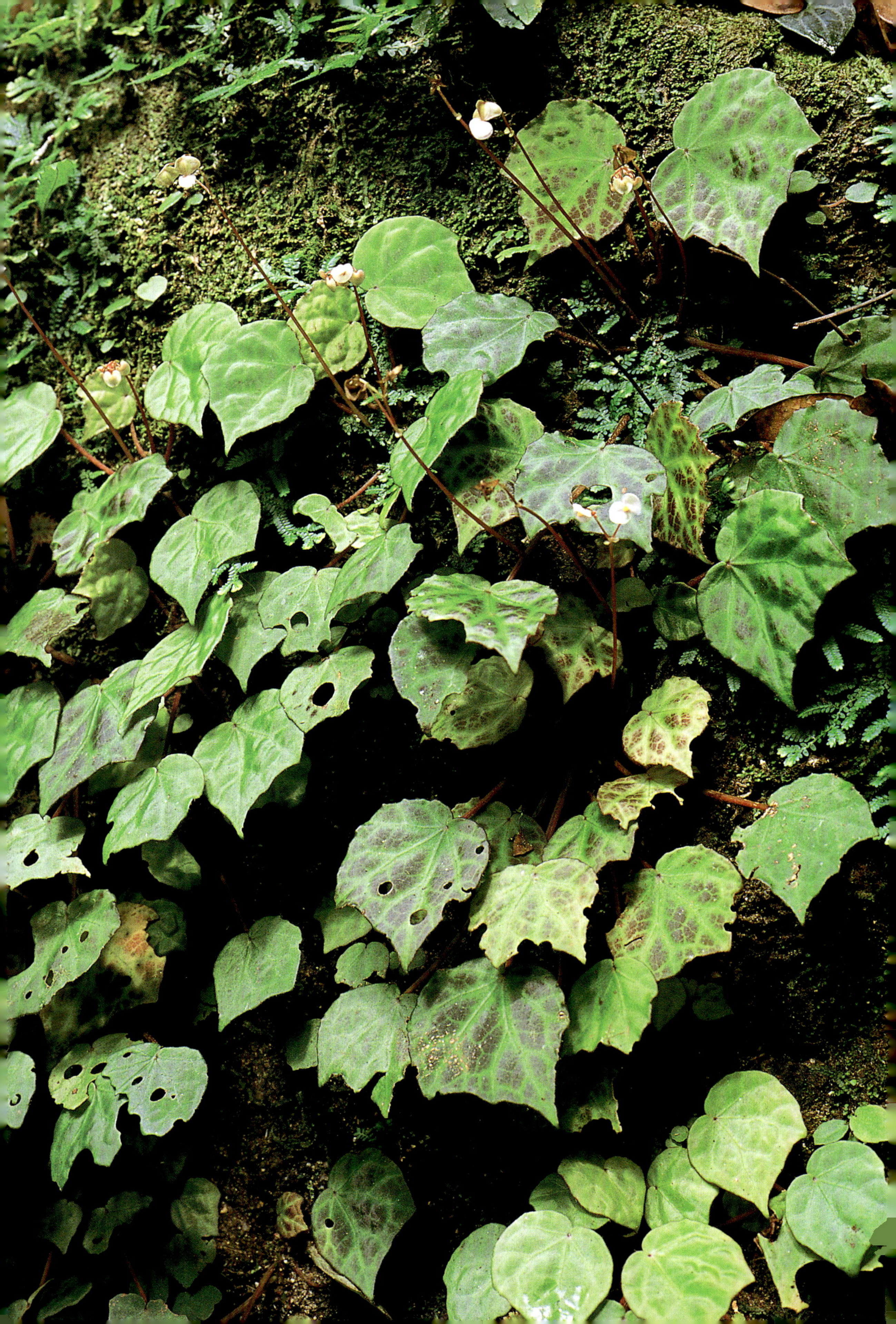

BEGONIA REGINULA

slightly oblique, glossy plain green or variegated between the veins, at first crimson then in mature leaves becoming brownish-red, veins the same colour as the blade, beneath green or dull red, margin red, blade slightly raised between the veins, thinly succulent in life, thinly leathery when dried, broadly ovate, asymmetric, 4–6.5 × 3.5–6.5 cm, broad side 2.25–3.5 cm wide, base rounded to slightly heart-shaped, basal lobes 2.5–8 mm long, margin red, sparsely hairy, not toothed, scalloped towards the tip, tip acute; venation palmate with 2–3 pairs of veins branching towards margin and another pair in the basal lobes, deeply impressed above, prominent and hairless beneath. **Inflorescences** axillary, crimson, without hairs, erect, longer than the leaves, (5.25–)8–11 cm long with two main branches 1.5–2.5 cm long, stalk (4–)6.5–9.5 cm long, male flowers many, female flowers up to 4, male flowers open first. Bract pair pale green, obovate, 3–4 × 1–3 mm, margin fringed by hairs, persistent. **Male flowers** with a deep rosy red stalk 5–8 mm long; tepals 2, hairless, margin not toothed, tip rounded to acute, outer two rosy red outside, deeper red towards base, sparkling white inside, rotund, 6–10 × 9–11 mm; stamens many, cluster globose, *c.* 2 mm across, scarcely stalked, stalk up to 0.5 mm long; filaments *c.* 0.5 mm long; anthers pale or deep yellow, broadly obovate, *c.* 0.5 mm long, tip slightly notched, opening by slits. **Female flowers** with a stalk 3–4 mm long; ovary pale green tinged pale red, 6–7 mm long, wings 3, equal, locules 3, placenta 1 per locule; tepals 2, deep pink, hairless, rotund, margin not toothed, tip acute, 5–7 × 6–8 mm; styles 3, styles and stigmas pale yellow, *c.* 2 mm long, stigmas spiral. **Fruits** dangling on a fine, hair-like stalk 8–10 mm long, capsule 6–9 × 13–16 mm, hairless, locules 3, wings equal, rounded, thinly fibrous, 4–5 mm wide, splitting between the locules and wings. **Seeds** barrel-shaped, *c.* 0.3 mm long, collar cells *c.* three quarters of the seed length.

DISTRIBUTION. Endemic in Peninsular Malaysia: Trengganu and Negri Sembilan.

HABITAT. On rocks on the banks of small waterfalls or streams above the flood zone in deep or semi-shade, occasionally on the base of mossy tree trunks at 30–100 m altitude.

OTHER SPECIMENS. NEGRI SEMBILAN—Pasoh F.R. *Saw & Mustafa FRI 37505* (KEP); Ulu Serting F.R. *Kiew s.n.* 16 Nov 1996 (SING), *RK 5175* (SING), *Saw & Mustafa FRI 37515* (KEP). TRENGGANU—Kuala Sungai Bok *Mohd Shah et al. MS 3511* (SING); Ulu Brang *Moysey & Kiah SFN 33803* (SING), Ulu Setui *Kiew RK 2259* (KEP), *RK 2278* (SING); *Ng FRI 22013* (KEP).

NOTES. The Queen Begonia was discovered in the foothills of Gunung Lawit (Kiew, 1989) in logged forest growing on rocks beside

---

(Opposite). The Queen Begonia, *Begonia reginula,* grows on rock faces or boulders by streams. (Top). Male flower. (Centre). Female flower. (Bottom). Fruit.

a small waterfall. Plants in this population were beautifully variegated with dark bronzy-green patches between the veins so I thought I had refound *Begonia rajah* (Kiew, 1989). The only material of *B. rajah* available was the watercolour painting in the Singapore Botanic Gardens, which showed it to have male flowers with four tepals, whereas the begonia I had found had only two. Mark Tebbitt checked *B. rajah* plants in cultivation in America and they all had four tepals. It was clear that the begonia I had found is a distinct species. I have called it 'reginula' because it is closely related to the Rajah Begonia but has smaller leaves. It is also different in a number of other characters:

Character	*Begonia rajah*	*Begonia reginula*
No. tepals in the male flower	4	2
No. tepals in the female flower	3 (rarely 4)	2
Leaf blade size (cm)	7–15 × 6–15	5.5–7.5 × 5.5–7.5
Leaf base	deeply heart-shaped	rounded
Leaf tip	abruptly pointed	tapered
Leaf surface between the veins	prominently raised	slightly raised
Length of inflorescence stalk (cm)	10–25	(4–)6.5–9.5
Fruit	longer than broad	broader than long
Fruit size (mm)	6–7 × 5–6	6–9 × 13–16

This difference in tepal number in the male flower in two closely related species mirrors the situation with *Begonia foxworthyi* (2 tepals) and *B. nurii* (4 tepals).

We returned to Trengganu in February 2000 to find the area clear-felled and the waterfall completely obliterated by bulldozing to make a road. Walking several hours deeper into the forest, we found a small population of a handful of plants by a small stream.

In 1992, Dr Saw Leng-Guan made an intriguing discovery of two other localized populations in Negri Sembilan—a large disjunction of distribution. There are only minor differences between the Trengganu and Negri Sembilan populations in the size of the outer male tepals that are larger (10 by 12 mm) in the Trengganu population compared with those from Negri Sembilan (6 by 9 mm). Unfortunately, the Negri Sembilan populations comprise mostly plants with plain green leaves rather than the striking bronzy-green ones of the Trengganu population. *Begonia reginula* is therefore not at all common and where found comprises only small populations and they are all in critical danger of being eliminated by habitat disturbance.

Young leaves of the Queen Begonia are brightly variegated.

39. THE ARING BEGONIA
Begonia yappii Ridl.
(R.H. Yapp, member of the Skeat Expedition to Gunung Tahan in 1899)

Ridley, Kew Bull. (1929) 258. **Type:** *Yapp 25*, Kuala Aring, Kelantan (not located).

Stem rhizomatous, brown, succulent, with 2 mm long hairs red on young stems, brown on old stems, little branched, slender, up to 8 cm long, 3–5 mm thick; without a tuber. Stipules pale red, densely hairy on midrib and margin, narrowly triangular, 5–9 × 1.5–2 mm, margin not toothed, tip ending in a long hair, soon falling. **Leaves** close together, up to 6 mm apart; stalk reddish becoming brown, (1.5–)3–5 cm long, 3–4 mm thick, densely hairy, hairs 1–1.5 mm long, round in cross-section; blade slightly oblique, plain dull mid-green above, pale green beneath, thin in life, papery when dried, broadly oval, slightly asymmetric, 5–6 × 3.5–5.5 cm, broad side 2.25–3.25 cm wide, base slightly heart-shaped, basal lobes 2.5–4 mm long, almost overlapping, margin not toothed, fringed by long hairs, undulate and sometimes with a single scallop on the broad side near the tip, tip pointed; venation palmate with 2 pairs of veins, branching more than halfway to margin, and 1 pair in the basal lobes, impressed above, beneath prominent with red hairs, the same colour as blade. **Inflorescences** axillary, reddish, without hairs, longer than the leaves, 7–9 cm long with two

The Aring Begonia, *Begonia yappii*.

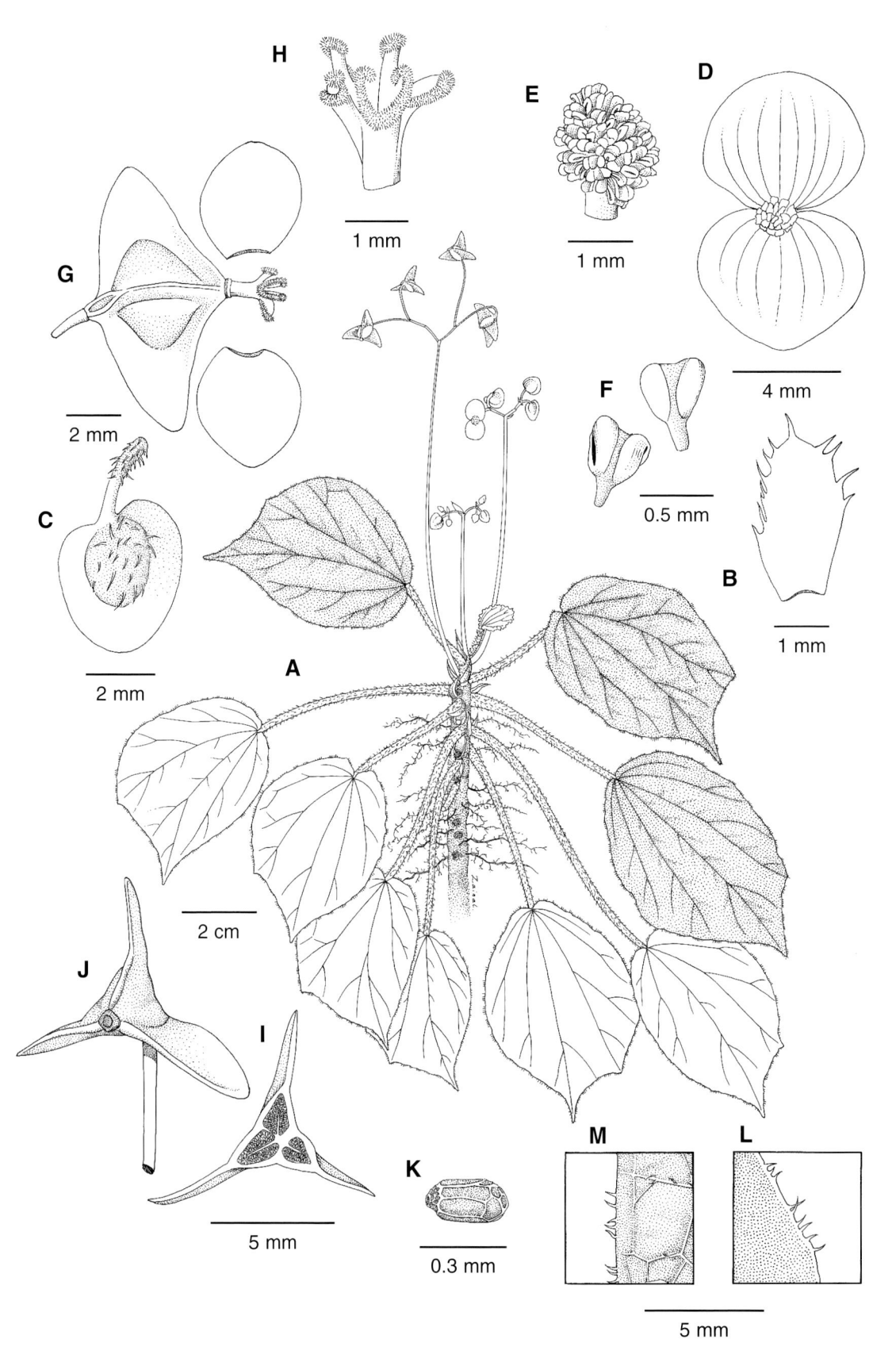

Begonia yappii Ridl. **A.** The plant. **B.** Stipule. **C.** Male bud. **D.** Male flower. **E.** Stamen cluster. **F.** Stamens. **G.** Female flower. **H.** Styles and stigmas. **I.** T.S. ovary. **J.** Fruit; **K.** Seed. **L.** The upper leaf surface. **M.** The lower leaf surface. (*RK 5198*)

BEGONIA YAPPII

The Aring Begonia grows on earth banks above the river.

main branches 0.5–2 cm long, stalk 5.5–8.5 cm long, lengthening in fruit to 13 cm, male flowers many, female flowers up to 4, male flowers open first. Bracts narrowly ovate, *c.* 2–3 × 1–1.5 mm, margin fringed by hairs, soon falling. **Male flowers** with a pale red stalk 7–8 mm long, without hairs; tepals 2, white, hairless or sometimes with short hairs outside, broadly oval, 4.5–7 × 4.5–7 mm, margin not toothed, tip rounded; stamens many, cluster globose, *c.* 1.5 mm across, stalk 0.3–0.5 mm long; filaments 0.5 mm long; anthers pale yellow, broadly ovate, 0.5–0.75 mm long, tip notched, opening by slits. **Female flowers** with a stalk 4–7 mm long, ovary pale green, 4–6.5 mm long, wings 3, equal, locules 3, placenta 1 per locule; tepals 2, white, hairless, rounded, margin not toothed, tip rounded, 4–5 × 4–6 mm; styles 3, styles and stigmas pale yellow, *c.* 2 mm long, stigmas U-shaped. **Fruits** dangling on a fine, hair-like stalk, 8–12 mm long, capsule 4–7 × 9–13 mm, hairless, locules 3, wings 3, equal, 3–5 mm wide, papery, splitting between the locules and wings. **Seeds** barrel-shaped, 0.25–0.3 mm long, collar cells *c.* half the seed length.

DISTRIBUTION. Endemic in Peninsular Malaysia—Kelantan, known only from Sungai Aring.

HABITAT. Not common, it grows in shade on steep river banks, usually on soil.

OTHER SPECIMENS. KELANTAN—Kuala Aring *Kiew RK 5198* (K, KEP, L, SING), *Yapp 36* (K).

NOTES. This little begonia was discovered during the Skeat Expedition, which attempted (but did not succeed) to reach from the north the summit of Gunung Tahan. It was probably discovered because transport in those days was by boat and this species, like *Begonia corneri*, grows in scattered patches on almost vertical river banks. Its rhizome grows vertically up the bank with the leaves spread against the bank.

It was previously placed by Doorenbos *et al.* (1998) in sect. *Diploclinium*, which is characterized by two placentas per locules. This is clearly in error as the Aring begonia has one placenta and so belongs in sect. *Reichenheimia*. It is distinctive among this group in its male and female flowers possessing two not four tepals and in its very small stamens.

Begonia yappii. (Left). Male flower. (Right). Female flower.

40. FOXWORTHY'S BEGONIA
Begonia foxworthyi Burkill *ex* Ridl.
(F.W. Foxworthy, Forester and Research Officer, Forest Department,
Federated Malay States, 1918–1932).

Ridley, FMP 5 (1925) 311; Irmscher, MIABH 8 (1929) 100; Chin, Gard. Bull. Singapore 30 (1977) 98. **Type:** *Haniff & Nur 10199*, 27 January 1923, Kuala Rek, Kelantan (holo K; iso SING).

Stem rhizomatous, red, densely hairy, unbranched, stout and wiry, up to 12 cm long, 9 mm thick; without a tuber. Stipules red, densely hairy, narrowly triangular, 7–12 × 2–5 mm, margin not toothed, tip ending in a hair, persistent. **Leaves** tufted, up to 5 mm apart; stalk pale red, woolly with red hairs, (2.5–)7–14(–25) cm long, grooved above; blade oblique, slightly glossy, plain dull green, sometimes with a slightly bronzy hue, thin in life, drying papery, almost rotund with a short pointed tip, asymmetric, (3–)5.5–12 × (2.5–)6–17.5 cm, broad side (1.5–)3.5–11.5 cm wide, base heart-shaped, rarely overlapping, basal lobes subequal, (0.3–)1.75–4.5 cm long, margin not toothed, wavy and densely fringed by red hairs, tip pointed, 5–7 mm long, sometimes rounded; venation palmate with 2–3 pairs and 1–2 veins in the basal lobes, veins and midrib branching

Foxworthy's Begonia, *Begonia foxworthyi*.

BEGONIA FOXWORTHYI

Growing on the same limestone hills as *Begonia nurii*, *B. foxworthyi* can be told apart by its larger leaves with a pointed tip and male flowers with only two tepals.

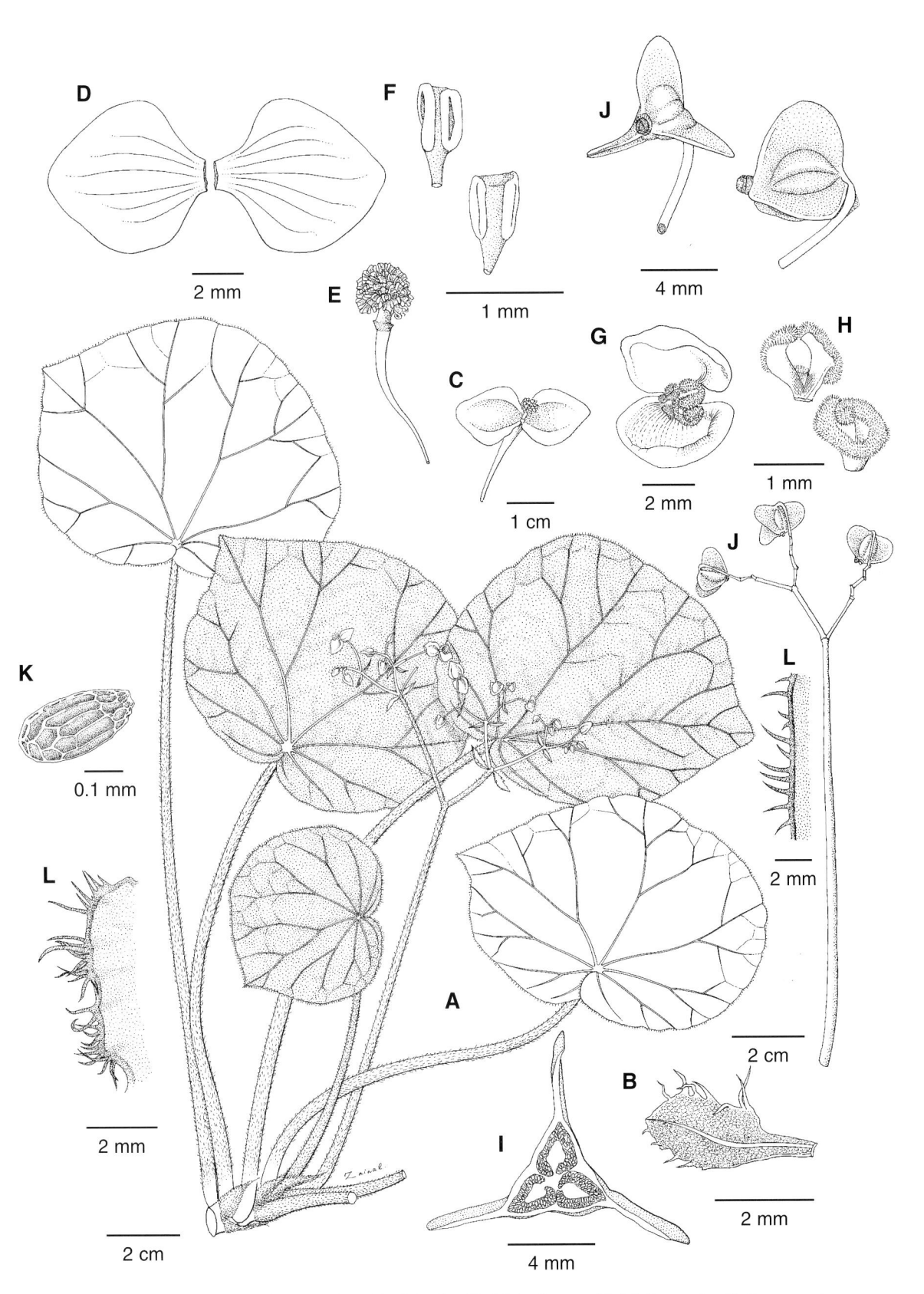

Begonia foxworthyi Ridl. **A.** The plant. **B.** Stipule. **C.** Male flower. **D.** Male tepals. **E.** Stamen cluster. **F.** Stamens. **G.** Female flower. **H.** Styles and stigmas. **I.** T.S. ovary. **J.** Fruit. **K.** Seed. **L.** The leaf margin. (*RK 5258*)

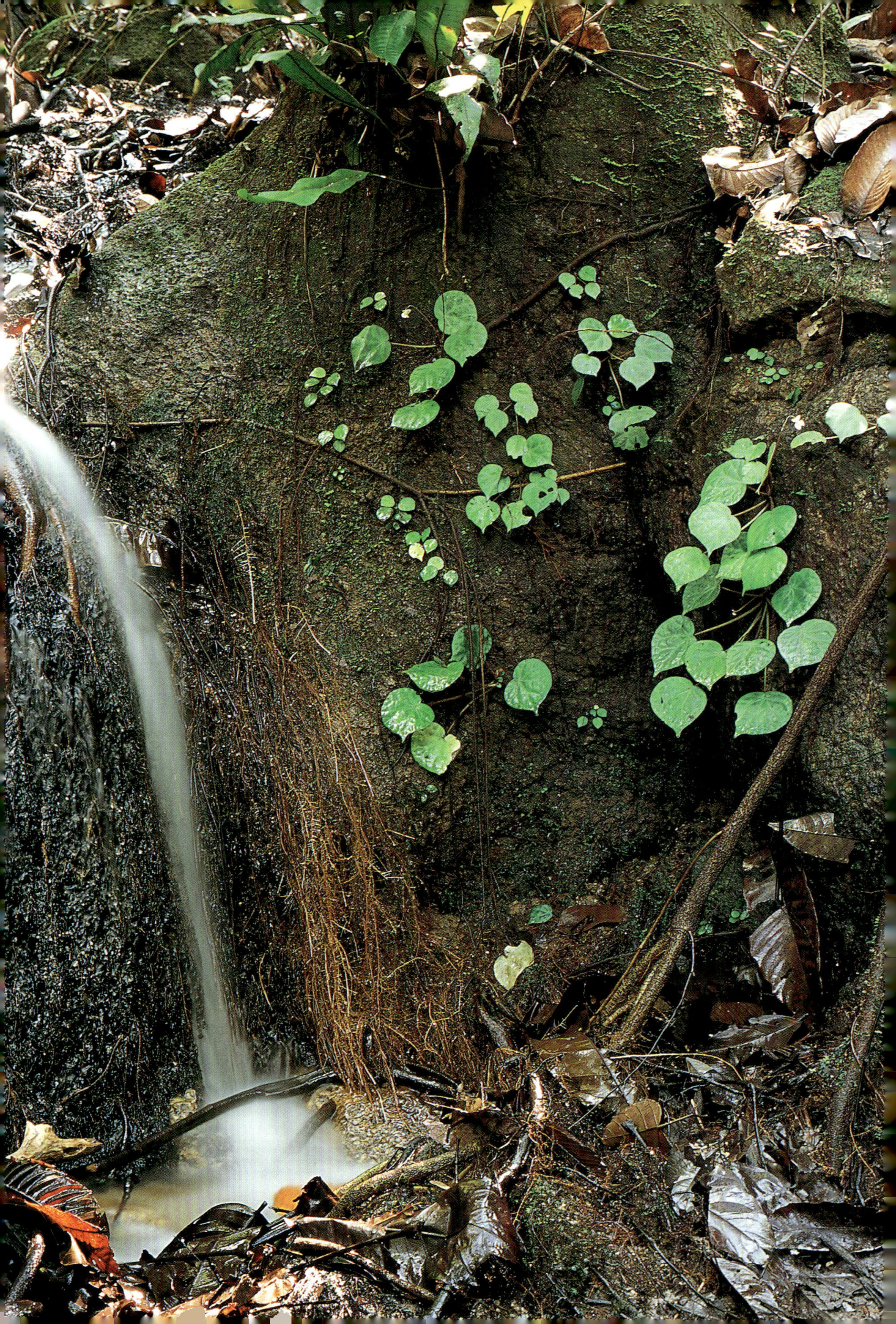

c. halfway to margin, plane above, beneath prominent, reddish and densely hairy. **Inflorescences** axillary, red, sparsely hairy, longer than the leaves, (5.5–)11–30(–41) cm, much branched, branches (0.5–)2–6(–11) cm long, stalk (5–)8–25(–30) cm long, male flowers many, female flowers 2–6, male flowers open first. Bract pair pale green, obovate, 4–6 × 2–4 mm, margin with long hairs, persistent. **Male flowers** with a densely hairy, red stalk, 3–10 mm long; tepals 2, rarely 4, white, sometimes pinkish outside, margin not toothed, tip rounded, outer two with a few red hairs on the outer surface, almost rotund, 4–6 × 4–6 mm, inner two (when present) narrowly obovate, *c.* 4 × 2.5 mm; stamens many, cluster globose, *c.* 2 mm across, stalk 0.5–0.75 mm; filaments 0.4–1–2 mm long; anthers pale yellow, obovate, 0.5–0.75 mm long, tip slightly notched, opening by slits. **Female flowers** with a hairy, red stalk 4–6 mm long; ovary brownish green, 2–4 mm long, wings 3, reddish, subequal, locules 3, placenta 1 per locule; tepals 2, white, reddish towards base, hairless, broadly ovate to rotund, margin not toothed, tip rounded, 3–6 × 4.5–6 mm; styles 3, styles and stigmas yellow, 2.5–4 mm long, stigmas U-shaped. **Fruits** dangling on a fine, hair-like stalk 6–13 mm long, capsule heart-shaped, 5–7(–9) × 11–13(–18) mm, hairless, locules 3, wings 3, subequal, triangular, tip blunt, papery, 3–6 mm wide, splitting between the locules and wings. **Seeds** barrel-shaped, *c.* 0.3 mm long, collar cells *c.* ⁴/₅ of the seed length.

DISTRIBUTION. Endemic in Peninsular Malaysia—Kelantan, Trengganu, Pahang and Negri Sembilan.

HABITAT. In the lowlands up to about 200 m altitude, on damp, shaded limestone cliff faces or on limestone-derived soils or on granite or shale rocks close to streams.

OTHER SPECIMENS. KELANTAN—Bertam, *UNESCO Exped. 29* (SING); Bukit Batu Papan, *Henderson SFN 29509* (SING); Ciku 5, *Kiew & Anthonysamy RK 3096* (KEP); Elephant Cave, *Kiew & Anthonysamy RK 2946* (SING); Gua Chawan, *Kiew RK 4917* (SING); Gua Ninik, *Kiew & Anthonysamy RK 2966* (KEP); Gua Sendok Utara, *Kiew & Anthonysamy RK 3061* (SING); Gua Teja, *Henderson SFN 29672* (SING); Gunung Brong, *Mohd Shah & Mohd Ali MS 2902* (SING); Gunung Panjang, *Kiew & Anthonysamy RK 2970* (KEP); Kampung Sta, *Kiew & Anthonysamy RK 2961* (SING); Kuala Betis, *Kiew RK 5258* (SING); Sungai Nenggiri, *Kiew & Anthonysamy RK 2921* (KEP). NEGRI SEMBILAN—Gua Pelangi, Pasoh 4 Felda Scheme, *Kiew s.n.* 16 Nov 1996

Begonia foxworthyi. (Left). Fruits. (Centre). Male flowers. (Right). Female flower. (Opposite). Foxworthy's Begonia grows both on limestone and granite. On granite, it is usually found close to streams.

(SING). PAHANG—Lancheng Recreation Forest, *Kiew RK 5078* (SING, KEP); Sungai Nering, *Henderson FMS 10599* (SING); Sungai Teku, *Kiah SFN 31781* (SING), *Seimund 526* (SING); TRENGGANU—Lata Renyok, *Carle AC 91232* (SING); Sekayu Recreation Forest, *Anthonysamy SA 650* (KEP), *Kiew RK 2695* (KEP), *Kiew RK 3788* (SING).

NOTES. Like *Begonia integrifolia* and *B. phoeniogramma*, *B. foxworthyi* is not restricted to limestone, but is also found on other rock types. In this respect it is unlike *B. nurii*, to which it is most closely related, as the latter is restricted to limestone. The most conspicuous characters that separate these two species are the leaf tip and leaf base—in *B. foxworthyi* the leaf tip is shortly pointed and the lobes of the heart-shaped leaf base are not overlapping, whereas in *B. nurii* the tip is rounded and the basal lobes are overlapping. Usually *B. foxworthyi* has much larger leaves and many-flowered inflorescences, but it flowers when it is as small as *B. nurii*. It has plain green leaves compared with the handsome variegated ones of *B. nurii*.

In addition, tepal number in the male flower also differs—*B. nurii* always has four tepals, whereas *B. foxworthyi* usually has only two. However, there are three populations where the male flowers have four tepals. All three are found at the southern limit of the geographical range of this species. The ones at Lancheng, Pahang, which grow on granite, and at Gua Pelangi, Negri Sembilan, that grows on limestone, have medium-sized leaves and, apart from tepal number, do not differ in any other respect. Their outer tepals have a few red hairs, a character typical of *B. foxworthyi* that is never seen in the 4-tepaled *B. nurii*, which has hairless tepals. The third population from Sungai Teku (not on limestone) at the foot of Gunung Tahan has very small leaves (up to 3.5 × 3.75 cm) and differs from other populations in that the leaves are spaced on the rhizome. This population may prove to be a different species when it is better known (its female flowers and fruits have not yet been collected).

Ridley (1925) recorded Foxworthy as the collector of the type specimen, although the specimen clearly states that the collectors were actually made by Mohd Nur and Haniff, who were taking part in one of Foxworthy's expeditions. Ridley also recorded the locality of Kuala Kek—it should be Kuala Rek.

Begonia nurii grows in cracks in limestone cliff faces.

41. THE DIMINUTIVE LIMESTONE BEGONIA
Begonia nurii Irmsch.
(Mohamed Nur bin Mohamed Ghous, Herbarium Assistant and Plant Collector, Botanic Gardens Singapore, 1911–1958)

Irmscher, MIABH 8 (1929) 95; Chin, Gard. Bull. Singapore 30 (1977) 99. **Type:** *Mohd Nur & Foxworthy 12026*, 9 February 1924, Sungai Keteh, Kelantan (holo SING, iso K).

Stem rhizomatous, reddish-brown, woody, densely hairy with red glandular hairs, unbranched, stout, up to 10 cm long, 5–7 mm thick; without a tuber. Stipules pale red, hairy, narrowly triangular, 5–10 × 3–4 mm, margin not toothed, tip ending in a long hair, persistent and becoming dry and papery. **Leaves** tufted, up to 7 mm apart; stalk pale or deep red, woolly with red hairs up to 3 mm long, (1.75–)4(–16.5) cm long, round in cross-section; blade slightly oblique, brownish purple when young, becoming variegated, bronze-green, brownish-red or purplish with green veins above, purplish and scintillating beneath, thin in life, papery when dried, orbicular, slightly asymmetric, (2–)4(–7) × (3–)4–6(–11) cm, broad side 2–3(–5.5) cm wide, base heart-shaped and often overlapping, basal lobes subequal 4–10(–20) mm long, margin not toothed, deep red, wavy and fringed by long hairs, tip rounded, rarely slightly pointed; venation palmate, 2(–3) pairs of veins with another pair in the basal lobes, veins and midrib branching about halfway to margin,

(Above). The Diminutive Limestone Begonia, *Begonia nurii*. (Following page). The Diminutive Limestone Begonia grows on cliff faces at the base of limestone hills.

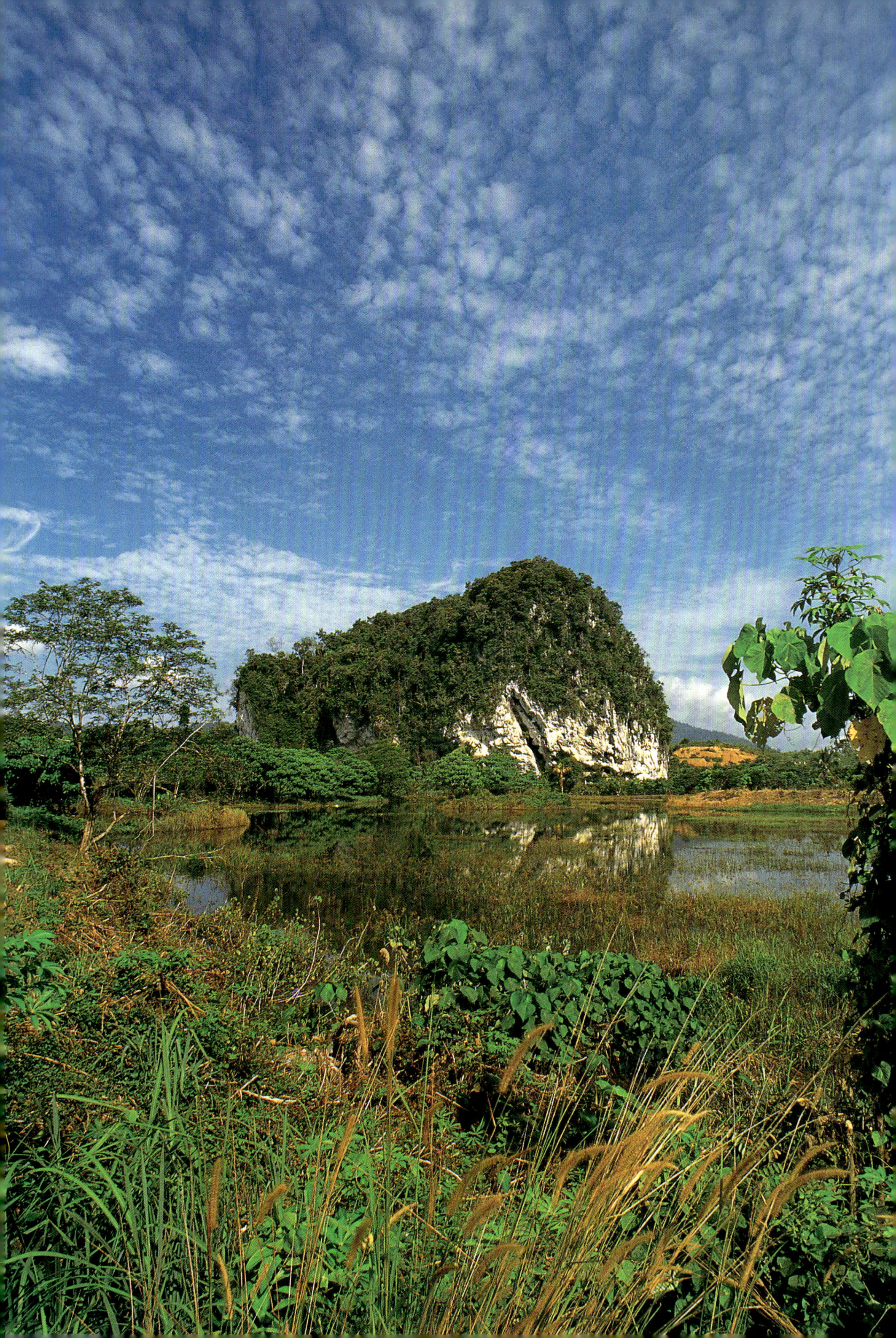

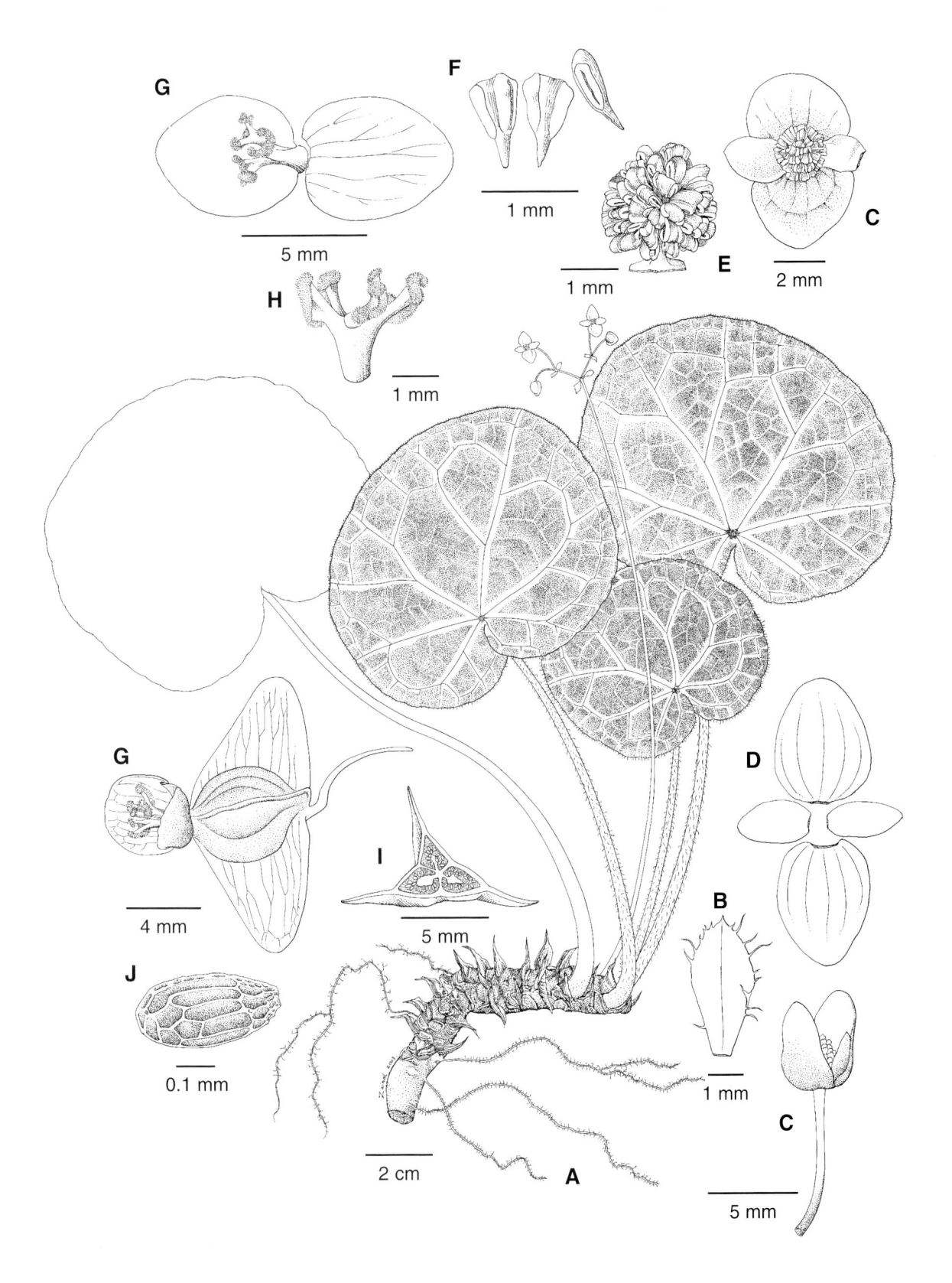

Begonia nurii Irmsch. **A.** The plant. **B.** Stipule. **C.** Male flower. **D.** Male tepals. **E.** Stamen cluster. **F.** Stamens. **G.** Female flower. **H.** Styles and stigmas. **I.** T.S. ovary. **J.** Seed. (*Piee & Yap s.n.*)

impressed above, beneath prominent, reddish and hairy. **Inflorescences** axillary, pale or deep red, sparsely hairy, longer than the leaves, 4–9(–19) cm long, unbranched or with two main branches 0.5–2.5 cm long, stalk 3.5–8(–16.5) cm long, male flowers many, female flowers 2–4, male flowers open first. Bract pair pale green, rotund to obovate, 2–4 × 0.75–2 mm, margin with long hairs, persistent. **Male flowers** with a reddish stalk 3–7 mm long; tepals 4, without hairs, margin not toothed, tip rounded, outer two white, sometimes tinged pink outside, broadly oval, 4–7 × 4–5 mm, inner two narrowly ovate, 3–4.5 × 1–2.5 mm; stamens many, cluster globose, *c.* 2 mm across, stalk *c.* 0.5 mm long; filaments *c.* 0.5 mm long; anthers pale lemon yellow, broadly obovate, 0.5–0.75 mm long, tip slightly notched, opening by slits. **Female flowers** with a hairy stalk 9–15 mm long; ovary pale green or reddish, 5–6 mm long, wings 3, equal, pale red, 4–6 mm wide, locules 3, placentas one per locule; tepals 2, completely white, hairless, broadly ovate to rotund, 4.5–6 × 3–5 mm, margin not toothed, tip rounded; styles 3, styles and stigmas golden yellow, 2.5–3 mm long, stigmas spiral. **Fruits** dangling on a fine, hair-like stalk 6–10 mm long, capsule 5–9 × 10–16 mm, hairless, locules 3, wings equal, tapering, thin and papery, 4–6 mm wide, splitting between the locules and wings. **Seeds** barrel-shaped, *c.* 0.4 mm long, collar cells *c.* $^2/_3$ of the seed length.

Begonia nurii. (Left). Female flowers. (Top right). Male flower. (Bottom right). Fruits. (Opposite). The Diminutive Limestone Begonia grows in crevices in the limestone cliff face.

BEGONIA NURII

DISTRIBUTION. Endemic in Peninsular Malaysia—Kelantan and northwest Pahang.

HABITAT. Growing in the lowlands up to 200 m altitude, it is restricted to limestone, where it grows on damp, shaded cliff faces below the tree canopy.

OTHER SPECIMENS. KELANTAN—Batu Lesong, *Kiew & Anthonysamy RK 2950* (KEP); Batu Machang, *Kiew & Anthonysamy RK 2859* (KEP); Batu Papan, *Kiew & Anthonysamy RK 2886A* (KEP); Bertam, *UNESCO Exped. 365* (SING); Bukit Pecah Kelubi, *Kiew & Anthonysamy RK 2976* (KEP); Bukit Tapah, *Boey 351* (KLU); Ciku 5, *Kiew & Anthonysamy RK 3108* (SING); Gua Ikan, *Kiew & Anthonysamy RK 2938* (SING); Gua Musang, *Kiew & Anthonysamy RK 2994* (KEP), *Ng FRI 5554* (KEP), *UNESCO Exped. 234* (SING). PAHANG—Gua Bekong, *Tan & Piee s.n.* 18 Oct 2001 (KEP, SING); Gua Tipus, *Henderson SFN 19382* (SING).

NOTES. This is one of the prettiest begonias in the Peninsula with dainty, small round leaves variegated with the same colouring as *Begonia rajah*, i.e., reddish brown between green veins. It is restricted to limestone in the Gua Musang area.

42. SCORTECHINI'S BEGONIA
Begonia scortechinii King
(Rev. Father B. Scortechini, Perak Government Botanist 1884–1886).

King, JASB 71 (1902) 62; Ridley, FMP 1 (1922) 860; Irmscher, MIABH 8 (1929) 133. **Type:** *Scortechini 1845*, Perak (lecto K, here designated). **Synonym:** *B. kunstleriana* King, JASB 71 (1902) 63, *B. scortechinii* var. *kunstleriana* (King) Ridl., FMP 1 (1922) 860; Irmscher, MIABH 8 (1929) 134. **Type:** *King's Collector 7194*, January 1895, Gunung Bujang Melaka, Perak (lecto K, here designated).

Stem rhizomatous, firmly rooted to rocks, scaly, unbranched, slender *c.* 5 cm long, *c.* 5 mm thick; without a tuber. Stipules without hairs, narrowly triangular, *c.* 7 × 2 mm, margin not toothed, tip acute ending in a hair, persistent. **Leaves** tufted, almost touching; stalk 5–20 cm long, covered in long brown hairs up to 1.5 mm long, grooved above; blade not oblique, densely hairy above, hairs *c.* 1–1.5 mm long, slightly hooked at tip, membranous when dried, narrowly oval to ovate, almost symmetric, 7–15 × 1.5–5 cm, broad side 0.9–3 cm wide, base narrowed, sometimes rounded, basal lobes not developed, margin toothed and fringed by hairs, tip elongate; venation pinnate with 3–4 pairs of veins along the midrib, not branching on the narrower side, major veins on the broader side branching *c.* halfway to margin, above with sparse brown hairs *c.* 1.5 mm long, beneath slightly prominent, densely covered in long brown hairs even on minor veins. **Inflorescences** axillary, without hairs, 13–18 cm long with two main branches *c.* 1–5 cm long, stalk 12–13 cm long, male flowers up to 7, male flowers open first. Bract pair ovate-oblong, *c.* 9 × 4 mm, sometimes with short hairs, margin not toothed, soon falling. **Male flowers** with a stalk 9–12(–20) mm long, without hairs; tepals 4, white tinged red or pink and green, hairless, margin not toothed, outer two ovate or elliptic, 10–25 × 6–13 mm, tip rounded, inner two oval; 7–9 × 2–4 mm; stamens many, cluster globose, *c.* 5 mm across, stalk short; filaments 1.5–2 mm; anthers narrowly oblong, 2–2.4 mm long, tip pointed, opening by slits. **Female flowers** [from King] ovary without hairs, locules 2; tepals 5, white tinged pink and green, oblong blunt, similar to male tepals; styles 2. **Fruits** [from King] capsule 2–2.5 cm wide, locules 2, wings 3, unequal, larger elongate, smaller two short and narrow. **Seeds** ellipsoid, brown, pitted.

DISTRIBUTION. Endemic in Peninsular Malaysia, known only from Gunung Bujang Melaka, Perak.

HABITAT. Growing on granite boulders in damp ravines at *c.* 300 m.

OTHER SPECIMENS. PERAK—*Scortechini s.n.* (K); Gunung Bujang Melaka, *Curtis 3123* (K, SING).

NOTES. Among the begonias with very narrow leaves, this species is distinct in its leaves being conspicuously hairy. King described *B. kunstleriana* as a species distinct from *B. scortechinii* because of its larger leaves and hairy leaf stalk. The fact that King's two species are both known only from Gunung Bujang Melaka and a specimen of each was collected by both Kunstler and Scortechini strongly suggests that they represent large and smaller plants within a single population.

Both Ridley and Irmscher treated *B. kunsterliana* as a variety of *B. scortechinii*. Irmscher had noted that leaf width alone is not a good character to separate the two. In fact, measurement of the leaves of the herbarium specimens shows that the measurements of leaf blade length, width and the ratio of width:length overlap and whether the leaf stalk is longer or shorter than the blade also does not separate the specimens. King's two species are therefore not considered to warrant even varietal status and so *B. kunstleriana* becomes a synonym of *B. scortechinii*.

The species has not been recollected since 1895, in spite of recent searches for it, so it is still poorly known, particularly the female flowers and fruits.

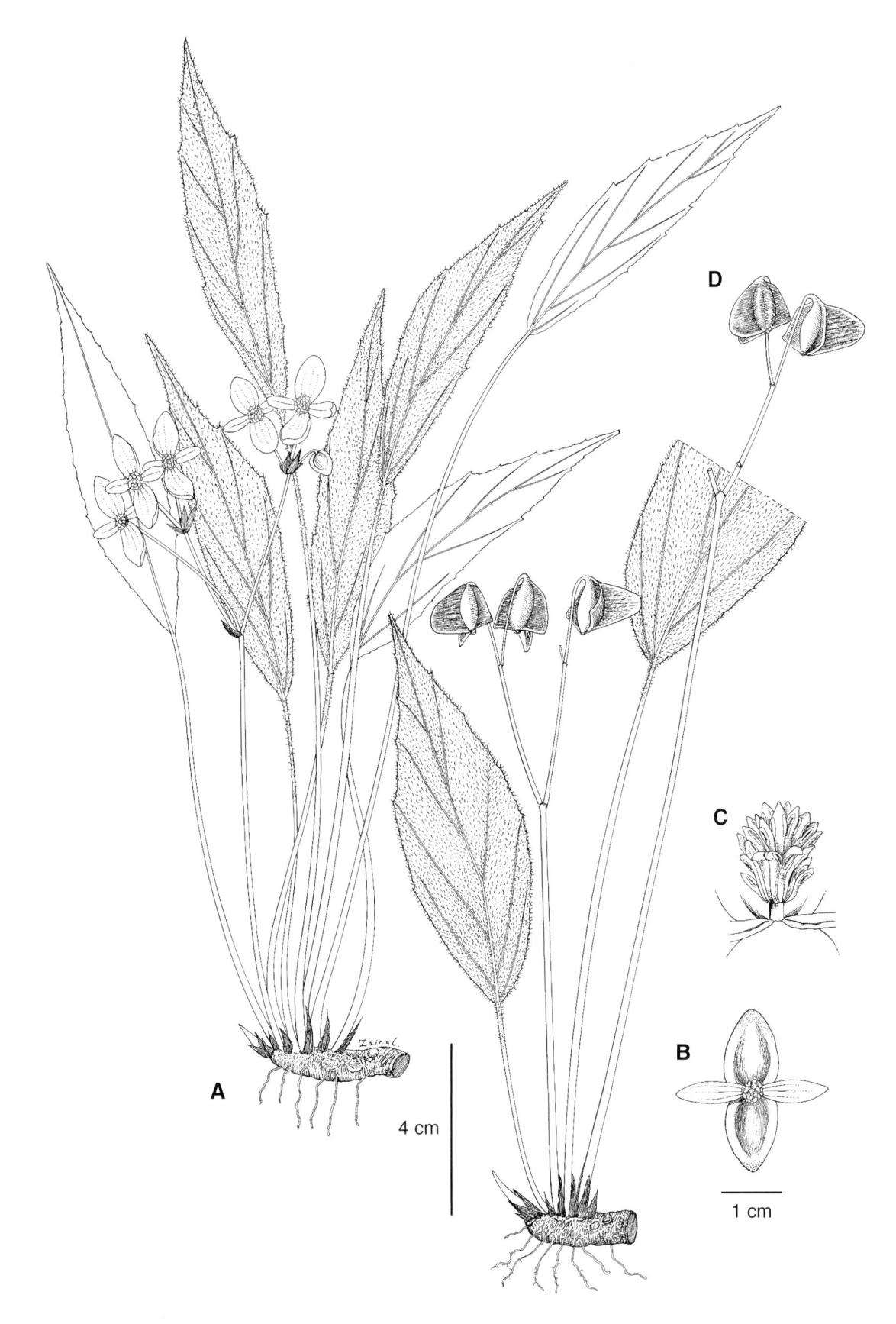

Begonia scortechinii King. **A.** Plant. **B.** Male flower. **C.** Stamen cluster. **D.** Fruit. (*Curtis 3132*)

43. THE PERAK BEGONIA
Begonia perakensis King
(Latin, *-ensis*=indicating origin, from Perak)

King, JASB 71 (1902) 64; Ridley, FMP 1 (1922) 861; Irmscher, MIABH 8 (1929) 129. **Type:** *King's Collector 10338*, June 1886, 'Kal', Selangor (lecto K, here designated).

Stem rhizomatous, clinging to rocks, unbranched, succulent, slender, up to 7 cm long, 2–5 mm thick; without a tuber. Stipules without hairs, narrowly triangular, 9–12 × 1.5–3 mm, margin not toothed, tip pointed, persistent. **Leaves** tufted, up to 3–4 mm apart; stalk grooved above; blade not oblique, plain green, thin in life, papery when dried, almost equal-sided, margin distantly toothed, tip elongate; venation pinnate, impressed above, prominent beneath. **Inflorescences** axillary, almost hairless, in flower shorter than the leaves, few-flowered, up to 15 cm long with two main branches 10–25 mm long, stalk to 13 cm long, in fruit elongating to 19 cm, male flowers 3–5, female flowers 1 or 2, male flowers open first. Bracts ovate, *c.* 12 × 5 mm long, without hairs. **Male flowers** with a hairless stalk 10–15 mm long, tepals 4, margin not toothed, tip rounded, outer two white tinged pink or sometimes completely pink, rotund-ovate to broadly oval, inner two white, oblong-oval, stamens many; filaments 1–2.5 mm; anthers golden yellow, narrowly obovate, 1.5–2 mm, tip rounded. **Female flowers** with ovary *c.* 5 mm long with 3 unequal wings, locules 2, placentas 2 per locule; tepals 5, styles 2, stigmas spiral. **Fruit** a splash cup pendent on a thin stiff stalk 15–20 mm long, capsule 12–15 × 22–35 mm, hairless, locules 2, wings 3, unequal, larger wing oblong, 12–20 mm wide, tip rounded, smaller two, rounded, 5–15 mm wide, thickly papery, splitting between the locules and wings. **Seeds** barrel-shaped, collar cells *c.* half the seed length.

The Perak Begonia, *Begonia perakensis* var. *conjugens*.

BEGONIA PERAKENSIS

DISTRIBUTION. Endemic in Peninsular Malaysia, on the Main Range in south Perak and Selangor.

HABITAT. On rocks in or besides small streams at 200–500 m altitude.

NOTES. This is a dainty begonia with a neat rosette habit, dark green leaves and deep pink flowers. Although it has narrow leaves that are not oblique, it is not a rheophyte as it does not grow in streams but on large boulders on stream banks and on rock faces close to small waterfalls. Irmscher (1929) distinguished two varieties based on leaf width and whether the leaf stalk was hairy or not. Thus, var. *perakensis* has very narrow leaves (3.5 to 5.25 times longer than wide) with a rounded, equal-sided base and hairy leaf stalks, while var. *conjugens* has broader leaves (usually less than three times, rarely 3.5 times, longer than wide) with a slightly unequal leaf base and leaf stalks without hairs.

Variation in leaf shape of the two varieties suggests that they are not distinct. Unfortunately, we have not been able to refind populations of var. *perakensis* to see how variable it is, so the two varieties are described here.

King was mistaken in calling this species 'perakensis' because in his time all collections were from Selangor. The mistake occurred because the specimens collected by H. Kunstler, King's Collector, were labelled as coming from Perak when in fact they came from Kal (or Kol) in the Kerling area of Selangor (Burkill, 1927). However, in trying to refind var. *perakensis*, one population of var. *conjugens* has been located in the extreme south of Perak.

(Left). *Begonia perakensis* var. *conjugens* only grows on boulders along streams. (Right). Male flowers.

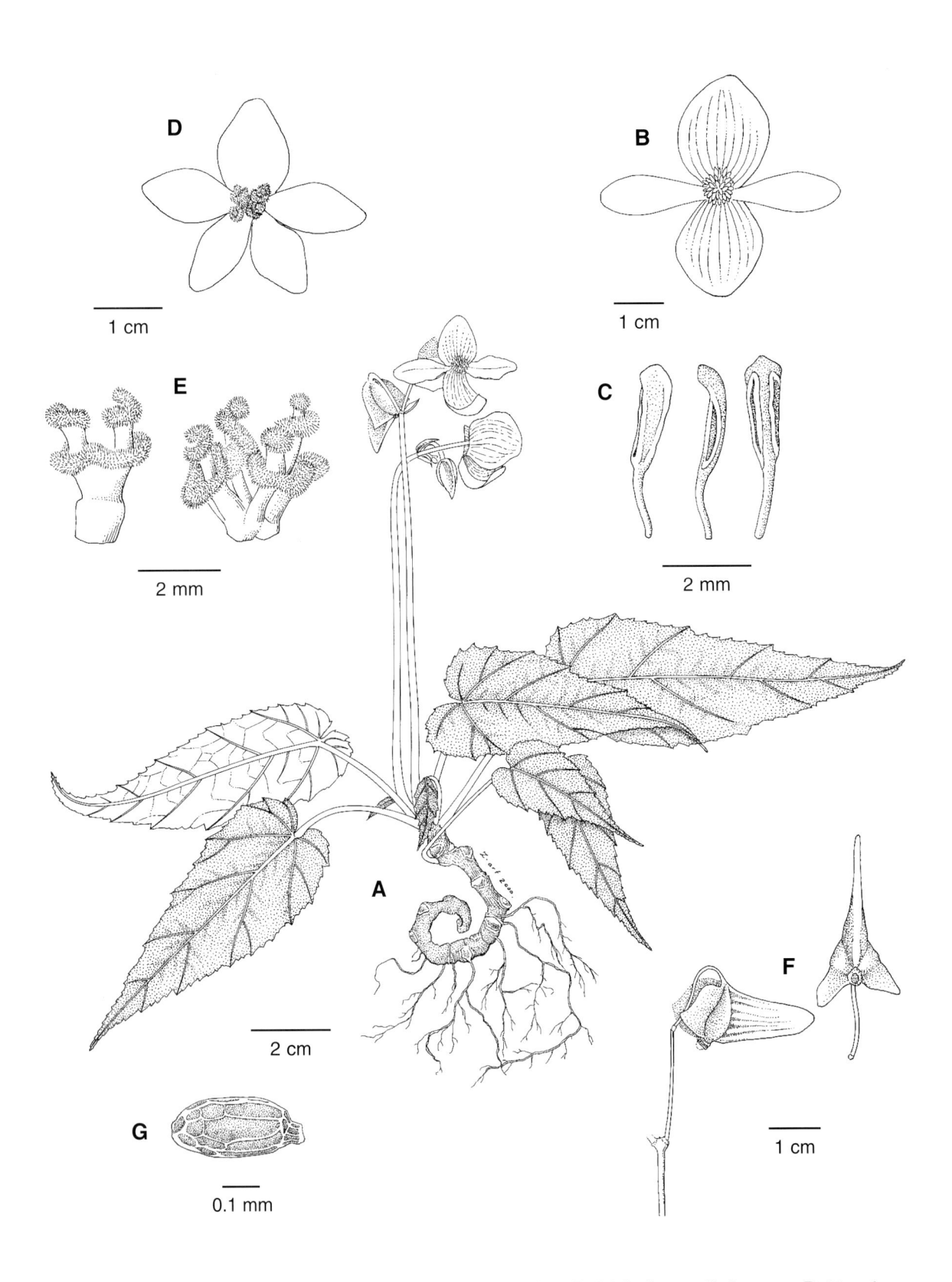

Begonia perakensis King var. ***conjugens*** Irmsch. **A.** The plant. **B.** Male flower. **C.** Stamens. **D.** Female flower. **E.** Styles and stigmas. **F.** Fruit. **G.** Seed. (*RK 4887*)

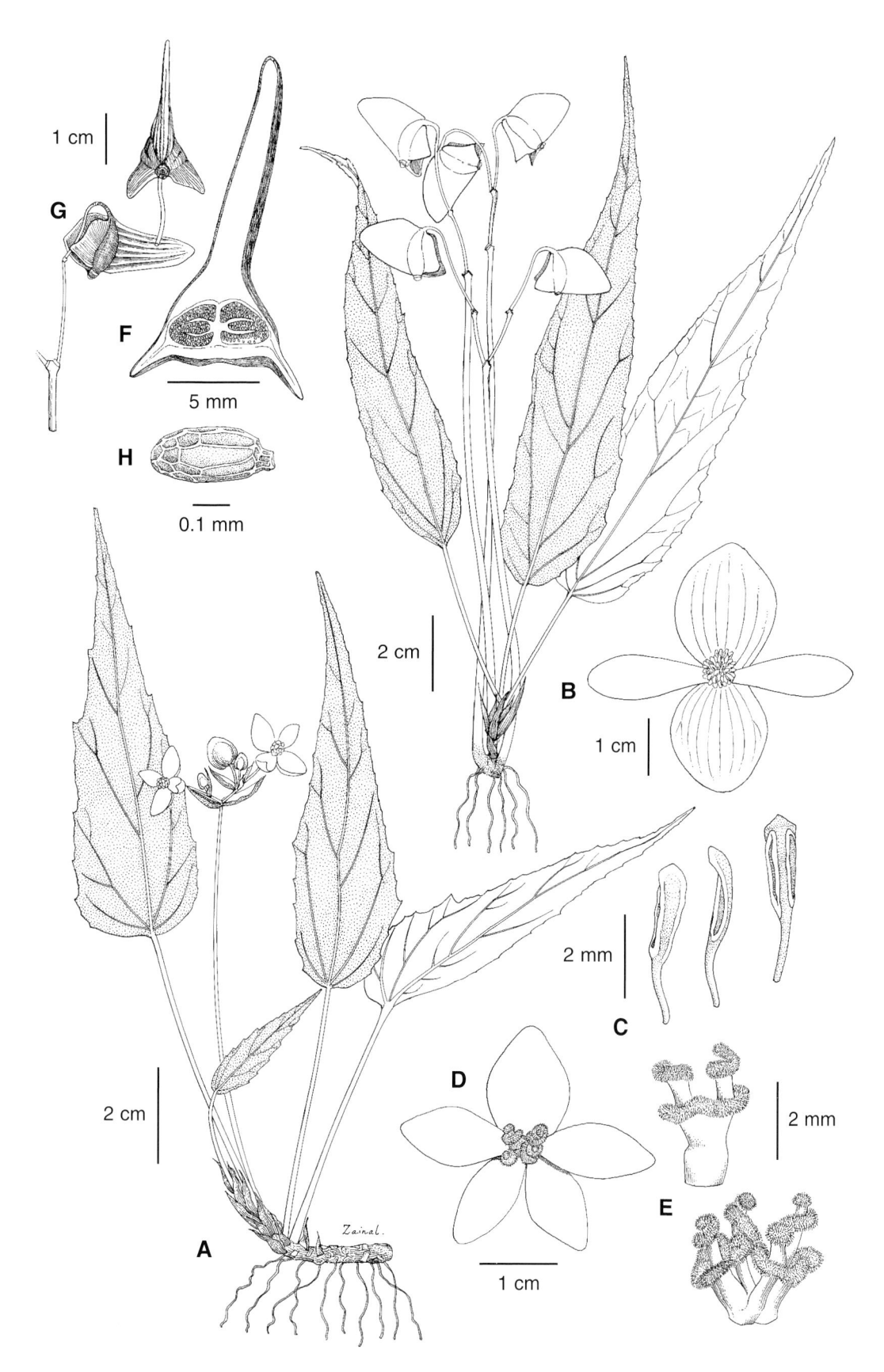

Begonia perakensis King var. **conjugens** Irmsch. **A.** The plant. **B.** Male flower. **C.** Stamens. **D.** Female flower. **E.** Styles and stigmas. **F.** T.S. ovary. **G.** Fruit. **H.** Seed. (*Piee s.n.*)

var. *perakensis*

Synonym: *B. perakensis* King var. *rotundata* Irmscher, MIABH 8 (1929) 129 & Fig. 6.

Leaves with a stalk 3.5–8 cm long, without hairs; blade very narrowly ovate, 8–14 × 1.75–4 cm, broad side 1–2.4 cm wide, base rounded, equal, veins 4–5 pairs, branching *c.* halfway to margin, hairless beneath. **Male flowers** with outer tepals *c.* 10 mm long, without hairs on the outer surface.

DISTRIBUTION. Known only from Selangor: Kal (or Kol) and Genting Bidai.

HABITAT. On rocks in or near streams.

OTHER SPECIMENS. SELANGOR—Kal, *King's Collector 10338* (K); Genting Bidai, *Ridley s.n.* 1896 (SING).

var. *conjugens* Irmsch.
Irmscher, MIABH 8 (1929) 129

Type: *Curtis s.n.*, May 1902, below the Gap, Selangor (lecto SING, here designated).

Stems deep red, hairs on stem, leaf stalk and undersurface of veins translucent and hooked at the tip. Stipules pale or deep red, almost translucent. **Leaves** tufted and held in a fan or tufted and up to 1 cm apart; stalk deep crimson, densely hairy, (3.5–)6–12 cm, blade dull dark green and sparsely hairy above, beneath green, reddish-green or crimson, narrowly ovate, 7.5–13.5 × 2.75–5 cm, broad side 1.5–3 cm wide, base usually unequal, slightly heart-shaped, basal lobe 4–8 mm long; veins 5–7 pairs, branching towards margin, densely hairy and deep crimson beneath. **Inflorescences** reddish brown to crimson. **Male flowers** with a pale pink stalk 12–15(–20) mm long; outer two dark pink outside, almost white inside, hairless, broadly oval, 15–26 × 8–19 mm, inner two white, oval, 12–25 × 7–10 mm; cluster ellipsoid, *c.* 6 × 4–7 mm, stalk 1–1.5 mm. **Female flowers** with a greenish brown to crimson stalk, 10 mm long; ovary reddish pale brown; tepals rosy pink, without hairs, outermost broadly obovate, *c.* 9 × 8 mm, inner tepals similar but slightly smaller.

DISTRIBUTION. Known only from south Perak and the Selangor Valley.

HABITAT. Near streams on large boulders on the downstream side, or on rocks beside streams or creeping up trees in the spray of waterfalls.

OTHER SPECIMENS. PERAK—Sungai Kelumpang, *Piee s.n.* 28 Feb 2004 (SING). SELANGOR—Fraser's Hill, path to Pine Tree Hill, *Addison SFN 37369* (SING); Kuala Kubu Baru-Gap Road, *Curtis s.n.* 1902 (SING), *Kiew RK 1108* (SING), *RK 1248* (SING), *RK 1249* (SING), *RK 4887* (SING), *Ridley s.n.* 1896 (SING).

Female flower of *Begonia perakensis* var. *conjugens*.

44. THE STREAM BEGONIA
Begonia rhoephila Ridl.
(Greek, *rhoe*=stream, *philos*=love, referring to the habitat)

Ridley, JSBRAS 75 (1917) 36, FMP 1 (1922) 860; Irmscher, MIABH 8 (1929) 137. **Type:** *Ridley s.n.*, 15 March 1915, Ulu Gombak, Selangor (lecto K, here designated; iso BM). **Synonym nova:** *Begonia collina* Irmscher, MIABH 8 (1929) 135. **Type:** *Kelsall s.n.*, 8 Jan 1891, Bukit Etam [Hitam], Selangor (holo K *ex* SING).

Stem rhizomatous, firmly rooted to rocks, deep red or brown, succulent, with long translucent slightly reddish hairs, unbranched, slender, up to 7 cm long, 4–7 mm thick; without a tuber. Stipules reddish, narrowly triangular with hairs along the midrib, 7–14 × 2–3 mm, margin not toothed, tipped by a long hair, soon falling. **Leaves** tufted, leaf bases touching or up to 2 mm apart; stalk deep red at the base, green at the top, becoming brown in mature leaves, sparsely to densely hairy, 4–6(–17) cm long, deeply grooved above; blade not oblique, dull plain deep green and sparsely hairy above, pale green beneath, thinly fleshy in life, papery when dried, narrowly lanceolate, often conspicuously parallel-sided, almost symmetric, 11.5–16.5(–21) × 2.25–3(–6) cm, broad side 1.25–3.5 cm wide, base narrowed sometimes unequal, basal lobes not developed, margin red, toothed, teeth tipped by a hair, sometimes undulate, tip elongate; venation pinnate, (4–)5–7 pairs of veins, not branching, impressed above, beneath slightly prominent, midrib densely hairy, lateral veins without hairs, red in young leaves, reddish green in adult leaves. **Inflorescences** axillary, reddish green or pale reddish brown, sparsely hairy, longer than the leaves, 4.75–7.5(–24)

The Stream Begonia, *Begonia rhoephila*.

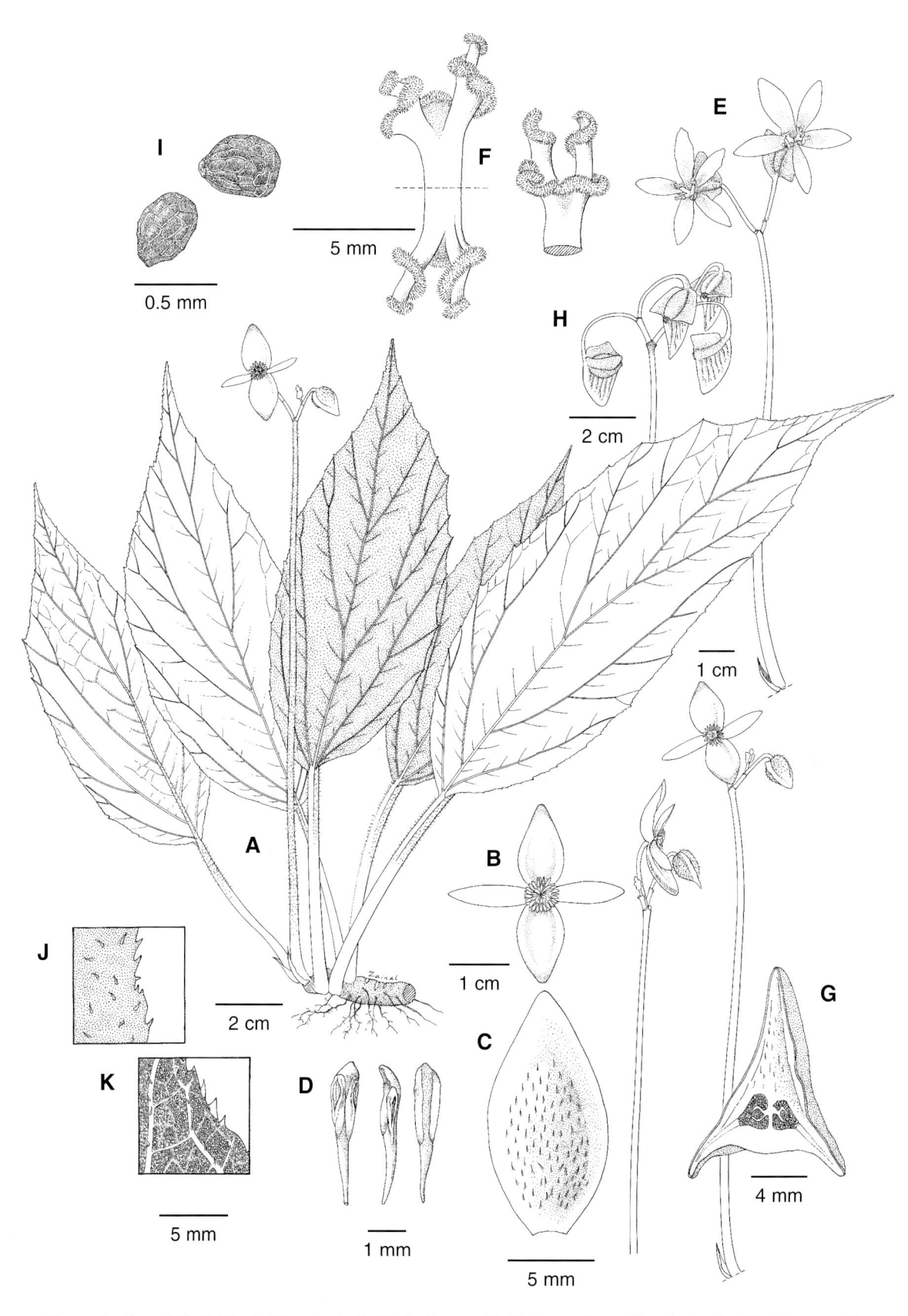

Begonia rhoephila Ridl. **A.** The plant. **B.** Male flower. **C.** Outer surface of male tepals. **D.** Stamens. **E.** Female flower. **F.** Styles and stigmas. **G.** T.S. ovary. **H.** Fruit. **I.** Seed. **J.** The upper leaf surface. **K.** The lower leaf surface. (*RK 5199*)

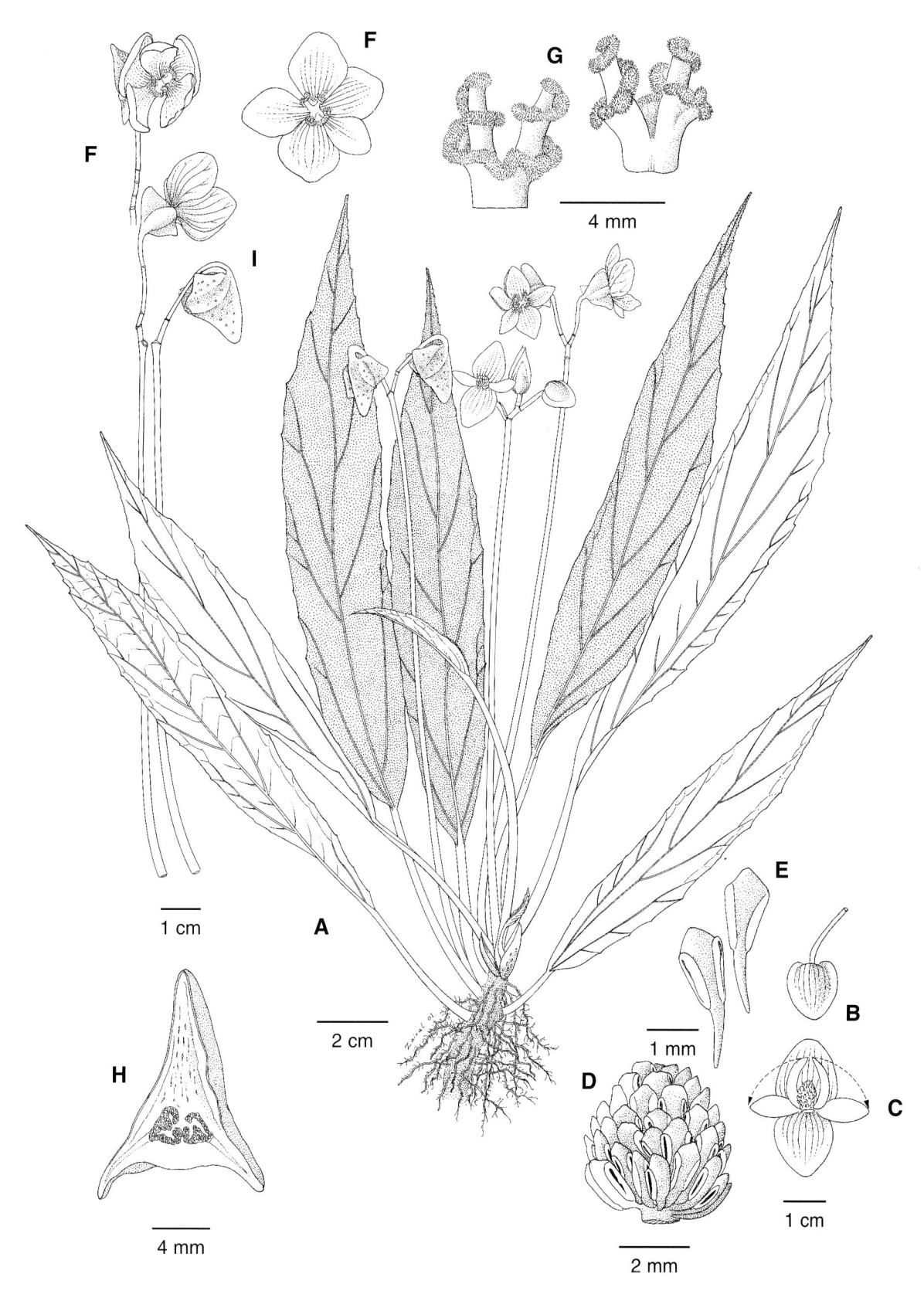

Begonia rhoephila Ridl. **A.** The plant. **B.** Male bud. **C.** Male flower. **D.** Stamen cluster. **E.** Stamens. **F.** Female flower. **G.** Styles and stigmas. **H.** T.S. ovary. **I.** Fruit. (*RK 5202*)

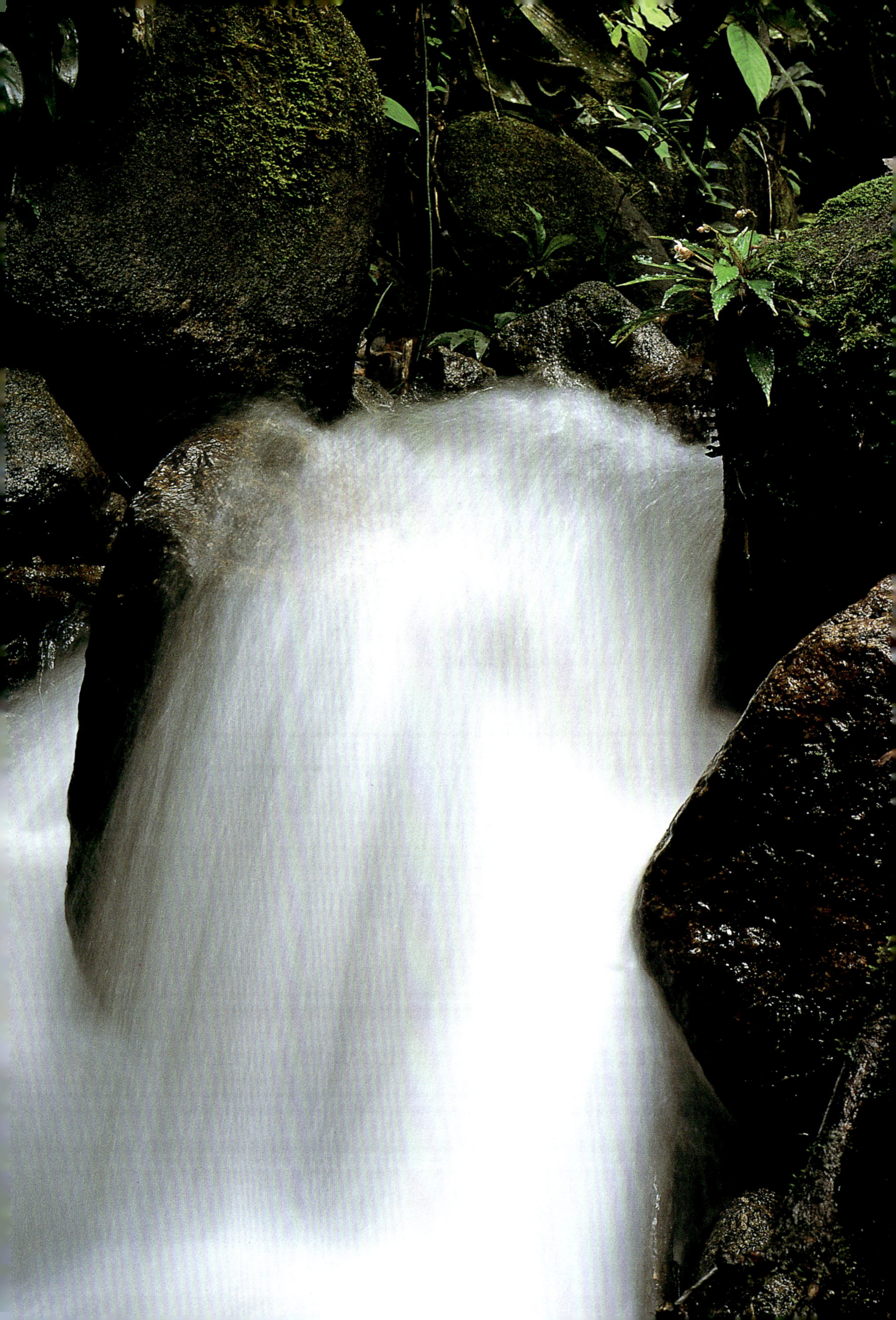

BEGONIA RHOEPHILA

cm with two main branches 0.3–1(–3) cm long, in fruit elongating to 30 cm long, few-flowered, male flowers 3–5, female flowers 1–4, male flowers open first. Bract pair reddish, hairy, narrowly ovate, *c.* 15 × 3 mm, soon falling. **Male flowers** with a pale pink or greenish white, hairless stalk 11–24 mm long; tepals 4, margin not toothed, tip rounded, outer two white tinged rosy pink, minutely hairy outside, broadly ovate, 11–23 × 10–16 mm, inner two pure white, oval, 9–18 × 5–8 mm; stamens many, cluster globose, 4–5 mm across, without a stalk; filaments 1–1.5 mm long; anthers golden yellow, narrowly oblong, 1.5–2 mm long, tip rounded, opening by slits. **Female flowers** with a greenish red, hairless stalk 5 mm long; ovary green, 6–10 mm long, wings 3, unequal, locules 2, placentas 2 per locule; tepals 5, rosy pink outside, paler inside, without hairs, outer rounded, margin not toothed, tip rounded, 6–10 × 6–8 mm, inner similar but smaller 5–9 × 4–8 mm; styles 2, styles and stigmas greenish yellow, 4–5 mm long, stigmas spiral. **Fruit** a splash cup pendent on a thin, stiff stalk, l5–23 mm long, capsule 10–14 × 17–28 mm, hairless, locules 2, wings 3, unequal, larger oblong, fibrous 10–16 mm wide, tip rounded, smaller two triangular, papery, 5–8 mm wide, splitting between the locules and wings. **Seeds** barrel-shaped, 0.3–0.5 mm long, collar cells *c.* half the seed length.

DISTRIBUTION. Endemic in Peninsular Malaysia, known only from Selangor in the foothills of the Main Range.

(Above). *Begonia rhoephila*. (Left). Female flowers. (top right). Male flower. (Bottom right). Fruit. (Opposite). The Stream Begonia grows within the flood zone of small torrential streams.

249

ALI IBRAHIM

BEGONIA RHOEPHILA

HABITAT. On rocks in or beside streams in light shade, about 15 cm above the normal water level and on the downstream side of boulders.

OTHER SPECIMENS. SELANGOR—Gombak Valley, *Kiew RK 1339* (SING), *RK 3250* (KEP), *RK 5199* (SING), *Wyatt-Smith KEP 94397* (KEP); Gunung Hitam, *Kelsall s.n.* 1891 (K *ex* SING), *Goodenough s.n.* 1897 (SING); Sungai Lewing, *Kiew RK 5202* (SING); Ulu Ampang Waterfall, *Shaharin Shah II* (KEP), *Shaharin s.n.* 1987 (KEP);

NOTES. At first sight, the Stream Begonia does not look like a begonia at all because its leaves are sleek and equal-sided. This narrow, streamlined leaf shape is typical of rheophytes (plants that grow in the flood zone of streams). This is one of the few begonias that grow in rocky streams just above the normal water level where they must be able to withstand strong water currents in times of flood. Its scientific name is therefore appropriate.

Irmscher did not see specimens of *Begonia rhoephila* but instead relied on Ridley's description. Of the specimens he saw that were annotated as *cf.* (=compare with) *Begonia rhoephila*, one had broad leaves, which Irmscher described as a new species, *B. aequilateralis*, and the other with narrow leaves he described as *B. collina*. He considered the latter distinct from *B. rhoephila* because it had a larger male flower (30 × 24 mm) and the leaf base was not decurrent (i.e., the blade does not extend down the petiole). In contrast, Ridley's *B. rhoephila* was described as having male flowers 20 × 10 mm and decurrent leaf bases. It was unfortunate that Irmscher did not see specimens as the leaf base of *B. rhoephila* is in fact not decurrent. Size of the male flower depends on its position in the inflorescence, the first flower to open is often much larger than the rest, so this character alone is not reliable.

Of the two specimens from Bukit Hitam, Irmscher described the Kelsall specimen as *B. collina*, and the Goodenough specimen, which had larger leaves, as *aff.* (=akin to) *B. collina*.

Now a wider range of specimens is available, not only from the type localities but also from a site intermediate between them, the size of the male flower no longer proves to be different as plants from a single site show greater variation than the difference between the two species and in both populations the outer male tepals are minutely hairy on the outer surface. For this reason, *B. collina* is treated as a synonym of *B. rhoephila*.

(Above). The Stream Begonia grows on the downstream side of boulders. (Opposite). The Stream Begonia is one of the rarest begonias in the Peninsula.

45. PIEE'S BEGONIA
Begonia abdullahpieei Kiew, sp. nov.
(pieei=Abdullah Piee, forest guide and explorer)

Begoniae rhoephilae Ridl. optime congruens, sed foliis latioribus (longitudine minus quam 3-plo longa quam lata) differt. **Typus:** *Kiew RK 4907*, 17 Feb 2000, Kelian Gunung, Perak (holo SING; iso K, KEP, L).

Stem rhizomatous, rooting at the nodes, succulent, with a few sparse hairs, slender, up to 9 cm long, 3–7 mm thick, branched with erect side shoots 4–9 cm long terminating in a small leaf and 1 or 2 inflorescences; without a tuber. Stipules sparsely hairy, narrowly triangular, 7–10 × 2–3 mm, margin not toothed, tip pointed, soon falling. **Leaves** tufted, up to 3 mm apart; stalk pale brown sometimes with a reddish tinge, almost without hairs, 7–11 cm long, round in cross-section; blade not oblique, plain bright fluorescent green, rarely dark green above, slightly paler or rarely reddish beneath with scattered erect hairs above and beneath, thin in life, thinly papery when dried, oval, almost symmetric, 10–13.5 × 4.5–6.5 cm, broad side 2.75–4 cm wide, base narrowed, rarely slightly rounded, basal lobes not developed, margin minutely toothed, teeth ending in a hair, tip elongate up to 2.5 cm long; venation pinnate, 4–7 pairs of veins, lower veins branching *c.* halfway to the margin, dark red and hairy above and beneath, impressed above, very prominent beneath. **Inflorescences** axillary from the rhizome or terminal on side shoots, reddish, at first minutely

Piee's Begonia, *Begonia abdullahpieei*.

BEGONIA ABDULLAHPIEEI

hairy, slightly longer than the leaf stalks, few-flowered, 6–11 cm long with two main branches 0.7–1.7 cm long, stalk 5.5–9 cm long, elongating to 14–27 cm in fruit, male flowers up to 5, female flowers 2, male flowers open first. Bract pair ovate, *c.* 12 × 8 mm, margin not toothed, soon falling. **Male flowers** with a stalk 13–15 mm long; tepals 4, margin not toothed, tip rounded, outer two pale pink, deeper pink towards the centre and base, veins translucent, minutely hairy outside, broadly oval to rotund, 11–17 × 9–16 mm, inner two narrowly oval, 10–14 × 6–9 mm; stamens many, cluster globose, *c.* 6 mm across, stalk *c.* 1 mm long; filaments 1.5–2 mm long, anthers golden yellow, narrowly obovate, *c.* 2 mm long, tip rounded, opening by slits. **Female flowers** with a deep pink stalk, 8–13 mm long; ovary reddish brown, minutely hairy, 7–9 mm long, wings 3, unequal, locules 2, placentas 2 per locule; tepals 4, pale pink to deep rosy red at base, without hairs, margin not toothed, outermost rotund, tip rounded, *c.* 12 × 12 mm, innermost similar but smaller *c.* 9 × 6 mm; styles 2, styles and stigmas golden yellow, *c.* 4 mm long, stigmas spiral. **Fruit** a splash cup pendent on a fine,

Begonia abdullahpieei. (Above). Habit. (Centre). Female flower. (Bottom). Fruits.

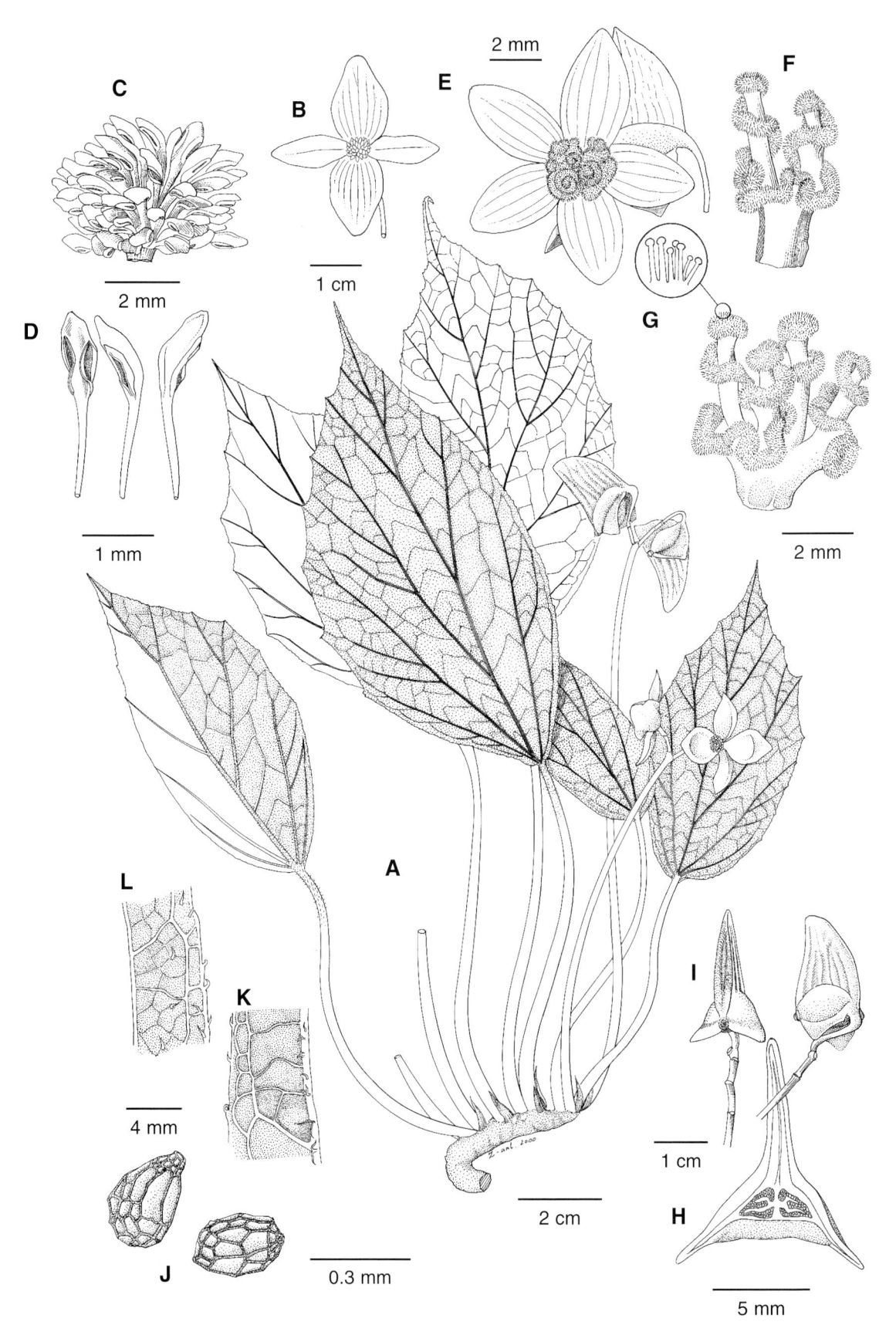

Begonia abdullahpieei Kiew. **A.** The plant. **B.** Male flower. **C.** Stamen cluster. **D.** Stamens. **E.** Female flower. **F.** Styles and stigmas. **G.** Papillose hairs of the stigma. **H.** T.S. ovary. **I.** Fruit. **J.** Seed. **K.** The upper leaf surface. **L.** The lower leaf surface. (*RK 4907*)

BEGONIA ABDULLAHPIEEI

stiff stalk (9–)16–17 mm long, capsule 11–14 × 22–25 mm, hairless, locules 2, wings 3, unequal, larger wing oblong with a rounded apex, thickly fibrous, 14(–17) mm wide, smaller two thinly fibrous, 4–6 mm wide, splitting between the locules and wings. **Seeds** barrel-shaped, *c.* 0.25–0.3 mm long, collar cells *c.* 0.75 of the seed length.

DISTRIBUTION. Endemic in Peninsular Malaysia, known only from the type locality in Perak.

HABITAT. Growing in deep shade on rocks in a small stream.

NOTES. Piee's Begonia is a very decorative species with its green fluorescent leaves, pink flowers and compact habit. It belongs to the group of stream-dwelling begonias that have almost equal-sided leaves that are not oblique. It is unique within this group in producing side shoots. In its narrowed leaf base (as opposed to a rounded or heart-shaped leaf base), it most resembles *Begonia rhoephila* but it is different from this species in its leaves having leaf stalks almost the same length as the blade and in its broader leaf, which is twice or less than twice as long as wide. In contrast, *B. rhoephila* has leaves with leaf stalks about three times as long as the blade and the leaf is more than three times longer than wide.

It belongs to section *Platycentrum* because it has a rhizome, more or less symmetrical leaves, male flowers with four tepals that open before the female flowers, and pendent fruits with three unequal wings and two locules each with two placentas.

Piee's Begonia grows on large boulders in a small stream.

46. THE WATERFALL BEGONIA
Begonia rhyacophila Kiew, sp. nov.
(Greek, rhyac=torrent; philos=loving)

Foliis *Begoniae alpinae* L.B. Smith & Wassh. optime congruens, sed foliis supra pilosis et lobos basales breviores gerentibus differt. **Typus:** *Piee RK 5103*, 6 Sept 2002, Kelian Gunung, Perak (holo SING; iso K).

Stem at first forming a compact rhizome attached to the rocks by a mat of fibrous roots, then producing a prostrate stem with internodes (3.5–)5–7 cm long, which root at the nodes, sparsely branched, reddish brown, succulent with long hairs slightly hooked at the tip, stout, 12–14 cm long, 4–8 mm thick; without a tuber. Stipules pale green, midrib densely bristly outside, narrowly triangular, 13–15 × 2–5 mm, margin not toothed, tip pointed, ending in a long hair, persistent. **Leaves** on the rhizome tufted and 4–6 mm apart, on the prostrate stem distant and 3.5–7 cm apart; leaf stalk deep red, bristly, hairs *c.* 1 mm long, 13–23.5 cm long, round in cross-section; blade not oblique, plain yellowish green above, slightly scintillating, veins darker green, beneath paler green, bristly above with long hairs *c.* 1.5–2 mm long as well as a fine covering of minute hairs, beneath with minute hairs only, thin in life, papery when dried, broadly ovate, almost symmetric, 8–11.5 × 6–8 cm, broad side 3–4.5 cm wide, base rounded to slightly heart-shaped, basal lobes equal, 7–10 mm long, margin minutely toothed, fringed by hairs, tip elongate; venation palmate-pinnate, 2 pairs of veins at the base, 2–3 pairs along the midrib, branching halfway to the margin, veins impressed above (blade almost bullate), beneath prominent, reddish and bristly, hairs *c.* 1 mm

Collecting the Waterfall Begonia requires special skills. (Opposite). *Begonia rhyacophila* grows on a rocky ledge in the waterfall.

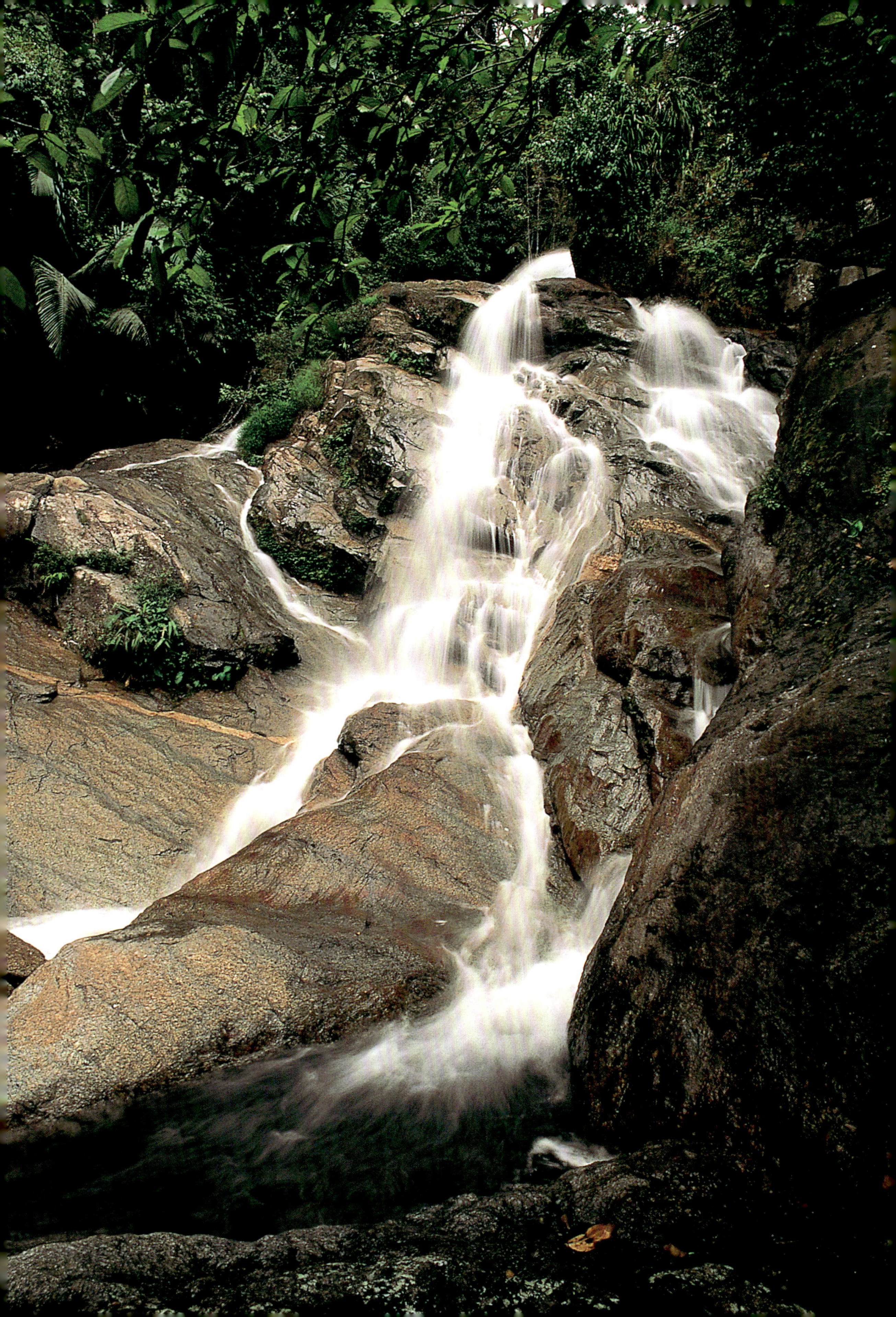

BEGONIA RHYACOPHILA

(Above). The Waterfall Begonia, *Begonia rhyacophila*. (Top). Male fowers.

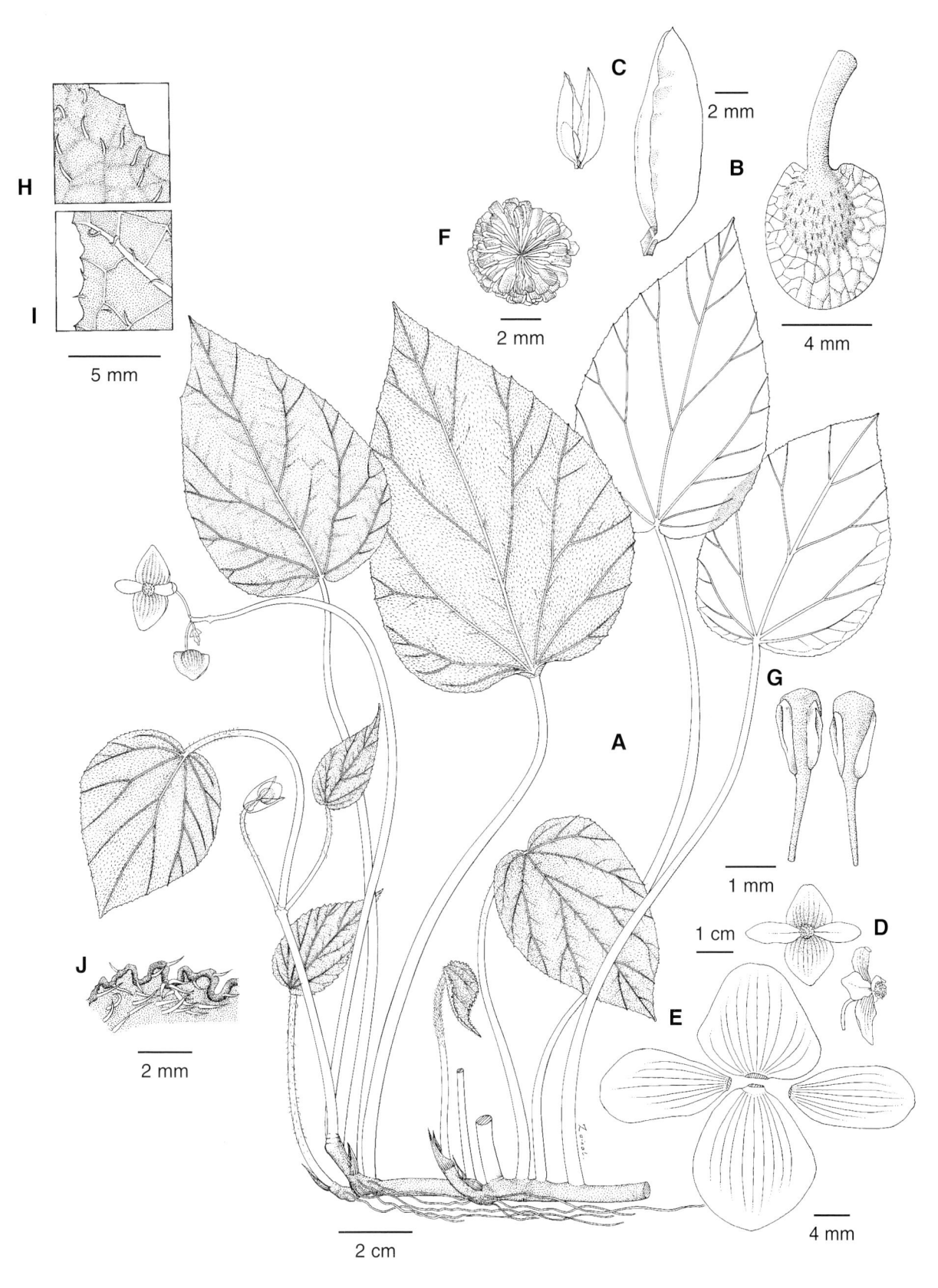

Begonia rhyacophila Kiew. **A.** Plant. **B.** Male bud. **C.** Bract. **D.** Male flower. **E.** Male tepals. **F.** Stamen cluster. **G.** Stamens; **H.** Upper leaf surface. **I.** Lower leaf surface. **J.** Leaf margin. (*RK 5103*)

BEGONIA RHYACOPHILA

Male flowers of *Begonia rhyacophila*.

long. **Inflorescences** axillary, reddish, hairy, in male flower shorter than the leaves, 13–22.5 cm long with branches 3–5 mm long, stalk 12.5–22 cm long, male flowers few, male flowers open first. Bract pair greenish yellow, narrowly ovate, 13–14 × 8–11 mm, margin not toothed, soon falling. **Male flowers** with a pale pink stalk 15–20 mm long; tepals 4, margin not toothed, outer two white with pink veins, outside with long hairs *c.* 0.5–0.75 mm long, broadly ovate, 16–18 × 7–18 mm, tip rounded, inner two without hairs, oval with a distinct midrib, 13–15 × 9–10 mm, tip acute or rounded; stamens many, cluster globose, *c.* 6 mm across, stalk *c.* 1 mm long filaments 1.25–2.25 mm long; anthers golden yellow, narrowly obovate, 1.25–1.5 mm long, tip rounded, opening by slits. **Female flower**, **fruit** and **seed** not known.

Distribution. Endemic in Peninsular Malaysia, known only from the type locality in Perak.

Habitat. It grows fully exposed to the sun in cracks on a sheer rock face beside the rush of water over a waterfall. In times of flood, it would suffer the full force of the torrential current.

Notes. The Waterfall Begonia is a remarkable species, not only in its habitat that is unique among Peninsular Malaysian begonias, but also because it does not appear to be closely related to any other species in the Peninsula. Among the group of begonias that have almost equal-sided leaves that are not oblique and in its leaf being widest as the base, it most resembles the Rhubarb-leaved Begonia, *Begonia rheifolia*, except that its leaf is very much smaller. In leaf size and shape (ovate) and in its long leaf stalks, it resembles the Mountain Begonia, *B. alpina*, but it is distinct from this species by its hairy, broader leaf and its leaf base that is rounded or scarcely heart-shaped.

It is a very rare plant. Only a few individuals were found. As yet the female flowers and fruits are not known and so the Waterfall Begonia cannot be placed with certainty in a section of *Begonia*. However, in other characters it most resembles begonias that are placed in section *Platycentrum*.

47. THE EQUAL-SIDED BEGONIA
Begonia aequilateralis Irmsch.
(Latin, *aequilaterus*=equal-sided, referring to the leaf)

Irmscher, MIABH 8 (1929) 134 & Fig. 7. **Type:** *Goodenough 10564*, October 1899, Sungai Buloh, Selangor (holo K *ex* SING).

Stem rhizomatous, tightly rooted to rocks, succulent, without hairs, unbranched, stout, 6–14 cm long, 3–5 mm thick; without a tuber. Stipules narrowly triangular, tip elongated, hairless, *c.* 10 × 3 mm, margin not toothed, persistent. **Leaves** tufted up to 5 mm apart; stalk reddish at the base, minutely hairy or without hairs, fleshy, 7–14.5 cm long, 1.5–2 mm thick, flattened above; blade not oblique, dull plain pale green and hairless above with minute hairs beneath, thin in life, membranous when dried, elliptic to slightly rhomboid, almost symmetric, 14–18 × 6–9.5 cm, broad side 3.5–5 cm wide, base narrowed, slightly extended down the leaf stalk, basal lobes not developed, margin deep red, minutely toothed, tip attenuated 0.75–2 cm long; venation pinnate, 3–4 pairs of veins along the midrib, branching towards the margin, impressed above, beneath prominent, deep red and minutely hairy. **Inflorescences** axillary, pale green or reddish, minutely hairy, erect, in male flower shorter than the leaves, 10–14 cm long with two main branches 0.5–2 cm long, stalk 11.5–13 cm long, elongating to 22.5 cm in fruit, few-flowered, male flowers 6–10,

Begonia aequilateralis is probabaly the most endangered begonia in the Peninsula, being known from a single small population in a forest area that is currently being cleared. (Following page). The Equal-sided Begonia, *Begonia aequilateralis*.

ALI IBRAHIM

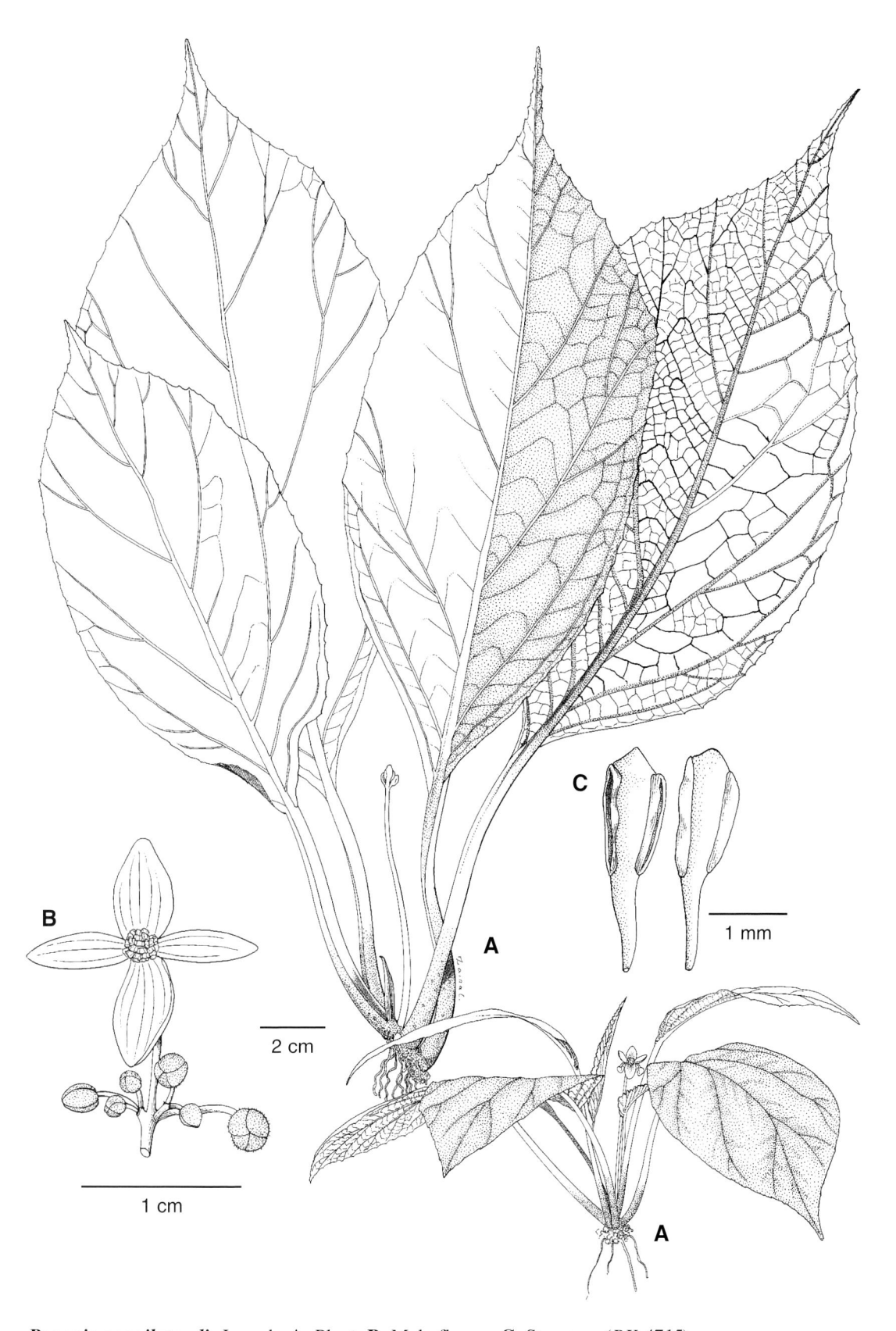

Begonia aequilateralis Irmsch. **A.** Plant. **B.** Male flower. **C.** Stamens. (*RK 4715*)

female flowers 2, male flowers open first. Bracts not known. **Male flowers** with a minutely hairy stalk 7–17 mm long; tepals 4, narrowly oval, tip acute, margin not toothed, outer two white tinged pink, minutely hairy outside, 10–12 × 4–8 mm; inner two white, without hairs, 7–15 × 2–6 mm; stamens many, cluster globose, 3–4 mm across, stalk 0.5–1 mm long; filaments 0.75–2 mm long; anthers narrowly oblong, 1–1.75 mm long, tip rounded, opening by slits. **Female flower** not known. **Fruit** a splash cup pendent on a slender, stiff stalk 20–30 mm long; capsule *c.* 12 × 30 mm, hairless, locules 2, placentas 2 per locule, wings 3, unequal, larger wing fibrous, 17–20 mm wide, tip rounded, smaller two thin, *c.* 8 mm wide, tip rounded, splitting between the locules and wings. **Seed** barrel-shaped, *c.* 0.3 mm long, collar cells three quarters of the seed length.

DISTRIBUTION. Endemic in Peninsular Malaysia, known only from the Sungai Buloh area in Selangor.

HABITAT. Growing on large boulders either on the banks of or in small rocky streams at about 30–70 m altitude. On rocks in streams, its rhizome clings tightly to the downstream side of rocks about 0.5–1 m above the water surface.

OTHER SPECIMENS. SELANGOR—Kepong, Sungai Kroh, *Wyatt-Smith 60831* (KEP); Sungai Buloh, *Kiew RK 1594* (KEP), *RK 1609* (SING), *RK 4715* (SING).

NOTES. The Equal-sided Begonia was first collected in 1899 and is extremely rare, currently being known from only one local population of less than 20 plants. The stream where it grows is rapidly being encroached on by development and the canopy opened up, making it one of the most critically endangered begonia species in Malaysia on the verge of extinction.

Among the species with symmetric leaves, it most resembles *Begonia tampinica* in its elliptic leaves but it is clearly distinct from this species in its smaller leaves.

Begonia aequilateralis. (Left) Male flowers. (Right). Lower surface of the leaf.

48. THE TAMPIN BEGONIA
Begonia tampinica Irmsch.
(Latinized name for Tampin)

Irmscher, MIABH 8 (1929) 130. **Type:** *Burkill SFN 3108*, 27 Jan 1918, Gunung Tampin, Negri Sembilan (lecto SING, here designated).

Stem rhizomatous, firmly rooted to rocks, reddish or green tinged red, succulent, with matted hairs, unbranched, stout, 20–30 cm long, 10–15 mm thick; without a tuber. Stipules green tinged red, densely brown hairy especially on the margin, narrowly triangular, 18–22 × 6–7 mm, margin not toothed, tip ending in a tuft of hairs, persistent. **Leaves** tufted, up to 5 mm apart; stalk green, when young sparsely covered in brown hairs, 12.5–34 cm long, 6–12 mm thick, at base flattened on the upper side, grooved towards top; blade not oblique, plain dull, yellowish green above with a red patch at the junction of the stalk and blade, bright magenta beneath, hairless above, thinly leathery in life, papery when dried, narrowly ovate, almost symmetric, 26–31.5 × 11.5–14.25 cm, broader side 6–7.5 cm wide, base rounded, basal lobes not developed, margin with minute widely spaced teeth, tip acute to slightly acuminate; venation pinnate, 5–6 pairs of veins along the midrib, branching towards the margin, prominent above and beneath, reddish above, beneath deep red and

The Tampin Begonia, *Begonia tampinica*.

Begonia tampinica Irmsch. **A.** Plant. **B.** T.S. leaf stalk. **C.** Fruit. **D.** T.S. fruit. **E.** Seed. (*RK 5151B*)

minutely hairy. **Inflorescences** axillary, coral pink to red, without hairs, in fruit longer than the leaves, 29–33 cm long with two main branches 1.8–2.5 cm long, stalk 27–30 cm long, male flowers many, female flowers 4, male flowers open first. Bracts not known. **Male flowers** with a hairless stalk 13–15 mm long; tepals 4, broadly ovate, without hairs, margin not toothed, tip rounded, outer two shell pink or white, 15–19 × 13–14 mm; inner two 12–16 × 5–6 mm; stamens many; filaments 1.25–2.5 mm long; anthers narrowly oblong, 1.75–2.25 mm long, tip notched, opening by slits. **Female flowers** with an ovary 6 mm long, wings 3, unequal, the longer wing 4–5 mm wide, the shorter two *c.* 2 mm wide, locules 2, placentas 2 per locule; tepals 5, without hairs, margin not toothed, tip rounded, outer oval, *c.* 8 × 5 mm, inner similar but smaller 6 × 4.5 mm; styles 2, styles and stigmas 3.3–5 mm long, stigmas spiral. **Fruit** a splash cup pendent on a fine stalk 12–15 mm long, capsule 10–13 × 22–25 mm, hairless, locules 2, wings 3, unequal, larger wing fibrous, 12–14 mm wide, tip rounded, smaller two papery, 7–8 mm wide, tip triangular, splitting between the locules and wings. **Seeds** barrel-shaped, 0.25–3 mm long, collar cells less than half the seed length.

DISTRIBUTION. Endemic in Peninsular Malaysia, known only from Gunung Tampin, Negri Sembilan.

HABITAT. Growing on large boulders at the headwaters of a small stream at *c.* 600 m altitude.

OTHER SPECIMENS. NEGRI SEMBILAN—Gunung Tampin, *Burkill SFN 2534* (SING), *Kiew RK 5181B* (SING).

NOTES. The Tampin Begonia and *Begonia aequilateralis* are some of the rarest begonias in the Peninsula, each being known from a single population of a few plants. Both grow on the West Coast in heavily populated areas and their habitats are fragile and particularly vulnerable to human disturbance. They are both on the brink of extinction.

(Left). *Begonia tampinica* grows on massive boulders in streams. (Right). Fruits.

49. THE HANGING-LEAF BEGONIA
Begonia praetermissa Kiew, sp. nov.
(Latin: *praetermissus*=overlooked)

Begoniae tampinicae Irmsch. affinis, sed laminis brevioribus (14–19.5 cm longis nec 26–32 cm longis) et tepalis masculis et feminis hirsutis differt. **Typus:** *Kiew RK 5275* 20 Feb 2003, E-W Highway, Kelantan (holo, SING).

Stem rhizomatous, succulent, unbranched, *c.* 3.5 cm long, *c.* 8 mm thick; without a tuber. Stipules reddish brown with long white hairs, narrowly triangular, *c.* 13 × 2 mm, margin not toothed, tip elongate ending in a hair, persistent. **Leaves** tufted, pendent and held in a downward position against the steep bank; stalk reddish brown, 11.5–20 cm long, densely hairy, grooved above; blade not oblique, plain, glossy light green above, sometimes with a bluish tinge, with scattered long hairs above, narrowly ovate, thin in life, thinly papery when dried, more or less symmetric to slightly asymmetric, 14–19.5 × 5.5–9 cm, broad side 3.25–5 cm wide, base rounded, basal lobes scarcely developed, margin toothed and densely hairy, tip elongate to *c.* 3 cm long; venation palmate-pinnate with 2 pairs at the base and 4–5 pairs along the midrib, branching towards the margin, with one vein in the basal lobe on the broader side, veins deeply impressed and hairy above, beneath prominent, densely hairy even on the fine veins, in young leaves red, becoming green in fully expanded leaves. **Inflorescences** axillary, pale reddish, minutely hairy, erect, shorter than the leaves, 13–23 cm long, with two main branches *c.* 1 cm long, male flowers 8–9, female

(Above). The Hanging-Leaf Begonia, *Begonia praetermissa*. (Opposite). The Hanging-Leaf Begonia grows on steep earth banks in forest.

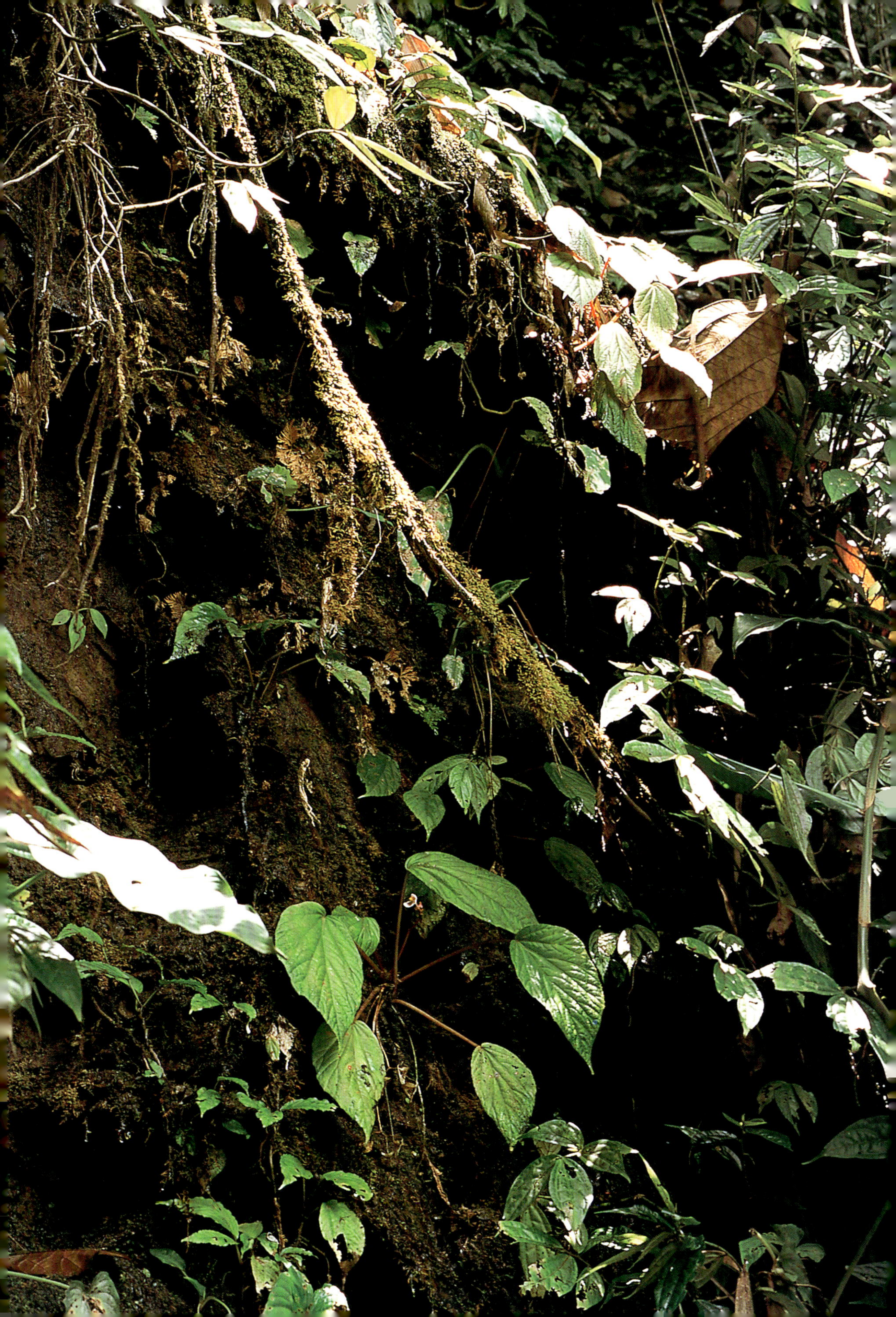

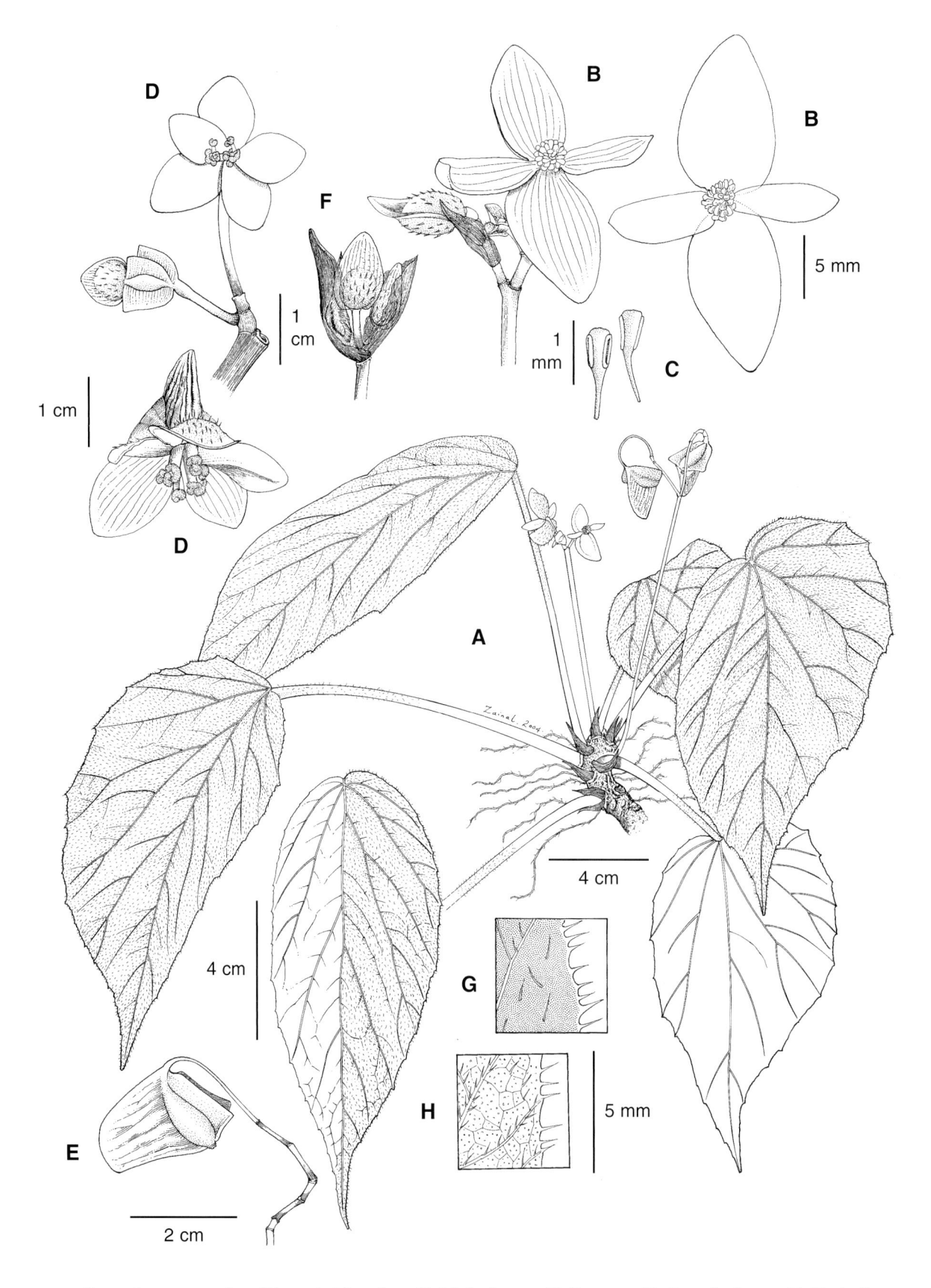

Begonia praetermissa Kiew. **A.** The plant. **B.** Male flower. **C.** Stamens. **D.** Female flower. **E.** Fruit. **F.** Bracts and flower buds. **G.** Upper leaf surface. **H.** Lower leaf surface.

flowers 4, male flowers open first. Bract pair reddish green, broadly ovate, *c.* 9 × 5 mm, margin entire, soon falling. **Male flowers** with pale red stalk, 10–11 mm long; tepals 4, margin entire, outer two white tinged pink, outside pink and densely hairy, oval, 15–16 × 10–13.5 mm, tip acute; inner two narrowly oval, white tinged pink, 11–15 × 3–4.5 mm, tip rounded; stamens many, cluster globose, 3.5–4 mm across, scarcely stalked, stalk *c.* 0.5 mm long; filaments *c.* 0.75 mm long, anthers golden yellow, narrowly obovate, 0.75 mm long, tip rounded, opening by slits. **Female flowers** with a reddish stalk, *c.* 20 mm long; ovary reddish green, *c.* 13.5 × 16 mm, wings 3, unequal, locules 2, placentas 2 per locule; tepals 5, pale pink with darker pink veins, margin not toothed, tip acute, outer two broadly oval, hairy outside, *c.* 13.5 × 13.5 mm, inner three oval, without hairs, *c.* 13.5 × 11 mm; styles and stigmas golden yellow, *c.* 5 mm long, stigmas spiral. **Fruit** a splash cup, pendent on a fine, stiff stalk *c.* 20 mm long, capsule 15 × 23 mm, hairless, locules 2, wings unequal, larger wing thickly fibrous *c.* 14 mm wide, smaller two thinly fibrous, *c.* 6 mm wide, splitting between the locule and wing. **Seed** not known.

DISTRIBUTION. Endemic in Peninsular Malaysia, in northern Kelantan.

HABITAT. In damp, shaded forest on steep earth banks.

NOTES. The Hanging-leaf Begonia is unique among Peninsular species in its leaves that have the tip of the leaf hanging downwards. It is most similar to *Begonia aequilateralis* and *B. tampinica* that have almost equal-sided leaves that are more than twice as long as wide. However, it differs from *B. aequilateralis* in its longer leaf stalk and rounded leaf base and from *B. tampinica* in its shorter leaf. In addition, it differs from both in the leaf blade and tepals being hairy. Among the begonias with asymmetric leaves, *B. klossii* has hairy leaves and tepals but, having heart-shaped leaves, it would not be confused with *B. praetermissa*.

Begonia praetermissa belongs to section *Platycentrum* because it has axillary inflorescences, male flowers that open before the female, and the ovary has two locules, each with two placentas, and the fruit has three very unequal wings.

There is a fruiting specimen from Gunung Setong, Kelantan (*Symington 37723*, KEP) that is rather similar and may belong to this species. However, its leaves are much more asymmetric.

Begonia praetermissa. (Left). Female flowers. (Right). Male flower.

50. HERVEY'S BEGONIA
Begonia herveyana King
(D.F.A. Hervey, Malayan Civil Service, 1870–1893,
Resident Councillor of Malacca 1882–1893)

King, JASB 71 (1902) 63; Ridley, FMP 1 (1922) 861; Irmscher, MIABH 8 (1929) 131. **Type:** *Hervey s.n.*, 1891, Malacca (lecto K—here designated).

Stem rhizomatous, firmly rooted on rocks, unbranched, brown, succulent, without hairs, to 20 cm long, in life *c.* 15 mm thick; without a tuber. Stipules brownish-green, hairless, broadly ovate, 5–7 × *c.* 4 mm, margin not toothed, tip acute, persistent. **Leaves** tufted and 10–15 mm apart; stalk brownish green, flecked with white, succulent, hairless, 10–18 cm long, in life 5–8 mm thick, shallowly grooved above; blade not oblique, plain dull pale green, succulent in life, papery when dried, with microscopic hairs beneath, symmetric, elliptic-oblong to ovate, 15–24.5 × 8.5–17 cm, base rounded, basal lobes 0.5–1 cm long, margin slightly undulate, tip elongate 2–2.5 cm long; venation palmate-pinnate with 2 pairs of veins at the base and 3–5 pairs along the midrib, branching towards margin, minor veins joining to form a marginal vein *c.* 2–3 mm from margin, impressed above, beneath major and minor veins deep red, prominent and without hairs. **Inflorescences** axillary, reddish brown, in male flower shorter than the leaves, 13–25 cm with two major branches, stalk 12–18.5 cm long, in the female and fruiting stages elongating to 19–30 cm and longer than the leaves, male flowers 7–13, female flowers 2–4, male flowers open first. Bract pair leaf-like, greenish, ovate, *c.* 15 × 7–10 mm, margin not toothed, soon falling. **Male flowers** with a pale pink to reddish stalk, 11–23 mm long; tepals 4, pale pink, whitish towards margin,

Hervey's Begonia, *Begonia herveyana*.

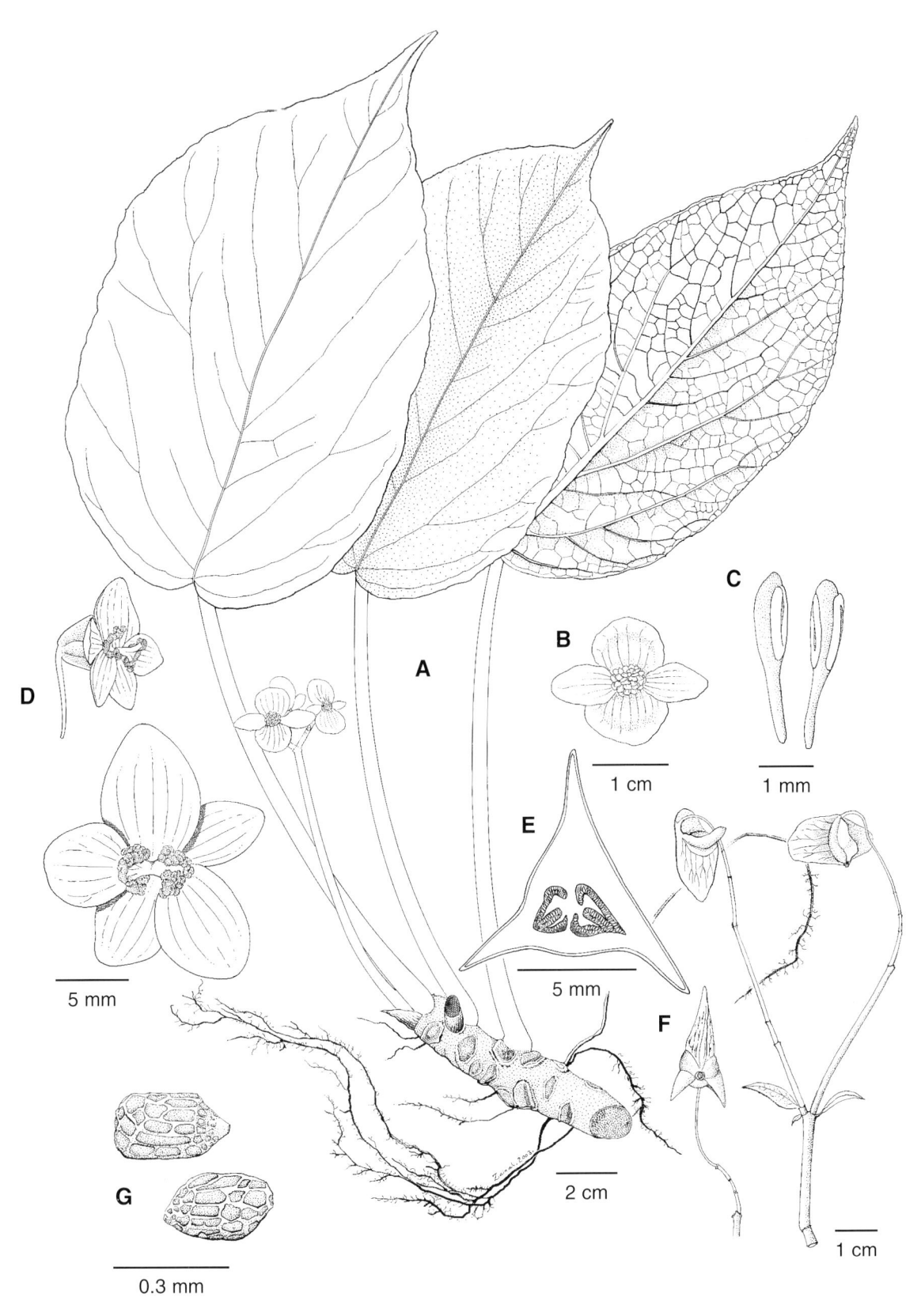

Begonia herveyana King. **A.** The plant. **B.** Male flower. **C.** Stamens. **D.** Female flower. **E.** T.S. ovary. **F.** Fruit. **G.** Seed. (*RK 5231*)

BEGONIA HERVEYANA

hairless, margin not toothed, tip rounded, outer two rotund, 12–18 × 12–18 mm, inner two oval, 12–19 × 5–7 mm; stamens many, cluster hemispherical, 6–7 × 7–9 mm, not stalked; filaments 1–1.5 mm long; anthers golden yellow, narrowly oblong, 1.5–3 mm long, tip rounded, opening by slits. **Female flowers** with a green 3-winged ovary, wings unequal, locules 2, placentas 2 per locule; tepals 5, deep pink, without hairs, subrotund, margin not toothed, tip rounded, inner similar but smaller; styles 2, styles and stigmas pale yellow, stigmas spiral. **Fruit** a splash cup pendent on a thin, thread-like stalk, 19–23 mm long, capsule 13–14 × 20–30 mm, hairless, locules 2, wings unequal, larger wing oblong, tip rounded, thickly fibrous, 10–17 mm wide, smaller two rounded, thinly fibrous, 5–7 mm wide, splitting between the locules and wings. **Seeds** barrel-shaped, *c.* 0.3 mm long, collar cells half the seed length.

DISTRIBUTION. Endemic in Peninsular Malaysia: Malacca and Johore (Pulau Tinggi).

HABITAT. Growing on boulders in streams in narrow valleys where it is extremely local.

OTHER SPECIMENS. MALACCA—Jeram Nyalas, *Derry 1130* (SING); Bukit Batu Lebah, *Kiew RK 5231* (SING). JOHORE—Pulau Tinggi, *Fielding s.n.* 1892 (SING), *Tan PT101* (SINU).

NOTES. Hervey's Begonia has a remarkable distribution. It is known only from the low hills inland from Malacca town and on Pulau Tinggi on the East Coast and nowhere in between. It was originally collected in 1892 from both these places and was only recently recollected, from Pulau Tinggi in 1992 and from a forest remnant in Malacca in 2003. The Malacca population is extremely vulnerable because logging activities in the vicinity place it in danger of extinction.

Begonia herveyana. (Above). Male flowers. (Bottom). Mature green leaf and immature red leaf.

51. THE CLOUD BEGONIA
Begonia nubicola Kiew, sp. nov.
(Latin: *nubicola*=dweller among clouds, referring to its habitat)

Begoniae herveyanae King affinis, sed petiolis et laminis longioribus, tepalis masculis majoribus, albis et hirsutis differt. **Typus:** *Kiew RK 5091*, 31 Aug 2000, Bukit Labohan, Trengganu (holo SING; iso K, KEP).

Stem rhizomatous, rooting tightly on rocks, stout and succulent, unbranched, *c.* 5 cm long, *c.* 12 mm thick; without a tuber. Stipules microscopically hairy outside, narrowly triangular, 14–21 × 7–12 mm, margin not toothed, tip pointed, persistent. **Leaves** about 6 per plant, tufted, up to 8–13 mm apart; stalk brown, hairs sparse, succulent, 15.5–36 cm long, flat to slightly grooved above; blade not oblique, dull, plain rice green or yellowish green, minutely hairy beneath, thinly succulent in life, papery when dry, symmetric, juvenile leaves lanceolate, *c.* 9.5 × 2.5 cm, adult leaves broadly oval to ovate, broadest just below midway, 22–33 × 12.5–20.5 cm, base rounded, margin with sparse hairs and widely spaced minute teeth, tip acute to pointed; venation palmate-pinnate with 2 pairs of veins at the base and 3–4 pairs along the midrib, branching towards the margin, impressed above, beneath prominent, pale fawn with sparse hairs. **Inflorescences** axillary, pinkish red, in male flower shorter than the leaves, 13–25 cm long with two main branches 5–8 cm long, stalk 12–22 cm long, elongating to 20.5–35 cm long and above the leaves in the female and fruiting stages, male flowers *c.* 14, female flowers up to 8, male flowers open first. Bract pair rosy

(Above). The Cloud Begonia grows on large boulders in small streams. (Following page). The Cloud Begonia, *Begonia nubicola*.

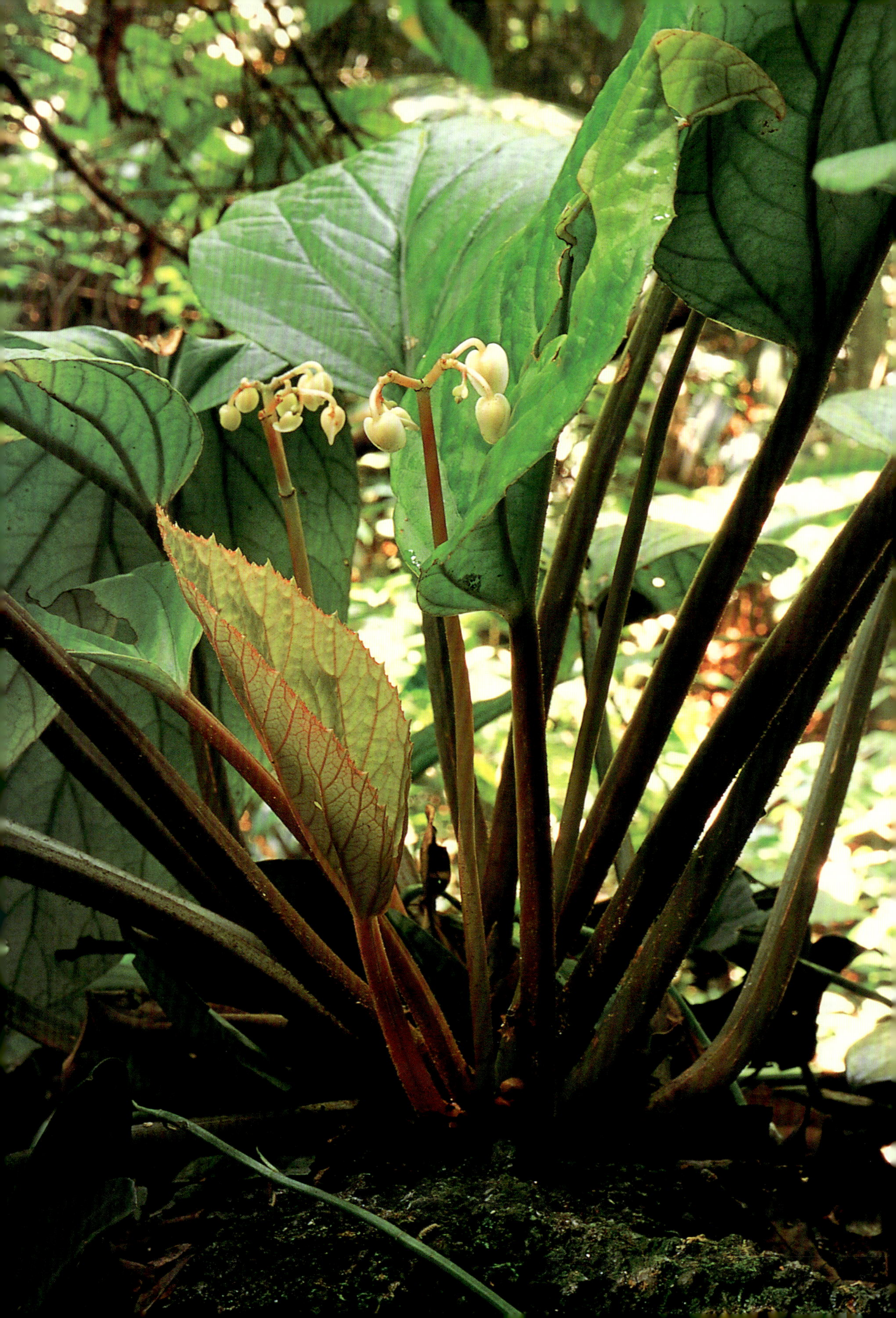

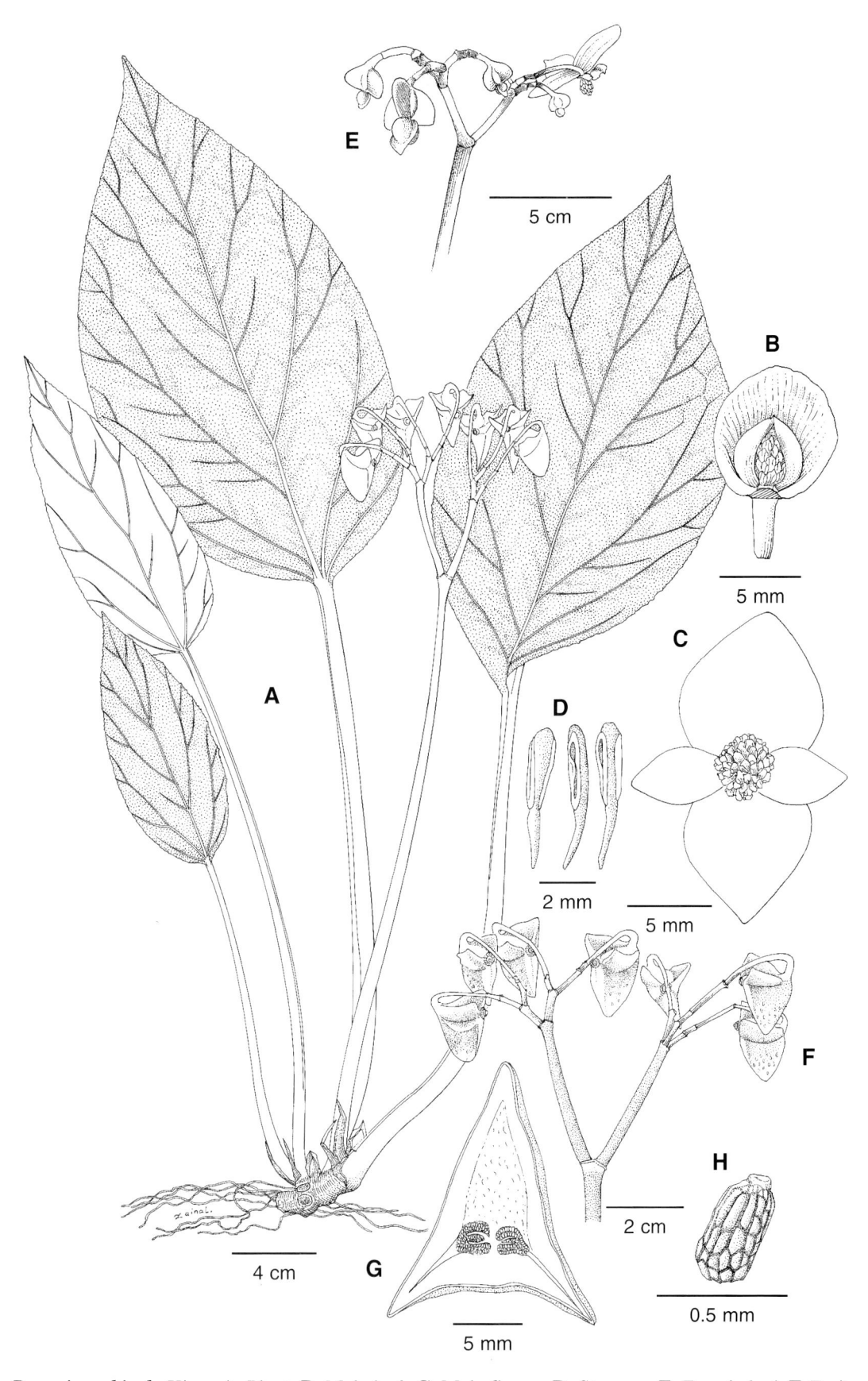

Begonia nubicola Kiew. **A.** Plant. **B.** Male bud. **C.** Male flower. **D.** Stamens. **E.** Female bud. **F.** Fruit. **G.** T.S. fruit. **H.** Seed. (*RK 5091*)

Begonia nubicola with male flowers. (Inset). Fruit.

pink, ovate, tip pointed, 20–21 × 10–14 mm, margin not toothed, soon falling. **Male flowers** with a white stalk, 12–20 mm long, microscopically densely hairy; tepals 4, waxy white, margin not toothed, tip rounded, microscopically hairy outside towards the base, outer two broadly oval, 16–20 × 11–17 mm, inner two slightly obovate, 13–15 × 6–9 mm; stamens many, cluster globose, 4–5 mm across, stalk *c.* 1 mm long; filaments 1–1.5 mm long; anthers yellow, narrowly oblong, *c.* 2 mm long, tip rounded, opening by slits. **Female flowers** (immature, tepals not open) with a stalk *c.* 13 mm long; ovary pale green, *c.* 8 mm long, wings 3, unequal, locules 2, placentas 2 per locule; tepals 5, microscopically hairy outside towards tip, margin not toothed, outer ovate, tip acute, *c.* 5 × 5 mm, inner two rotund, tip rounded, slightly larger, *c.* 7 × 7 mm; styles 2, styles and stigmas *c.* 4 mm long, stigmas spiral. **Fruit** a splash cup pendent on a thin stalk 12–25 mm long, capsule 13–17 × 18–25 mm, without hairs, locules 2, wings 3, unequal, larger wing oblong, tip rounded, fibrous, 10–16 mm wide, smaller two broadly rounded, papery, 6–8 mm wide, splitting between locule and wing. **Seeds** barrel-shaped, *c.* 0.3–0.4 mm long, collar cells *c.* half the seed length.

DISTRIBUTION. Endemic in Peninsular Malaysia, known only from two hills on the East Coast—Bukit Labohan, Trengganu, and Bukit Pelindung, Pahang.

HABITAT. Growing on large boulders in light shade in gullies at the headwaters of intermittent streams in coastal forest at *c.* 230 m altitude.

OTHER SPECIMENS. Peninsular Malaysia: PAHANG—Bukit Pelindung, *Kiew RK 5291* (SING); TRENGGANU—Bukit Labohan, *Davison BL 1* (KEP).

NOTES. Bukit Labohan and Bukit Pelindung are low, isolated granite hills overlooking the sea. *Begonia nubicola* grows on large boulders in gullies in the upper reaches of intermittent streams, where it was locally common. This begonia occupies a very specific habitat being confined to the cloud line, the altitude where cloud settles below the peak of the hill in the late morning, hence its name. Presumably the cooler and moister conditions within the cloud enables this begonia to survive on the dry coastal hill because it is not found on boulders lower down even though the streams carry more water.

Begonia nubicola most resembles *B. herveyana* in leaf shape but differs in its longer petiole (15.5–36 cm long) and larger blade (22–33 × 12.5–20.5 cm), the larger male flowers, which are waxy white, hairy outside and longer than wide (32–40 × 22–34 mm). In contrast, *B. herveyana* has a leaf stalk 10–18 cm long, the blade is 15–24.5 × 8.5–17 cm, the male flowers are smaller, pink, hairless outside and as wide as long (24–36 × 24–36 mm).

Like the Rhubarb-leaved Begonia, *Begonia rheifolia*, the juvenile leaves have a very different shape from ones on flowering plants, being comparatively much narrower.

Male flower of *Begonia nubicola*.

52. THE RHUBARB-LEAVED BEGONIA
Begonia rheifolia Irmsch.
(*Rheum*=scientific name for rhubarb; Latin, *folius*=leaf)

Irmscher, MIABH 8 (1929) 132 & Fig. 6. **Type:** *Ridley s.n.*, Aug 1891, Tahan Valley, Pahang (holo K *ex* SING). **Synonym:** *B. herveyana* King var. *robusta* Ridl., FMP 1 (1922) 861. **Type:** *Wray & Robinson 5546*, Aug 1905, Gunung Tahan, Pahang (lecto K, here designated; iso SING). **Synonym nova:** *Begonia tiomanensis* Ridl., Kew Bull. (1928) 73; Irmscher, MIABH 8 (1929) 133. **Type:** *Mohd Nur SFN 18796*, Pulau Tioman, Pahang (holo K).

Robust begonia. **Stem** rhizomatous, firmly rooted to rocks, reddish brown, fleshy with felty brown hairs, unbranched, stout, up to 8 cm long, 17–18 mm thick; without a tuber. Stipules deep red, midrib hairy, narrowly triangular, *c.* 25 × 13–15 mm, margin not toothed, tipped by a long hair, soon falling. **Leaves** tufted or distant and up to 2.5 cm apart; stalk pale green or reddish brown, when young with dense matted dark brown hairs, 23–47.5 cm long, fleshy, 10–17 mm thick, grooved or flat above; blade not oblique, dull, plain green above and beneath, sometimes reddish beneath, fleshy in life, papery when dried, without hairs, juvenile leaves narrowly elliptic, *c.* 14 × 3 cm and narrowed to base and apex, becoming broadly oval, 21–25 × 13–14 cm and rounded at the base, then broadly ovate, 26–30 × 18–26 cm with a truncate base in mature plants, symmetric,

(Above). The Rhubard-leaved Begonia, *Begonia rheifolia*. (Opposite). The Rhubard-leaved Begonia usually lives on rocks in small streams but in very damp conditions it can grow on tree trunks.

TAN JIEW HOE

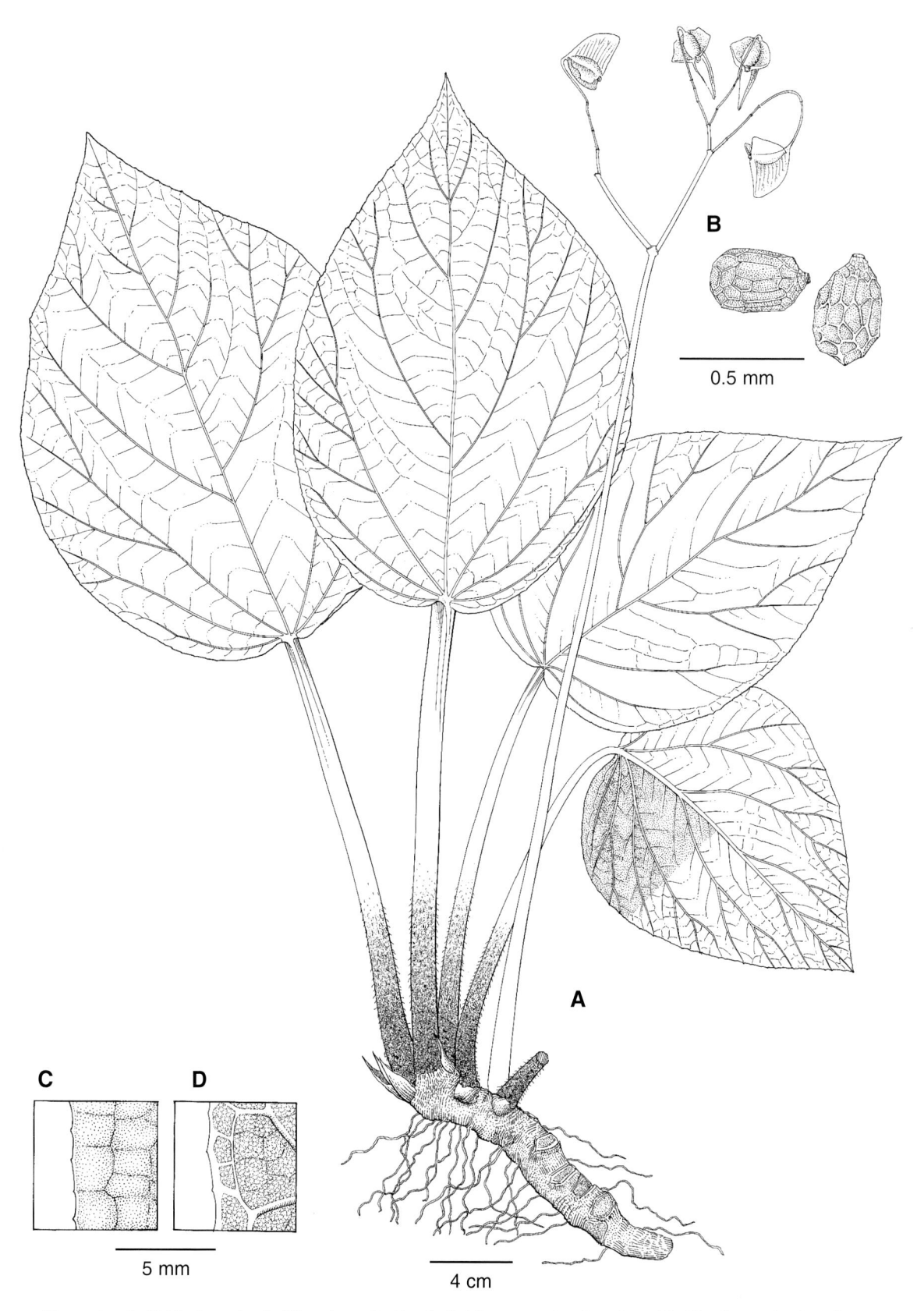

Begonia rheifolia Irmsch. **A.** The plant. **B.** Seed. **C.** The upper leaf surface. **D.** The lower leaf surface. (*RK 5206*)

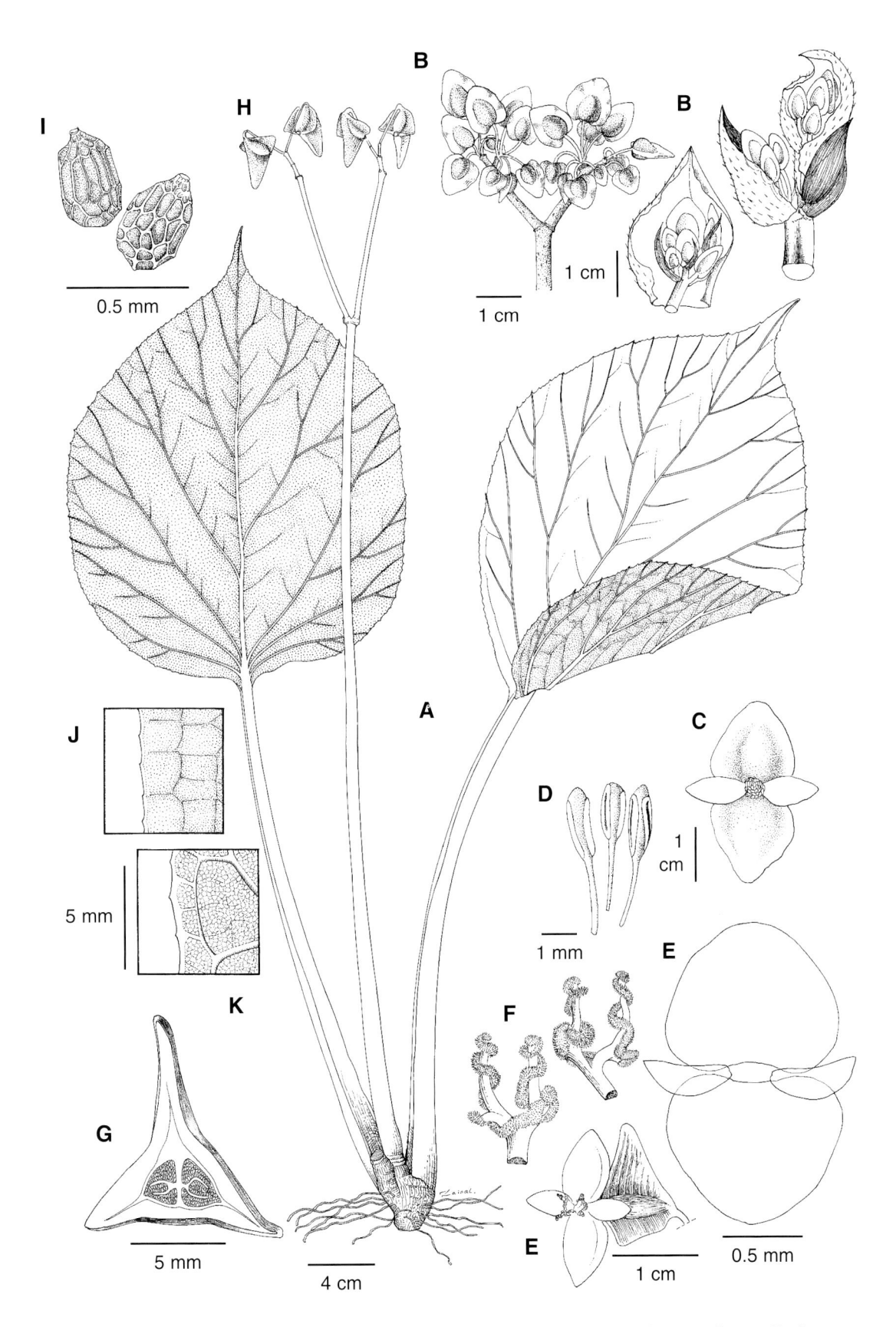

Begonia rheifolia Irmsch. **A.** The plant. **B.** Inflorescence with male buds. **C.** Male flower. **D.** Stamens. **E.** Female flower. **F.** Styles and stigmas. **G.** T.S. ovary. **H.** Fruit. **I.** Seed. **J.** The upper leaf surface. **K.** The lower leaf surface. (*RK 5093*)

broad side 10–13.5 cm wide, basal lobes not developed, margin slightly toothed, tip acute to acuminate; venation palmate-pinnate, 2–3 pairs of veins at the base and 4–7 pairs along the midrib, branching less than halfway to the margin, veins impressed above, beneath prominent, reddish with densely matted pale fawn hairs. **Inflorescences** axillary, reddish or brownish, without hairs, at first shorter than the leaves, 15.5–32 cm long with two main branches 1.5–8 cm long, stalk 14–29 cm long, lengthening to 38–60 cm in fruit and held above the leaves, male flowers many, female flowers up to 8, male flowers open first. Bract pair deep red, without hairs, broadly ovate, 1.5–3.5 × 0.6–2.5 cm, margin not toothed, enclosing the developing flowers, falling once the flowers are mature. **Male flowers** with a rosy pink stalk 13–18 mm long; tepals 4, white or pink, hairless, margin not toothed, tip rounded, outer two ovate, 12–18 × 8–17 mm, inner two oblong-obovate, 10–15 × 6–7 mm; stamens many, cluster globose, *c.* 6 mm across, not stalked; filaments 1–1.5 mm long; anthers lemon yellow, narrowly oblong, 1.5–2 mm long, tip rounded, opening by slits. **Female flowers** with a bright pink or red stalk 8–25 mm long; ovary rosy pink, oval, *c.* 9 mm long, wings 3, unequal, longer wing 5–10 mm long, shorter two 3–5 mm long, locules 2, placentas 2 per locule; tepals 4(?5), pale pink in bud, pure white when open, hairless, oval, margin not toothed, tip rounded, outermost 22–23 × *c.* 7 mm, inner similar but smaller; styles 2, styles and stigmas pale lemon yellow, 5–7 mm long, stigmas spiral. **Fruit** a splash cup pendent on a thin and stiff stalk 20–27 mm long, capsule 14–19 × 27–42 mm, hairless, locules 2, wings 3, unequal, larger thickly fibrous, oblong, tip rounded, 15–24 mm wide, smaller two rounded, 5–10 mm wide, splitting between locules and wings. **Seeds** barrel-shaped, *c.* 0.4 mm long, collar cells *c.* half the seed length.

DISTRIBUTION. Endemic in Peninsular Malaysia: Kelantan, Pahang and Trengganu.

HABITAT. Growing in the lowlands, in deep shade in small rocky side streams either on boulders on the downstream side or on earth banks, between 180–280 m altitude, except on Pulau Tioman where it grows on the top of massive boulders near the summit at 550 m.

Male flowers of *Begonia rheifolia*.

BEGONIA RHEIFOLIA

OTHER SPECIMENS. KELANTAN—Gunung Ayam, *Saw & Kamarudin FRI 37602* (KEP). PAHANG—Sungai Tahan, *Corner s.n.* 1937 (SING), *Kiew RK 2414* (KEP), *RK 2424* (SING), *Yong G.C. s.n.* 1986 (KEP); Sungai Teku, *Kiah SFN 31784* (SING); Pulau Tioman, *Mohd Nur SFN 18796* (K), *Kiew RK 5206* (K, KEP, SING); Tekam F.R., Jeruntut, *Sam et al. FRI 44495* (KEP); Telom Valley, Sungai Lano, *Strugnell & Kiah SFN 23970* (SING). TRENGGANU—Batu Biwa, *Kiew RK 2354* (KEP), Kenyir Dam Waterfall, *Kiew RK 5093* (SING).

NOTES. This robust, fleshy begonia is remarkable for its large leaves with prominent veins and broad base reminiscent of rhubarb leaves, hence the scientific name. It has the largest leaf of any begonia in the Peninsula.

The Rhubarb-leaved Begonia is also remarkable for the change in leaf shape as the plant grows. As a juvenile plant, the leaves are very narrow—very like those of *Begonia rhoephila*—successive leaves become broader and oval with a rounded base until they reach the mature size and shape, ovate with a very broad base. This change in leaf shape is only seen in this species and *B. nubicola*, another equal-sided begonia.

Among the begonias with more or less equal-sided leaves, it is distinct by its broad base. Ridley's *B. tiomanensis* matches *B. rheifolia* not only in leaf shape but also in venation and fruit characters. It is therefore reduced to synonymy with *B. rheifolia*. A notable difference between the Pulau Tioman population and the one on the mainland is habitat. On the mainland, *B. rheifolia* is always associated with streams in the lowlands (up to 280 m), while the small population of Pulau Tioman is found near the summit at 550 m altitude, where it grows on the top of enormous boulders 5–6 m high.

On the mainland, the Rhubarb-leaved Begonia is found in the centre of the Peninsula, from Trengganu (Kenyir Dam), to the Tahan valley in Taman Negara, to Gunung Ayam east of the Main Range, to the Telom valley below Cameron Highlands on the west of the Main Range. *Begonia longifolia* is another species found both in the Cameron Highlands area and on Pulau Tioman. Another equal-sided begonia, *B. herveyana*, is also found on an east coast island, Pulau Tinggi south of Pulau Tioman, as well as on the mainland in Malacca.

Begonia rheifolia. (Left). Fruits. (Right). Female flower.

53. THE WAX FLOWER BEGONIA
Begonia cucullata Willd.
(Latin, *cucullatus*=hooded, referring to the leaves)

Willd., Sp. Pl. 4 (1805) 414.

Stem pale green or brownish-red, nodes swollen, succulent, glossy, without hairs, erect shoots produced from the short prostrate rhizome, flowering at *c.* 15 cm tall, ultimately growing to *c.* 60 cm tall and 1 cm thick at the base, shoots unbranched or with a few branches; without a tuber. Stipules in pairs, semi-transparent, pale green or reddish green, without hairs except for the margin, triangular, 8–12 × 5–6 mm, margin fringed by hairs, tip pointed, frequently persistent. **Leaves** distant and up to 12 cm apart; stalk pale green or pale reddish brown, without hairs, 0.75–7.5 cm long, the lower ones much longer than the upper, shallowly grooved above; blade not oblique, plain deep green, pale beneath, in life succulent, papery when dried, glossy above, broadly ovate to heart-shaped, almost symmetric, 5.5–7.25 × 5.75–7.5 cm, broad side 3.5–4.25 cm wide, base angular or broadly rounded, basal lobes shallow, *c.* 1 cm long, margin strongly wavy and minutely toothed, teeth tipped by a hair, tip acute; venation palmate, 2–3 pairs of veins branching *c.* halfway to margin, with another 1(–2) veins in the basal lobe, impressed above, slightly prominent beneath, same colour as blade, without hairs. **Inflorescences** axillary, green or pale red, without hairs, shorter than the leaves, few-flowered, *c.* 1.5–3.75 cm long lengthening to 3–4 cm in fruit, male flowers 3, female flowers 2, male flowers open first. Bract pair broadly ovate, 4–6 × 4–6 mm, margin fringed by hairs, persistent. **Male flowers** with a stalk 5–20 mm long, white tinged rosy pink; tepals 4, deep rosy red, without hairs, margin not toothed, outer two broadly rounded, 10–14 × 14–17 mm, inner two narrowly oval, 4–11 × 2–4 mm; stamens about 20, stamen cluster globose, 4–5 mm across, stalk *c.* 1 mm long; filaments *c.* 1.5 mm long; anthers yellow, narrow, slightly obovoid, *c.* 2.5–3 mm long, tip rounded, opening by slits. **Female flowers** with a rosy red stalk 12–14 mm long; ovary green with white locules, 10–11 × 18–19 mm, wings 3, unequal, white or bright pink, locules 3, white, placentas 2 per locule; tepals 5, deep rosy red, white at base, without hairs, outer oval, margin not toothed, tip rounded, 6–9 × 5–5.5 mm, inner similar

Begonia cucullata. (Left). Female flower. (Right). Male flower. (Opposite). Habit.

but smaller 5–7 × 3–4 mm; styles 3, styles and stigmas golden yellow, 3.5–6 mm long, stigmas spiral. **Fruits** pendent on a thin, stiff stalk 6–17 mm long, capsule 9–16 × 24–26 mm, hairless, locules 3, wings unequal, drying papery, larger wing 15–17 mm wide with a pointed tip, smaller two, 4–7 mm wide with a rounded tip, splitting between the locules and wings, stigmas persistent. **Seeds** barrel-shaped, *c.* 0.4 mm long, collar cells about half the seed length.

DISTRIBUTION. An exotic from Brazil, it is now a garden escape both in the highlands, e.g., at Cameron and Genting Highlands, and in the lowlands.

HABITAT. On the banks of and in wet concrete storm drains. It spreads by seed and detached pieces of stem.

SPECIMENS. PAHANG—Cameron Highlands *Soeriaatma 0041* (KLU), Fraser's Hill, *Neal et al. SN2* (KEP); Genting Highlands, *Khair Talib s.n.* (UKMB), *Stone 14040* (KLU).

NOTES. Its lush green leaves and waxy deep rosy pink flowers make this an attractive ornamental plant. It was first introduced into the Singapore Botanic Gardens in 1893. It is now widely cultivated, especially at hill resorts where it is used as a bedding plant. Compared with the bearded begonia, *Begonia hirtella*, it is a more recent garden escape with the earliest records dating from the 1970s. It belongs to the semperflorens-cultorum group of hybrids that in temperate regions are important horticulturally, where the dwarf forms, including the hybrids with copper-coloured leaves, are commonly used *en masse* as bedding plants or in floral displays.

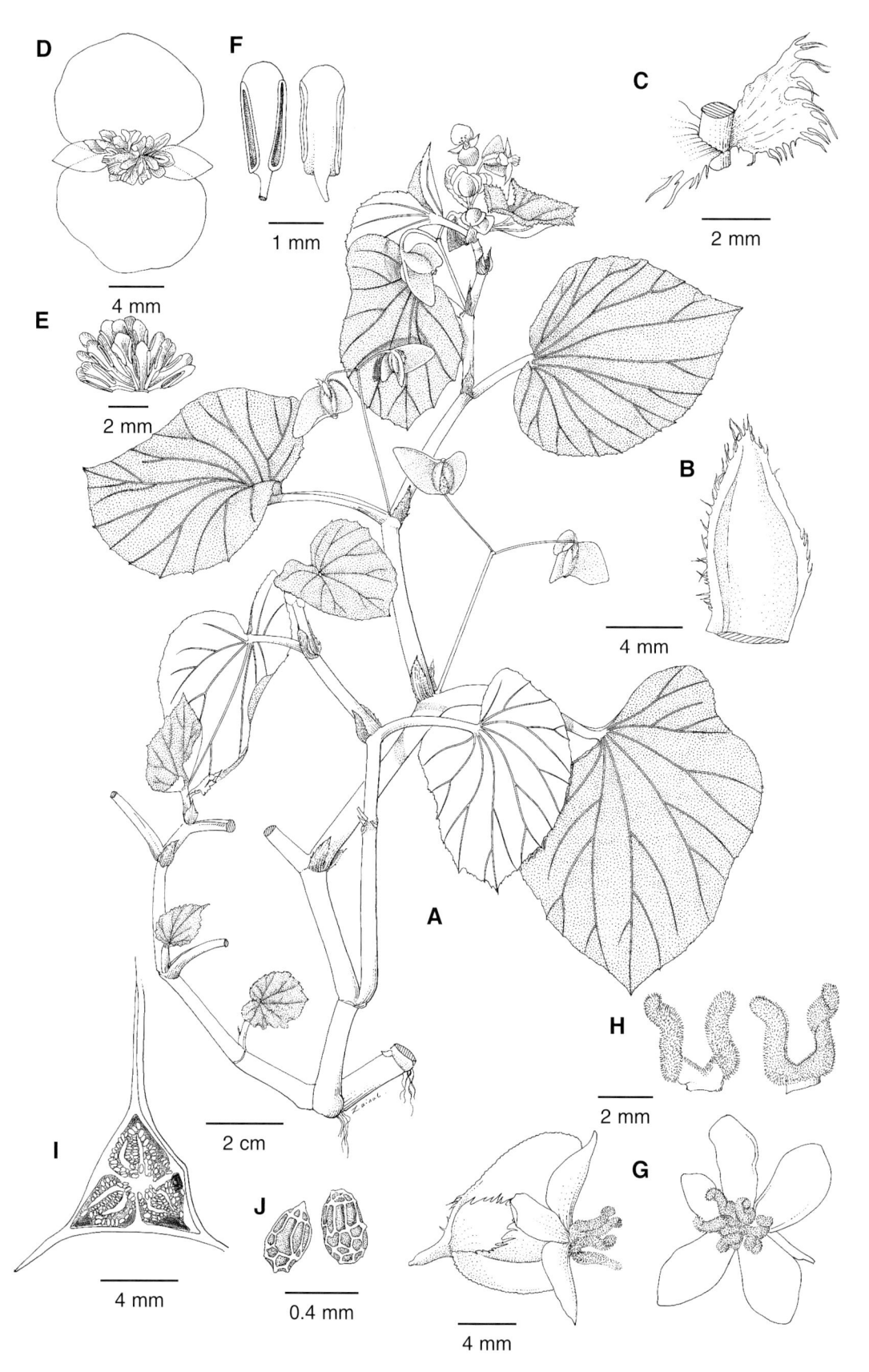

Begonia cucullata Willd. **A.** The plant. **B.** Stipule. **C.** Bracteole. **D.** Male flower. **E.** Stamen cluster. **F.** Stamens. **G.** Female flower. **H.** Styles and stigmas. **I.** T.S. ovary. **J.** Seed. (*Yap s.n.*)

54. THE BEARDED BEGONIA
Begonia hirtella Link
(Latin, *hirtellus*=minutely hairy)

Link, Enum. Hort. Berol. 2 (1822) 396; Ridley, FMP 1 (1922) 864; Henderson, MWF Dicot (1959) 165 & Fig 158.

Stem pale green, succulent, erect, up to 25 cm tall and 8–9 mm thick, flowering at *c.* 3 cm tall, little or unbranched, hairs soft, white *c.* 4 mm long, nodes not swollen; without a tuber. Stipules pale green, densely hairy, obliquely ovate, *c.* 9 × 8 mm, margin not toothed, tip rounded, persistent. **Leaves** distant, 0.75–8 cm apart; stalk green or translucent red, almost woolly, 1–6 cm long, round in cross-section; blade oblique, plain dull rice green above with a crimson spot at the junction with the stalk, paler beneath, hairy above, soft and thin in life, papery when dried, ovate, asymmetric, 3–7.5 × 2.75–6.75 cm, broad side 2–4.75 cm wide, base slightly heart-shaped, basal lobes rounded, 0.75–3 cm long, margin minutely toothed and fringed with hairs, tip acute or elongate; venation palmate, 1–2 pairs of veins branching towards margin with another 2(–3) veins in the basal lobe, veins deeply impressed above, prominent and minutely hairy beneath, the same colour as blade or sometimes crimson towards the stalk. **Inflorescences** axillary, reddish, hairy, longer than the adjacent leaf stalk, unbranched, 2–2.25 cm long, lengthening to 4 cm in fruit, few-flowered, male flowers 2–3, female flowers 2–3, female flowers open first. Bract pair pale green tinged reddish,

The Bearded Begonia, *Begonia hirtella*. (Following page). In hill stations in the Peninsula, *Begonia hirtella* has become established along roadside banks and on rocks along the forest margin.

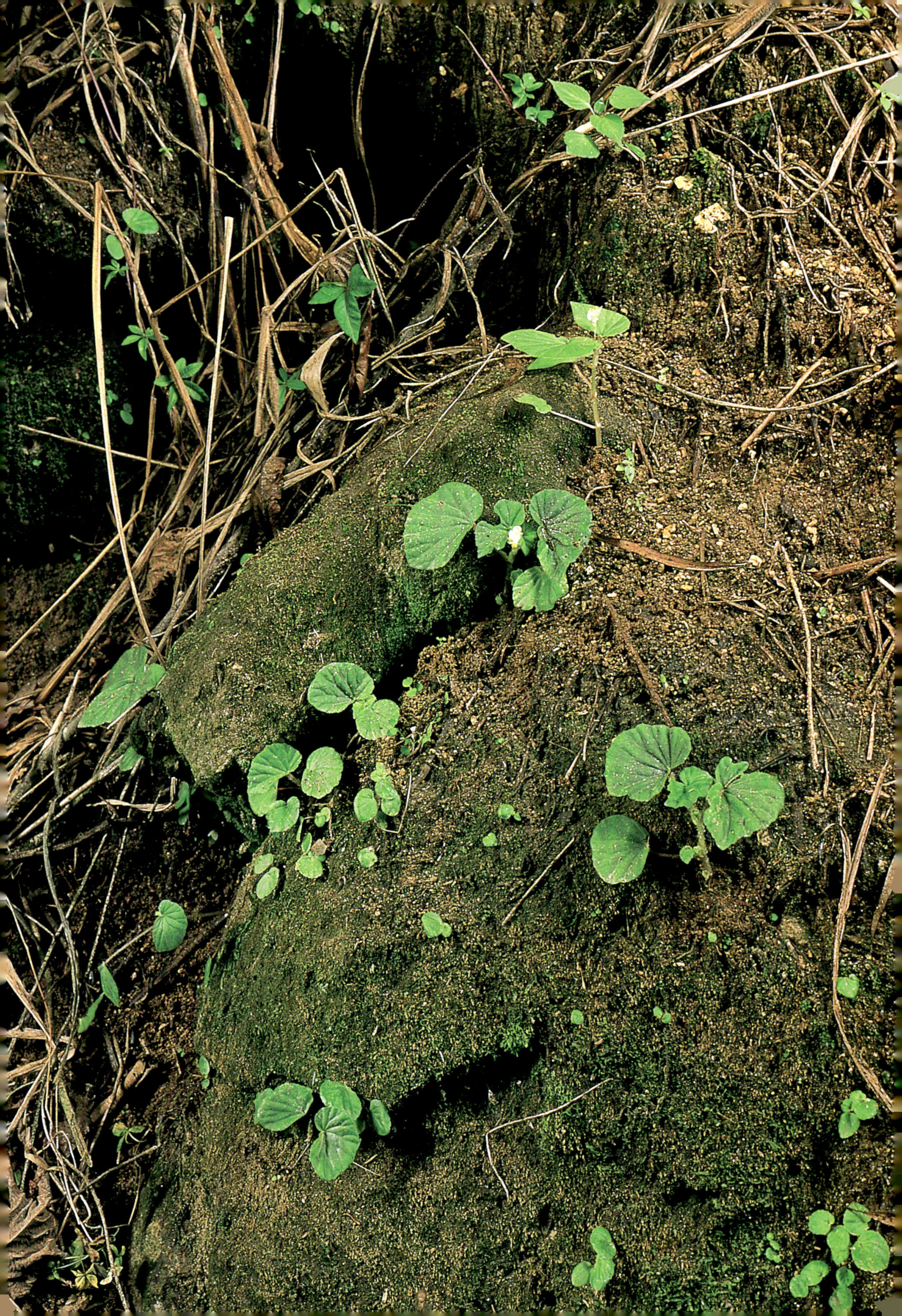

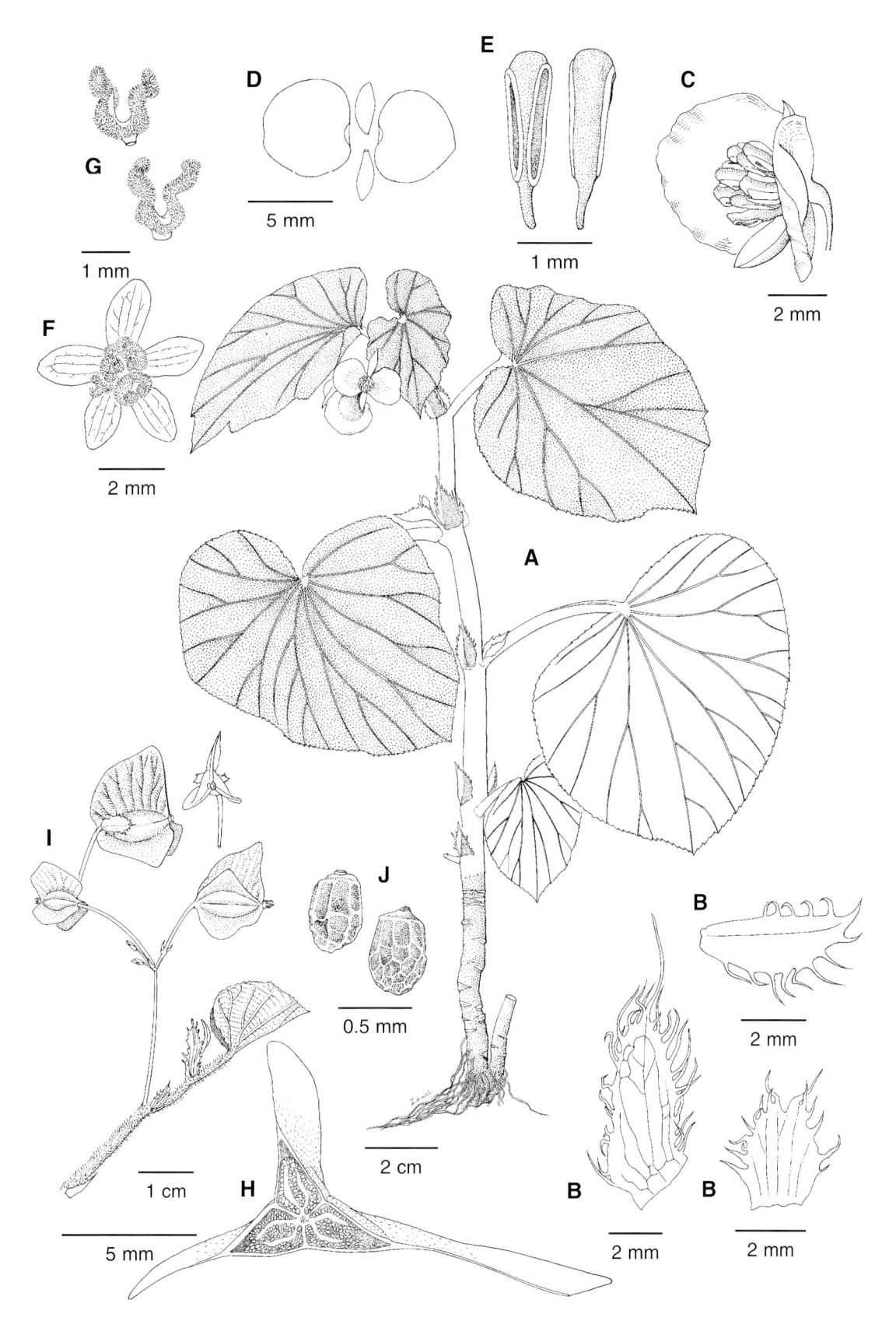

Begonia hirtella Link. **A.** The plant. **B.** Stipule. **C.** Male flower. **D.** Tepals of male flower. **E.** Stamens. **F.** Female flower. **G.** Styles and stigmas. **H.** T.S. ovary. **I.** Fruit. **J.** Seed. (*RK 4908*)

BEGONIA HIRTELLA

Begonia hirtella. (Left). Male flower and fruit. (Right). Male flowers and a young female flower.

narrowly triangular, 2.5 × 1–2 cm, margin not toothed, fringed by hairs, persistent. **Male flowers** with a white stalk *c.* 11–12 mm long, sparsely hairy; tepals 4, white, without hairs, margin not toothed, outer two almost rotund, 4.5–5.5 mm diam., tip rounded, inner two narrowly oval, *c.* 3 × 1 mm; stamens few, *c.* 8, cluster forming a narrow fan, *c.* 2 × 2.5 mm; filaments *c.* 0.75 mm long, free at the base except for the centre three that are joined halfway; anthers pale yellow, narrowly oblong, *c.* 2 mm long, tip rounded, opening by slits. **Female flowers** with a reddish, sparsely hairy stalk 4–5 mm long; ovary white, hairy, *c.* 3 × 4 mm, wings 3, unequal, locules 3, placentas 2 per locule; tepals 5, white, without hairs, oval, margin not toothed, tip rounded, outer *c.* 3.5 × 1.3 mm, inner similar but smaller *c.* 3 × 1 mm; styles 3, styles and stigmas golden yellow, 1.5–2 mm long, stigmas spiral. **Fruits** dangling on a fine and hair-like stalk, 7–22 mm long, capsule 10–11 × 7–19 mm long, hairless, locules 3, wings unequal, larger wing squarish 4–11 mm wide, smaller two rounded, 3–5 mm wide, thin and papery, splitting between the locules and wings. **Seeds** barrel-shaped, *c.* 0.6 mm long, collar cells *c.* one third to half the seed length.

DISTRIBUTION. An exotic species that grows in Brazil, Peru and the West Indies, it is naturalised on Kedah Peak (Gunung Jerai), Penang Hill and Maxwell Hill and in Singapore.

HABITAT. Growing on lightly shaded roadside earth banks and on forest margins.

SPECIMENS. KEDAH—Kedah Peak, *Burkill HMB 3310* (SING); PENANG—Penang Hill, *Burkill SFN 748* (SING), *Kiew RK 1595* (KEP), *RK 1596* (KEP), *Kok Choy s.n.* (SING), *Nauen s.n.* (SING), *Ridley 9346*, 1898 (SING), *Stone 6152* (KLU), Waterfall Garden, *Nauen s.n.* (SING); PERAK—Maxwell Hill, *Burkill & Holttum SFN 12811* (SING), *SFN 12997* (SING), *Kiew RK 4908* (SING), *Kloss s.n.*, 1900 (SING), *Stone 11763* (KLU); SINGAPORE—Tanglin, *Ridley 6929*, 1896 (SING).

NOTES. The Bearded Begonia must have been introduced very early on as a garden plant, because it was reported as a garden escape in Singapore in 1896, on Penang Hill in 1898 and on Maxwell Hill in 1900. Until today it persists on Penang Hill and Maxwell Hill. Indeed, among all the locally cultivated begonias, it is the only one that self-seeds very freely and is dispersed efficiently. It can become quite a weed in gardens. The naturalized form has white flowers, but pale pink-flowered cultivars are available from nurseries.

POSTSCRIPT

But this is not the end of the story. There are still new species that cannot be named because they are not completely known. These include the one-leaf begonia shown below, certainly distinct from *Begonia sinuata* because it does not have star-like hairs. However, it cannot be described as a new species without flowers and fruits.

Another is the variety that Irmscher (1929) described as *Begonia herveyana* var. *barnesii* Irmsch. based on a collection made by Barnes from Kluang Terbang, Pahang, in 1900. It certainly does not belong to *B. herveyana*, which has equal-sided leaves, because the Barnes specimen has one side of the leaf considerably larger and the base is much broader. The specimen has male flowers but no fruits.

Yet another enigmatic specimen was collected by Alvins in 1885 from Bukit Sutu on the Negri Sembilan-Selangor border. Ridley identified this as *B. herveyana* but it is very different in its leaves being oblique and asymmetric. It is without flowers and fruits. Alvins recorded its local name as *asam susu* and mentioned that its sour leaves were used in curries.

It is also likely that the Teku plant that has here been provisionally included in *Begonia foxworthyi* represents a new species.

In addition to these, there must be several more species yet to be discovered as large areas of the Peninsula remain unexplored botanically. This includes most peaks in the Main Range, especially those north of Cameron Highlands, as well as the east side of the Main Range and the west side of the Trengganu Hills.

I hope that this book will encourage naturalists to take notice of begonias and to look out for new and rare species.

A new, unnamed begonia known only from sterile plants.

ACKNOWLEGEMENTS

I am deeply indebted to Tan Jiew-Hoe, who originally suggested that I write this book and who arranged for the field work, illustrations and photography and to Dr Tan Wee-Kiat and Dr Chin See-Chung for their positive support throughout the project. The attractiveness of the book is due to the fine quality of Zainal Mustapha's detailed botanical drawings, Wendy Gibbs's delicate watercolour paintings and Yap Kok-Sun's stunning photography, all enhanced by Datuk C.L. Chan's outstanding skills in natural history publication. Many friends have been involved in the project—Abdullah Piee, Elisabeth Eber-Chan, Geoffrey W.H. Davison, Saw Leng-Guan, Soong Wye-Ping and Yap Kok-Sun all shared materials of new species they had found; Saw Leng-Guan, Lim Swee-Yian and Sam Yen-Yen rediscovered species that had not been seen for many years and were thought extinct; Francis Seow-Choen identified the stick insect; Ali Ibrahim, John Dawn, J.K. Sew and Tan Jiew-Hoe trekked to distant places to take photographs for the book; and Abdullah Piee, Tan Jiew-Hoe and Yap Kok-San, the field team for the book, gave dedicated support over all terrains and in all weathers and succeeded in tracking down 47 of the 52 wild species. This book has been many years gestating during which time several of my students, Noraini Munasib, Christine C.R. Sheue and Teo Lai-Lai, carried out research on begonias; many friends have shared with me the excitement of botanical exploration, in particular S. Anthonysamy, Geoff Davison, John Dawn, Kiew Bong-Heang and Dennis G.C. Yong. I also thank the curators of the herbaria at BM, CGE, K, KEP, KLU, SINU and UKMB for allowing me to examine specimens in their care, Mark J.E. Coode for correcting the Latin diagnoses, Andrea Kee and Norhayati Mohd Din for their dedicated care in maintaining the SBG living collection of begonias, and Elisabeth Eber-Chan for her meticulous proof reading.

APPENDIX: BEGONIA SECTIONS IN PENINSULAR MALAYSIA

With over 1,500 species of begonias worldwide, botanists have grouped the species into smaller, more manageable units called sections. Doorenbos *et al.* (1998) have provided a recent review of the sections currently used.

There are eight sections represented in Peninsular Malaysia and below the species are grouped into their respective sections. Some species Doorenbos *et al.* (1998) were not able to place with certainty because complete data were not available. In a few cases, the species were placed in the wrong section because the information then available was inaccurate. For example, *Begonia carnosula* and *B. yappii* were placed in sect. *Diploclinium* whereas the former belongs to sect. *Parvibegonia* and the latter to sect. *Reichenheimia*, and *B. herveyana* was placed in sect. *Petermannia* whereas it belongs in sect. *Platycentrum*, and *B. integrifolia* should be moved from sect. *Platycentrum* to sect. *Parvibegonia*.

In Peninsular Malaysia, the largest section by far is sect. *Platycentrum* with 23 species (Table 1). Two sections, *Heeringia* and *Ridleyella*, are endemic in Peninsular Malaysia.

Table 1. Size of *Begonia* sections in Peninsular Malaysia.

Section	No. Species
Platycentrum	23
Reichenheimia	10
Parvibegonia	8
Petermannia	5
Ridleyella	2
Diploclinium	2
Heeringia	1
Sphenanthera	1

Section *Diploclinium* (Lindl.) A.DC.

Stem rhizomatous or erect; tubers absent or present; leaf symmetric or asymmetric, not peltate; venation palmate or palmate-pinnate; inflorescence axillary (or terminal); male flowers basal and female flowers distal, protandrous; placenta with 2 branches; capsule pendulous, 3-loculate, wings 3, equal or unequal.

B. *jayaensis*
B. *lowiana*

Section *Heeringia* Irmsch.

Stem erect; tubers present; leaf symmetric, not peltate, sometimes opposite; venation palmate; inflorescence terminal; placentas with 2 branches; capsule 2-loculate, wings 3, unequal.

B. *sibthorpioides*

Irmscher (1929) erected a new section *Heeringia* for *Begonia sibthorpioides* based on three characters: the scaly, rhizomatous (cylindric) tuber, opposite leaves and anthers with pores. Doorenbos *et al.* (1998) stated that, in addition, the stipule is toothed but it is actually entire and fringed by large hairs. However, this species produces globose as well as cylindric tubers and its anthers dehisce by slits, not pores. Its anthers are unique among Malaysian begonias in the slits being positioned on the inner face (instead of laterally on the side of the anther) and in its outer face being covered by minute pimples.

This section is closely similar to sect. *Parvibegonia* differing only in possessing opposite leaves and anthers which split on the inner face and have a pimply outer face. Among all the Malaysian begonias, it is most similar to *B. sinuata* of this section in its symmetric leaves, which are not oblique, the small number of stamens (about 15), and the U-shaped stigmas.

Section *Parvibegonia* A.DC.

Stem erect or rarely rhizomatous; tubers present or absent; leaf asymmetric (rarely symmetric), not peltate; venation palmate or palmate-pinnate; inflorescence terminal; male flowers basal and female flowers distal, protandrous; placentas with 2 branches; capsule pendulous, 2-loculate, wings 3, unequal.

This section reaches its southern limit in Malaysia and indeed several Malaysian species, *B. integrifolia*, *B. martabanica* and *B. sinuata*, are widespread in Asia.

B. *carnosula*
B. *elisabethae*
B. *integrifolia*
B. *phoeniogramma*
B. *martabanica* var. *pseudoclivalis*
B. *variabilis*

These six species form a very natural group. They are all soft, fleshy plants with a small tuber. Doorenbos *et al.* (1998) stated that they are perennial but in the north, where there is a regular dry season, they die down in dry weather and behave like annuals.

APPENDIX

Doorenbos *et al.* (1998) mentioned the irregular splitting of the locules to release the seeds as characteristic of this section. In actual fact, the locule does split neatly along its edge but because the locule wall is thin, it becomes ragged with age.

B. thaipingensis

This species is unusual for the section in possessing a creeping rhizome.

B. sinuata

The sparkling begonia is anomalous for this section in having very few leaves, which are symmetric and not oblique, and in being an annual plant. It is unique among Malaysian begonias in having star-like hairs and leaves that sometimes produce bulbils.

Section *Petermannia* (Klotzsch.) A.DC.

Stem erect (rarely rhizomatous); tubers absent; leaf asymmetric, not peltate; venation palmate to pinnate; inflorescence axillary or terminal; male flowers distal, female flowers basal, protogynous; placentas with 2 branches; capsule pendulous (rarely erect), 3-loculate, wings 3, equal to unequal.

B. holttumii
B. isopteroidea
B. jiewhoei
B. wrayi

These four species are all cane-like begonias. Doorenbos *et al.* (1998) recorded the stem as herbaceous but in fact it becomes woody with age. Among this group, only *B. jiewhoei* has variegated leaves.

B. barbellata

This species produces a short, hairy erect stem and its leaves have very short stalks and are not oblique. Species with these characters are common in Borneo (Kiew, 2001), where sect. *Petermannia* is very well represented.

Section *Platycentrum* (Klotzsch.) A.DC.

Stem rhizomatous or erect; tuber absent; leaf symmetric or asymmetric, not peltate; venation palmate or palmate-pinnate; inflorescence axillary; male flowers basal and female flowers distal, protandrous; placentas with 2 branches; capsule pendulous, 2-loculate, wings 3, very unequal.

This large section, which includes almost half the species in Peninsular Malaysia, separates into several groups based on whether the leaf is oblique and asymmetric versus not oblique and more or less symmetric. These informal groups can be further subdivided based on whether they are semi-erect or have exceptionally narrow leaves.

APPENDIX

(a) leaves oblique and asymmetric
- (i) plant rhizomatous
  - B. alpina
  - B. decora
  - B. klossii
  - B. koksunii
  - B. paupercula
  - B. pavonina
  - B. vallicola
  - B. wyepingiana

- (ii) plant rhizomatous, but also producing weak, semi-erect stems
  - B. fraseri
  - B. longicaulis
  - B. maxwelliana
  - B. venusta

(b) leaves symmetric and not oblique
- (i) leaves extremely narrow
  - B. perakensis
  - B. rhoephila
  - B. scortechinii
- (ii) leaves broad
  - B. abdullahpieei
  - B. aequilateralis
  - B. herveyana
  - B. nubicola
  - B. praetermissa
  - B. rheifolia
  - B. rhyacophila
  - B. tampinica

Section *Reichenheimia* (Klotzsch.) A.DC.

Stem rhizomatous (rarely erect); tuber absent or present; leaf symmetric or asymmetric, sometimes peltate; venation palmate; inflorescence axillary; male flowers basal and female flowers distal, protandrous; placenta unbranched; capsule pendulous, 3-loculate, wings 3, equal or subequal.

These ten species form a natural group of small rhizomatous plants with small, rounded leaves with palmate venation. *Begonia corneri* is unusual in producing a creeping stem with widely spaced leaves and only *B. ignorata* and *B. tigrina* have peltate leaves.

B. corneri
B. forbesii
B. foxworthyi

APPENDIX

B. ignorata
B. lengguanii
B. nurii
B. rajah
B. reginula
B. tigrina
B. yappii

Section *Ridleyella* Irmsch.

Stem rhizomatous; tubers absent; leaf asymmetric, peltate; venation palmate; inflorescence axillary; male flowers basal and female flowers distal, protandrous; placentas unbranched; capsule pendulous, 2-loculate, wings 3, unequal.

Doorenbos *et al.* (1998) doubtfully included a Thai species, *B. pumila*, in this section. Although it has peltate leaves, it is unlike the other two species in its upright habit and symmetric leaves. Excluding this species, sect. *Ridleyella* is endemic in Peninsular Malaysia.

B. eiromischa
B. kingiana

Section *Sphenanthera* (Hassk.) Warb.

Stem erect or rhizomatous; tuber absent; leaf asymmetric, not peltate; venation palmate or palmate-pinnate; inflorescence axillary; male flowers basal and female flowers distal, protandrous; placentas with 2 branches; berry erect or pendulous, locules 3 (or 4), without wings, not splitting. This section groups together all the species with a berry-like fruit. It is not a natural group as it includes obviously unrelated species (Tebbitt, 2003). In its cane-like habit, *B. longifolia* is closest to species in sect. *Petermannia*.

B. longifolia

Exotic species

The two garden escapes belong to **Section *Begonia*** (*B. cucullata*) and **Section *Doratometra*** (*B. hirtella*).

GLOSSARY

Berry – A fleshy fruit with many seeds that does not split

Bract – A leaf-like structure associated with inflorescences

Bracteole – A small bract, usually attached above the bract

Bulbil – A special bud, often produced on the leaves, which readily detaches and grows into a new plant

Capsule – A dry fruit that splits to release the seeds

Cultivar – Cultivated plants with distinguishing features that are retained when the plant is propagated or reproduced

Defoliate – To remove all the leaves

Endemic – Confined in its distribution to a specific area that can be small, e.g. a single hill or large, like a country or continent

Endosperm – The cellular food reserve in the seeds of flowering plants

Exotics – Not native, plants originating from outside the area they are now found

Genus – A group of species recognized as sharing characters in common that distinguish them from other genera

Herb – A soft plant without woody tissue

Herbivory – The eating of plants

Hybrid – A cross between different entities, usually between species

Inflorescence – The flowering axis comprising bracts, bracteoles and flowers

Lanceolate – Spear-shaped, i.e., widest in the lower half

GLOSSARY

Locule – A chamber or cavity, in begonias referring to the cavities in the ovary (see page 27)

Node – The point on the stem where structures like leaves are attached

Obovate – Widest in the upper half

Palmate venation – Veins radiating from the point of attachment of the leaf blade with the stalk (see page 28)

Palmate-pinnate venation – In begonias this means veins at the base are palmate and the ones above are pinnate (see page 28)

Papillose – Covered in soft superficial glands

Peltate – Of leaves, where the leaf stalk is attached above the base of the leaf blade (see page 28)

Pinnate venation – Veins arise from the midrib (see page 28)

Placenta – The tissue to which the ovules are attached in the ovary. In begonias, the placenta is a vertical plate which projects into the locule cavity (see page 27)

Prostrate – Lying flat on the ground or rocks

Rheophyte – A plant that grows in streams or rivers where it is periodically assailed by strong water currents

Rhizome – A prostrate or underground stem

Rice green – The bright green of young rice or grass leaves

Shrub – A low woody plant with many stems or branches

Species – A group of individuals (plants) recognized as distinct which under normal conditions reproduce with one another but not with individuals of a different species

Stamen – The male part of the flower that consists of a stalk called the filament and the anther that produces the pollen

Stigma – The surface at the top of the style that receives pollen

Stipule – In begonias it is a leaf-like structure attached at the base of the leaf stalk

Style – The columnar structure between the ovary and the stigma

Tepals – The flower parts where the petals and sepals are similar. In the case of begonias the tepals are petaloid, i.e. the outer ones are not small and green like sepals

GLOSSARY

Terrarium – An enclosed transparent structure for growing plants that need high humidities

Turgid – Swollen, i.e., the cells are filled to capacity with water

Understorey – The layer of the forest below the tree canopy where the herb and shrub layer grows

REFERENCES

Berg, R.G. van den (1985). Pollen morphology of the genus *Begonia* in Africa. In: J.J.F.E. de Wilde (ed.). *Studies in Begoniaceae* II. Agric. Univ. Wageningen Papers 84–3: 5–94.

Boesewinkel, F.D. and A. de Lange (1983). Development of ovule and seed in *Begonia squamulosa* Hook.f. *Acta Bot. Neerl.* 32: 417–425.

Burkill, I.H. (1927). Botanical collectors, collections and collecting places in the Malay Peninsula. *Gard. Bull. Str. Sttl.* 4: 113–202.

Clarke, C.B. (1879). Begoniaceae. In: J.D. Hooker (ed.). *Flora of British India* 2: 635–656.

Corner, E.J.H. (1960). The Malayan Flora. In: R.D. Purchon (ed.). *Proc. Centen. & Bicenten.. Cong. Biol., Singapore.* Pp. 21–24.

Doorenbos, J., M.S.M. Sosef and J.J.F.E. de Wilde (1998). The sections of *Begonia. Studies in Begoniaceae* VI. Agric. Univ. Wageningen Papers 98-2: 1–266.

Gagnepain, F. (1921). *Begonia integrifolia.* In: F. Gagnepain (ed.). *Flora Generale de l'Indochine* 2: 1114.

Henderson, M.R. (1959). *Malayan Wild Flowers Dicotyledons.* Malayan Nature Society, Kuala Lumpur.

Irmscher, E. (1929). Die Begoniaceen der Malaiischen Halbinsel. *Mitt. Inst. Allg. Bot. Hamburg* 8: 86–160.

Kiew, R. (1988). Herbaceous flowering plants. In: E. Cranbrook (ed.). *Key Environments Malaysia.* Pergamon Press, U.K. Pp. 56–76.

Kiew, R. (1989). Lost and found—*Begonia eiromischa* and *B. rajah. Nature Malaysiana* 14: 64–67 & front cover.

Kiew, R. (1991). The Limestone flora. In: R. Kiew (ed.). *The State of Conservation in Malaysia.* Malayan Nature Society, Kuala Lumpur. Pp. 42–50.

Kiew, R. (1997). The Malaysian highlands and limestone hills: threatened ecosystems. In: *State of the Environment in Malaysia.* Consumers Assoc. Penang, Penang. Pp. 66–73.

Kiew, R. (2001). The limestone begonias of Sabah, Borneo—Flagship species for conservation. *Gard. Bull. Singapore* 53: 241–286.

Kiew, R. and C. Geri (2003). Begonias from the Bau limestone, Borneo, including a new species. *Gard. Bull. Singapore* 55: 113–123.

REFERENCES

Kiew, R. and A. Kee (2002). Begonias galore. *Gardenasia* 13: 12–14, 30–31.

Kiew, R., L.L. Teo and Y.Y. Gan (2003). Assessment of the hybrid status of some Malesian plants using amplified fragment length polymorphism. *Telopea* 10: 225–233.

King, G. (1902). Begoniaceae. *J. Asiat. Soc. Bengal* 71: 56–68.

Lee, D.W. (2001). Leaf colour in tropical plants: Some progress and much mystery. *Malay. Nat. J.* 55: 117–131.

Ridley, H.N. (1909). *Botanic Gardens, Singapore*. Annual Report on the Gardens and Forests Department, Straits Settlements for 1908.

Ridley, H.N. (1909). *Begonia robinsonii. J. Fed. Mal. St. Mus*. 4: 22.

Ridley, H.N. (1911). *Begonia clivalis. J. Str. Br. Roy. As. Soc*. 57: 49.

Ridley, H.N. (1922). Begoniaceae. *Flora of the Malay Peninsula* 1: 853–864.

Ridley, H.N. (1925). *Begonia foxworthyi. Flora of the Malay Peninsula* 5: 311.

Savile, D.B.O. and H.N. Hayhoe (1978). The potential effect of drop size on efficiency of splash-cup and springboard dispersal devices. *Can. J. Bot*. 56: 127–128.

Sheue, C.R., V. Sarafis, H.Y. Liu, Y.P. Yang and R. Kiew (2003). Structural study on leaves of super shade plants. *Proc. of the 24th Republic of China Symposium on Microscopy*. B-0-13-14.

Stapf, O. (1892). *Begonia decora. Gard. Chron*. III, 12 (19 Nov): 621.

Tebbitt, M.C. (2003). Taxonomy of *Begonia longifolia* Blume (Begoniaceae) and related species. *Brittonia* 55: 19–29.

Teo, L.L. and R. Kiew (1999). First record of a natural begonia hybrid in Malaysia. *Gard. Bull. Singapore* 51: 103–118.

Thompson, M.L. and E.J. Thompson (1981). *Begonias, the Complete Reference Guide*. New York Times Book, New York.

Watson, W. (1892). *Begonia decora. Garden and Forest* 5 (26 Nov): 561.

INDEX
Complied by Kay Lyons

Note: Numbers in **bold** refer to illustrations

A
Abrosoma festinatum, 18
Alvins, 293
angel-wing begonia, 3
Anisophyllaceae, 26
Apis cerana, 13
Aring begonia, 223–6, **223–6**
asam riang, 74
asam susu, 293

B
Barnes, 293
bearded begonia, 3, 287, 289–92, **289–92**
bee pollination 13, **13**, 138
Begon, Michel, 25
Begonia, 26–9; sections, 27, 295–9
    abdullahpieei, 32, 252–5, **252–5**
    aequilateralis, 23, 32, 251, 261–4, **261–4**, 271
    alpina, 32, 33, 149–52, **149–52**, 260; habitat, 19; leaves, 5, 8
    barbellata, 31, 128, 129–32, **129–32**; distribution, 20; habitat, 19; habit, 5; stick insect, 18
    baturongensis, 128
    berhamanii, 132
    carnosula, 31, 70–3, **70–3**, 84; habitat, 19; leaves, 8
    cathayana, 138
    clivalis, 29, 62, 68
    coccinea, 3
    collina, 29, 251
    corneri, 26, 29, 30, 106, 202–5, **202–5**, 226; distribution, 20; habitat, 19; habit, 5
    cucullata, 3, 31, 32, 286–7, **286–8**
    curtisii, 29, 84
    debilis, 29, 84
    decora, 3, 30, 133–8, **133–8**, 144, 148; flowers, 10, **13**; habitat, 19; hybrid, 13, 138; leaves, 8, **9**, 10; pollination, 13; seeds, 16; vegetative propagation, 17, **17**
    decora × *B. venusta*, 138–9, **139**
    deliciosa, 138
    eiromischa, **22**, 26, 29, 30, 35, **36**, 42;
        extinction, 22; leaves, 5
    elisabethae, 32, 84, 98–101, **98–101**; distribution, 20; habitat, 19; vegetative propagation, 17
    forbesii, 29, 33, 206–10, **206–10**; distribution, 21; habitat, 19, 23
    foxworthyi, 32, 222, 227–32, **227–31**, 293; fruits, 14; habitat, 19; pollination, 13
    fraseri, 33, 178, 184–7, **184–7**, 191; habitat, 19
    goegoensis, 3, 217
    guttata, 25, 29, 84
    haniffii, 29, 84
    hatacoa, 138
    heracleifolia, 3
    herveyana, 32, 272–4, **272–4**, 279, 285, 293; distribution, 20
    herveyana var. *barnesii*, 293
    hirtella, 3, 31, 287, 289–92, **289–92**
    holttumii, 31, 112–15, **112–15**, 116, 122, 128; distribution, 20; Endau-Rompin State Park, 22; flowers, 10; fruits, 13; habit, 5; habitat, 19; pests, 18, **18**; reproduction, 17
    ignorata, 26, 30, 42, 43–9, **43–9**, 53; distribution, 20; flowers, 11; habitat, 18; leaves, 5; Taman Negara, 22; vegetative propagation, 17
    integrifolia, **8**, 29, 31, 68, 73, 74–84, **74–82**, 91, 97, 101, 232; discovery, 25; distribution, 19–20; flowers, 11; habit, 5; habitat, 19; leaves, 8; Perlis State Park, 22; seasonality, 17
    isopteroidea, 29, 31, 116, **117**; habit, 5
    jayaensis, **viii**, 26, 30, 196, 197–201, **197–201**; distribution, 20; flowers, 11; habitat, 18; leaves, 10
    jiewhoei, **ii**, 31, 123–8, **123–8**; distribution, 20; habit, 5; habitat, 18
    kingiana, **i**, **x**, **1**, 26, 30, 35, 37–42, **38–42**, 49; distribution, 20; flowers, **9**, 11; habitat, 18; leaves, 5, 10; name, 25; Perlis State Park, 22
    klossii, 29, 33, 164–9, **164–9**, 271; habitat, 19; pests, 18

INDEX

koksunii, **24**, 33, 153–7, **154–7**
kunstleriana, 238
lengguanii, 33, 211–15, **211–15**; Bukit Rengit Wildlife Reserve, 22; distribution, 20; habitat, 19
leucantha, 29, 84
longicaulis, 33, 178, 187, 188–91, **188–91**; distribution, 20; Taman Negara, 22
longifolia, 27, 31, 107–11, **107–11**, 285; distribution, 19, 20; flowers, 12; habit, 5; habitat, 19; pests, 18; seeds, 13–14, 16
lowiana, 192–6, **192–6**, 201; flowers, 11, 26, 30; habit, 5; habitat, 19; leaves, 10
martabanica, 31; distribution, 19–20
martabanica var. *pseudoclivalis*, 29, 68, **69**
masoniana, 3, 138
maxwelliana, 33, 169, 173, 174–8, **174–8**, 187; distribution, 20; leaves, 10
monticola, 152
nubicola, 32, 275–9, **275–9**, 285; distribution, 20; habitat, 19; leaves, 6
nurii, 32, 222, 232, 233–7, **232–7**; distribution, 20; habitat, 18; leaves, 5
paupercula, 29, 33, 170–3, **170–3**; pests, 18; seedlings, **16**
pavonina, **7**, 29, 33, 73, 157, 158–63, **158–63**; habitat, 19; leaves, 8, **9**; vegetative propagation, 17
perakensis, 240; habitat, 19; leaves, 5
perakensis var. *conjugens*, 32, 152, **240–4**, 244
perakensis var. *perakensis*, 30, 244
phoeniogramma, 84, 85–91, **87–91**, 97, 232; flowers, 11, 26, 31; habitat, 19; leaves, 8; seasonality, 17
praeclara, 138
praetermissa, 32, 268–71, **268–71**
rajah, 3, **4**, 29, 33, 216–17, **217**, 222, 237; cultivation, 23, 217; distribution, 20; hybrids, 217
reginula, 33, 216, 218–22, **218–22**; distribution, 20; habitat, 19; leaves, 6; threats to, 23
rex, 1, **2**, 3; cultivars, 138
rheifolia, 29, 32, 111, 260, 279, 280–5, **280–5**; distribution, 20; habit, 5; habitat, 19; leaves, 5; Taman Negara, 22
rhoephila, 27, 29, 30, 245–51, **245–51**, 285; habitat, 19; leaves, 5
rhyacophila, 27, 32, 256–60, **256–60**; habitat, 18
robinsonii, 29, 163, 191
scortechinii, 29, 30, 238, **239**
sibthorpioides, 26, 31, 54–7, **54–7**; habit, 5; leaves, 5
sinuata, 26, 31, 56, 58–63, **58–64**, 122, 293; discovery, 25; distribution, 20; Endau-Rompin State Park, 22; habit, 5; habitat, 19; leaves, 8, **9**, 10; seasonality, 17; Taman Negara, 22; vegetative propagation, 17
sinuata var. *pantiensis*, 26, 64, **65–7**; leaves, 6
sinuata var. *sinuata*, **60**, 63–4
tampinica, 29, 32, 264, 265–7, **265–7**, 271; threats to, 23
thaipingensis, 26, 30, 102–6, **102–6**, 205; habit, 5; habitat, 19; leaves, 8, **9**, 10
tigrina, 26, 30, 42, 50–3, **50–3**; distribution, 20; habitat, 18; leaves, 5
tiomanensis, 29, 285
tricornis, 111
vallicola, 30, 145–8, **146–8**
variabilis, 26, 32, 84, 91, 92–7, **93–6**, 101; flowers, 11; habitat, 19
venusta, 33, 116, 173, 178, 179–83, **179–83**, 187, 191; flowers, 11; habit, 5; habitat, 19; hybrid, 13; leaves, 10; pollination, 13; seeds, 16
wrayi, 26, 31, 115, 116, 118–22, **118–22**, 128; distribution, 20; flowers, 10, 11; fruits, 137; habit, 5; habitat, 19; reproduction, 17
wyepingiana, 30, 140–4, **140–4**; flowers, 11; leaves, **6**, 10; vegetative propagation, 17
yappii, 29, 33, 215, 223–6, **223–6**; distribution, 20; habitat, 19
Begoniaceae, 26
begonias: acids, 3; conservation, 20–23; cultivated species, 1–3; cultivation, 23; distribution, 19–20; flowers, 10–13, **11**, **12**; fruits, 13–14, **14**, **27**; habit, 5; habitats, 18–19; history, 25; identification, 28–9; leaf shapes, **28**; leaves, 5–10; pests, 18; seasonality, 17; seed dispersal, 13–14; seeds, 14–16, **14**, **15**; uses, 3, 293; vegetative propagation, 17
berry begonia, 107–11, **107–11**; distribution, 19; flowers, 11; seeds, 13–14, 16
Berumban begonia, 116, **117**
Big Decora hybrid, 3
Bolbitis heteroclita, 101
bristly begonia, 128, 129–32, **129–32**; distribution, 20; habit, 5; habitat, 19; stick insect, 18
Bukit Rengit Wildlife Reserve: begonias, 22

INDEX

Burkill, I.H., 84
Burmese begonia, 68, **69**

C
cabbage-leaved begonia, 179–83, **179–83**; flowers, 11; habit, 5; hybrid, 13; leaves, 8, 10; seeds, 16
caterpillars, 18, **18**
cave begonia, **viii**, 197–201, **197–201**; flowers, 11; habitat, 18; leaves, 10
Clarke, C.B., 84
cloud begonia, 275–9, **275–9**; habitat, 19; leaves, 6
common Malayan begonia, 115, 118–22, **118–22**; distribution, 20; flowers, 10
Corner's begonia, 202–5, **202–5**
Cucurbitaceae, 26
Cucurbitales, 26
Curtis, Charles, 138, 148

D
Datiscaceae, 26
Dawn, John, 62
decorative begonia, 3, 133–8, **133–8**; flowers, 10; hybrid, 13; leaves, 8, 10; seeds, 16
diminutive limestone begonia, 233–7, **232–7**
Doorenbos, J., 73, 84, 163, 226

E
Eber-Chan, Elisabeth,101
Elisabeth's begonia, 98–101, **98–101**
Endau-Rompin State Park: begonias, 22
equal-sided begonia, 261–4, **261–4**

F
Foxworthy, F.W., 232
Foxworthy's begonia, 227–32, **227–31**
Fraser's Hill begonia, 184–7, **184–7**
fringed limestone begonia, 43-9, **43-9**

G
Gagnepain, F., 84
Gopeng begonia, 170–3, **170–3**
Gunung Tahan begonia, 188–91, **188–91**

H
hanging-leaf begonia, 268–71, **268–71**
Haniff, 232
Henderson, M.R., 25
Hervey's begonia, 272–4, **272–4**
Hillebrandia, 26
Holttum's begonia, 112–15, **112–15**; flowers, 10

I
Irmscher, Edgar, 25, 57, 62, 68, 84, 91, 116, 148, 183, 191, 238, 241, 251, 293
iron-cross begonia, 3

J
Jiew-Hoe's begonia, **ii**, 123–8, **123–8**

K
King, Sir George, 25, 84, 116, 173, 210, 238, 241
King's begonia, 49
Kloss, C.B., 169
Kloss's begonia, 164–9, **164–9**
Kok-sun's begonia, **24**, 153–7, **154–7**
Kunstler, H., 173, 238, 241

L
Larut begonia, 206–10, **206–10**
Leng-Guan's begonia, 211–15, **211–15**
Lim Swee-Yian, 210
Low's begonia, 192–6, **192–6**; flowers, 11; leaves, 10
Loxonia hirsuta, 111
lush-leaved begonia, 70–3, **70–3**

M
maple-leaf begonia, 64, **65–7**
Maxwell Hill begonia, 174–8, **174–8**
Mohamed Nur bin Mohamed Ghous, 232
mountain begonia, 149–52, **149–52**, 260

O
Octomeles sumatrana, 26

P
painted-leaf begonia, 3
peacock begonia, **7**, 73, 158–63, **158–63**; leaves, 8
Pentaphragma begoniifolia, 5
Perak begonia, 240–4, **240–4**
Perlis State Park: begonias, 22
Piee's begonia, 252–5, **252–5**
Plumier, Charles, 25
pokok riang, 3

Q
queen begonia, 218–22, **218–22**

R
rajah begonia, 216–17, **217**
red-haired begonia, 140–4, **140–4**; leaves, 10

307

INDEX

red striped begonia, 85–91, **87–91**; flowers, 10; leaves, 8
Rex begonia, 1
rhubarb-leaved begonia, 260, 279, 280–5, **280–5**; habit, 5; leaves, 5, 6
riang batu, 118
Ridley, Henry N., 25, 49, 62, 68, 73, 84, 91, 111, 116, 138, 152, 163, 191, 232, 238, 251, 285, 293

S

Sam Yen-Yen, 217
Saw Leng Guan, 215, 222
Scortechini, B., 238
Scortechini's begonia, 238, **239**
shilling begonia, 54–7, **54–7**; habit, 5; leaves, 5
Sonerila begoniifolia, 5
Soong Wye Ping, 144
sparkling begonia, 56, 58–63, **58–64**; distribution, 20; habit, 5; habitat, 19; leaves, 8, 10; seasonality, 17; vegetative propagation, 17
spotted begonia, 74–84, **74–82**; distribution, 19–20; flowers, 10; habit, 5; habitat, 19; leaves, 8; seasonality, 17
Stapf, O., 135
star-leaf begonia, 3
stick insect, 18
stream begonia, 245–51, **245–51**
Strugnell, 144

T

Taiping begonia, 102–6, 102–6; leaves, 10
Taman Negara: begonias, 22
Tampin begonia, 265–7, **265–7**
Tebbitt, Mark, 111, 222
Tetrameles nudiflora, 26
tiger begonia, 50–3, **50–3**
tortoise shell begonia, **i**, **x**, 37–42, **38–42**; flowers, 11; leaves, 10
Trigona sp. bees, 13, **13**, 138

V

valley begonia, 145–8, **146–8**
variable begonia, 92–7, **93–6**; flowers, 11

W

Wallich, Nathaniel, 25
waterfall begonia, 256–60, **256–60**
Watson, W., 138
wax flower begonia, 3, 286–7, **286–8**
woolly-stalked begonia, 35, **36**
Wray, L., 116, 138

Y

Yap Kok Sun, 18, 157

Z

Zippelia begoniifolia, 5

Titles by Natural History Publications (Borneo)
For more information, please contact us at

NATURAL HISTORY PUBLICATIONS (BORNEO) SDN. BHD.
A913, 9th Floor, Phase 1, Wisma Merdeka
P.O. Box 15566, 88864 Kota Kinabalu, Sabah, Malaysia
Tel: 088-233098 Fax: 088-240768 e-mail: info@nhpborneo.com
website: www.nhpborneo.com

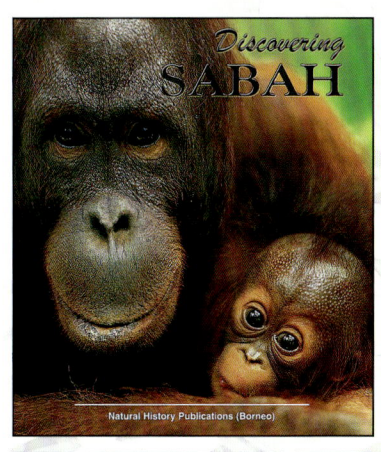

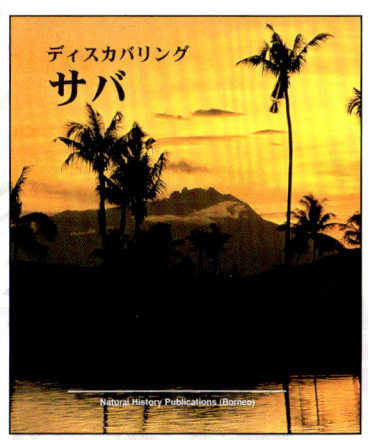

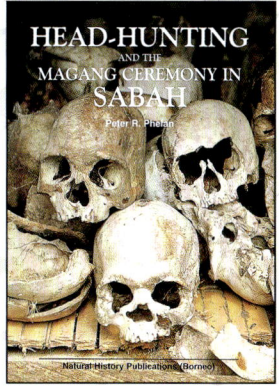

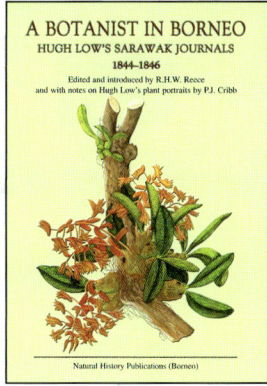

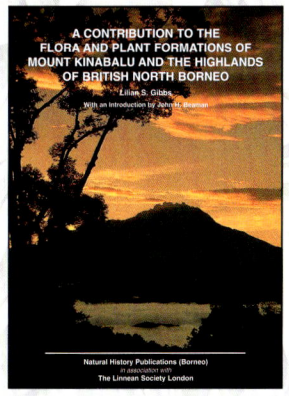

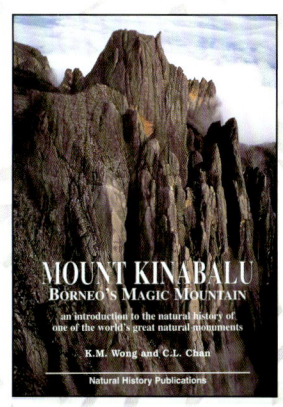

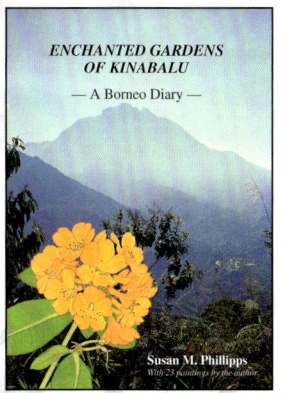

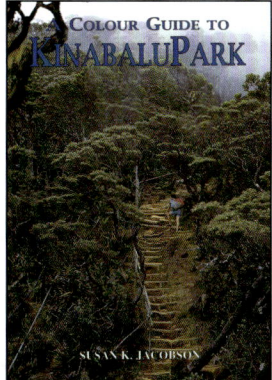

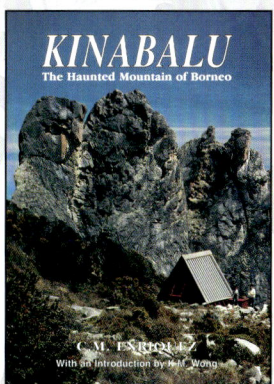

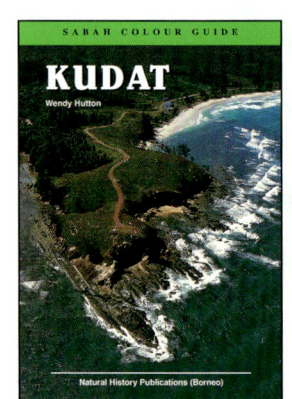

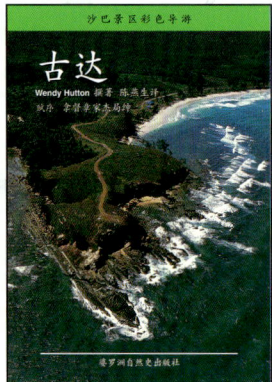

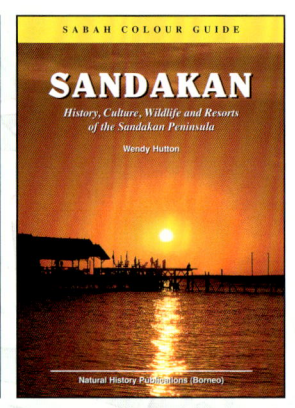

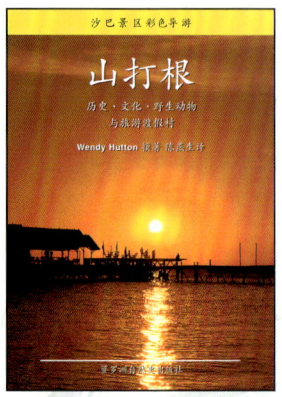

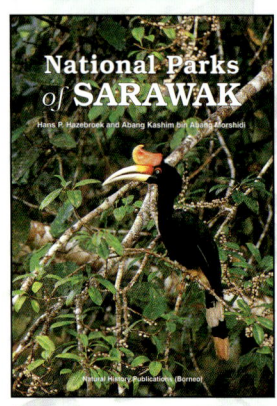

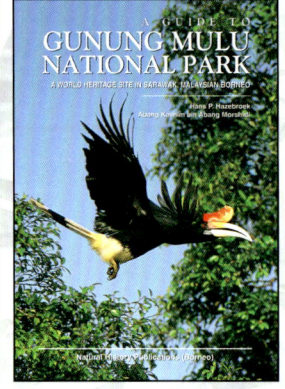

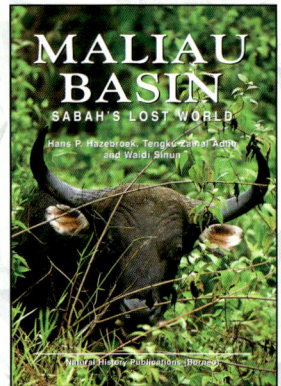

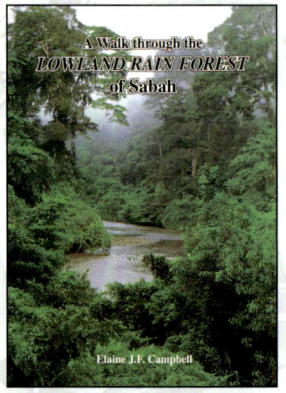

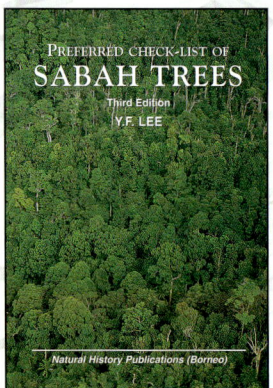

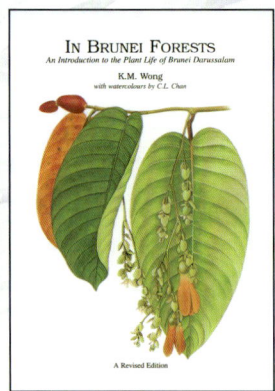

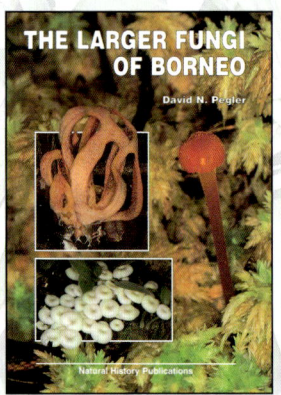

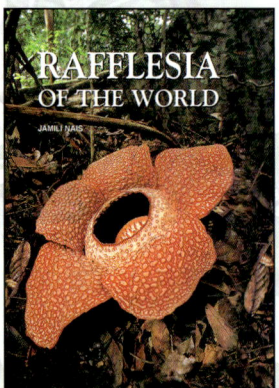

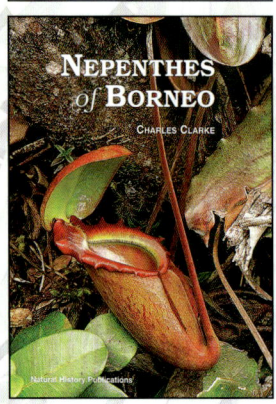

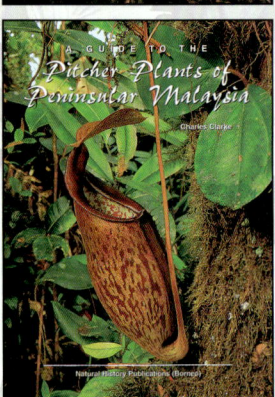

SABAH COLOUR GUIDE **KUDAT** Wendy Hutton	沙巴景区彩色导游 **古达** Wendy Hutton 撰著 陈燕生译	**SABAH COLOUR GUIDE** **SANDAKAN** History, Culture, Wildlife and Resorts of the Sandakan Peninsula Wendy Hutton	沙巴景区彩色导游 **山打根** 历史・文化・野生动物与旅游度假村 Wendy Hutton 撰著 陈燕生译
National Parks of SARAWAK Hans P. Hazebroek and Abang Kashim bin Abang Morshidi	A GUIDE TO **GUNUNG MULU NATIONAL PARK** A WORLD HERITAGE SITE IN SARAWAK, MALAYSIAN BORNEO Hans P. Hazebroek Abang Kashim bin Abang Morshidi	**MALIAU BASIN** SABAH'S LOST WORLD Hans P. Hazebroek, Tengku Zainal Adlin and Waidi Sinun	A Walk through the *LOWLAND RAIN FOREST* of Sabah Elaine J.F. Campbell
PREFERRED CHECK-LIST of **SABAH TREES** Third Edition Y.F. LEE	**In Brunei Forests** An Introduction to the Plant Life of Brunei Darussalam K.M. Wong with watercolours by C.L. Chan A Revised Edition	**THE LARGER FUNGI OF BORNEO** David N. Pegler	**RAFFLESIA OF THE WORLD** JAMILI NAIS
PITCHER-PLANTS OF BORNEO Anthea Phillipps and Anthony Lamb with watercolour paintings by Susan M. Phillipps and a Foreword by Tan Jiew Hoe	**NEPENTHES OF BORNEO** CHARLES CLARKE	A GUIDE TO THE *Pitcher Plants of Sabah* Charles Clarke	A GUIDE TO THE *Pitcher Plants of Peninsular Malaysia* Charles Clarke

NEPENTHES
OF SUMATRA
AND PENINSULAR MALAYSIA
Charles Clarke
Natural History Publications (Borneo)

BORNEO
HUGO STEINER
ITS MOUNTAINS AND LOWLANDS
WITH THEIR PITCHER PLANTS
TREKKING FROM 1992 TO 2002

ORCHIDS
of SARAWAK
TEOFILA E. BEAMAN, JEFFREY J. WOOD,
REED S. BEAMAN AND JOHN H. BEAMAN
Natural History Publications (Borneo)
in association with
The Royal Botanic Gardens, Kew

ORCHIDS
of SUMATRA
J.B. Comber
Natural History Publications (Borneo)
in association with
The Royal Botanic Gardens, Kew

DENDROCHILUM
OF BORNEO
Jeffrey J. Wood
Natural History Publications (Borneo)
in association with
The Royal Botanic Gardens, Kew

THE PLANTS of
MOUNT KINABALU
3. GYMNOSPERMS AND NON-ORCHID MONOCOTYLEDONS
John H. Beaman and Reed S. Beaman
Natural History Publications (Borneo)
in association with
The Royal Botanic Gardens, Kew

THE PLANTS of
MOUNT KINABALU
4. DICOTYLEDON FAMILIES ACANTHACEAE TO LYTHRACEAE
John H. Beaman, Christiane Anderson and Reed S. Beaman
Natural History Publications (Borneo)
in association with
The Royal Botanic Gardens, Kew

The Genus
PAPHIOPEDILUM
Second Edition
Phillip Cribb
Natural History Publications (Borneo)
in association with
The Royal Botanic Gardens, Kew

SLIPPER ORCHIDS
OF BORNEO
Phillip Cribb
Natural History Publications

THE GENUS
COELOGYNE
A SYNOPSIS
Dudley Clayton
Natural History Publications (Borneo)
in association with
The Royal Botanic Gardens, Kew

ORCHID
CONSERVATION
Editors
Kingsley W. Dixon, Shelagh P. Kell,
Russell L. Barrett, Phillip J. Cribb
Natural History Publications (Borneo)

GINGERS
of PENINSULAR MALAYSIA
AND SINGAPORE
K. Larsen, H. Ibrahim, S.H. Khaw and L.G. Saw
Natural History Publications (Borneo)

MOSSES AND LIVERWORTS
of MOUNT KINABALU
Jan Peter Frahm, Wolfgang Frey,
Harald Kürschner and Mario Menzel

PHASMIDS
of BORNEO
Philip E. Bragg
Natural History Publications (Borneo)

An Illustrated Guide to the
STICK AND LEAF INSECTS
of Peninsular Malaysia and Singapore
FRANCIS SEOW-CHOEN
Natural History Publications (Borneo)

A GUIDE TO THE
DRAGONFLIES
OF BORNEO
THEIR IDENTIFICATION AND BIOLOGY
A.G. ORR
including photographs by
M. Hämäläinen
Natural History Publications (Borneo)

A Guide to Australian
GRASSHOPPERS AND LOCUSTS
D. C. F. Rentz, R. C. Lewis, Y. N. Su and M. S. Upton
Natural History Publications (Borneo)

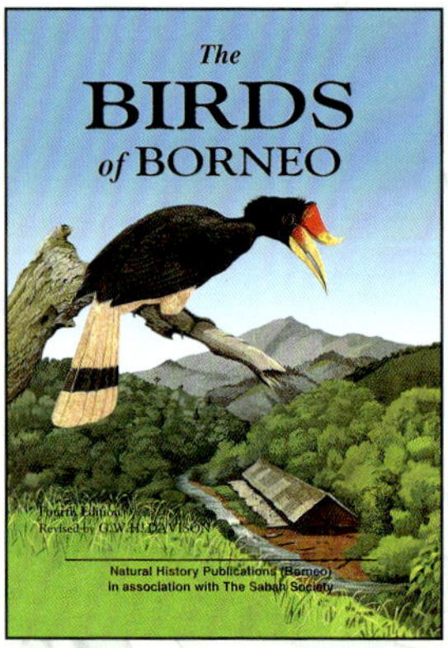
The
BIRDS of BORNEO
Natural History Publications (Borneo)
in association with The Sabah Society

The
BIRDS of BURMA
Fourth Edition
Bertram E. Smythies
Natural History Publications (Borneo)

Swiftlets of Borneo
BUILDERS OF EDIBLE NESTS
Lim Chan Koon and Earl of Cranbrook
Natural History Publications (Borneo)

The Natural History of
ORANG-UTAN
Elizabeth L. Bennett
Natural History Publications (Borneo)

PROBOSCIS MONKEYS OF BORNEO
Elizabeth L. Bennett and Francis Gombek

A Field Guide to the
FROGS OF BORNEO
Robert F. Inger
Robert B. Stuebing

Panduan Lapangan
KATAK-KATAK BORNEO
Robert F. Inger
Robert B. Stuebing
Disunting oleh
Anna Wong
Shahrul Anuar Mohd. Sah
Natural History Publications (Borneo)
Jabatan Muzium Sabah

The Natural History of
AMPHIBIANS AND REPTILES
IN SABAH

An Introduction to the
Amphibians and Reptiles of Tropical Asia
Indraneil Das
Natural History Publications (Borneo)

A FIELD GUIDE TO THE
SNAKES of BORNEO
Robert B. Stuebing and Robert F. Inger
Natural History Publications (Borneo)

Man-eating
CROCODILES of Borneo
JAMES RITCHIE
with JOHNSON JONG
Natural History Publications (Borneo)

TURTLES of BORNEO
AND PENINSULAR MALAYSIA
Lim Boo Liat and Indraneil Das
Natural History Publications (Borneo)

A POCKET GUIDE
LIZARDS OF BORNEO
Indraneil Das
Natural History Publications (Borneo)

A POCKET GUIDE
PITCHER PLANTS OF SARAWAK
Charles Clarke and Ch'ien Lee
Natural History Publications (Borneo)

MARINE FOOD FISHES
AND FISHERIES
OF SABAH
Chin Phui Kong
Natural History Publications

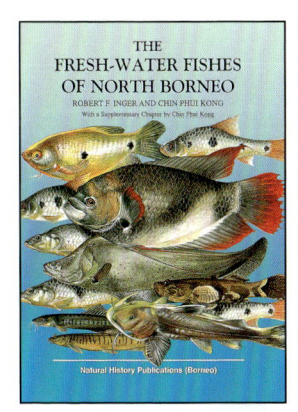

THE FRESH-WATER FISHES OF NORTH BORNEO
ROBERT F. INGER AND CHIN PHUI KONG
With a Supplementary Chapter by Chin Phui Kong

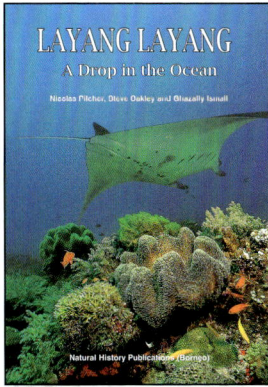

LAYANG LAYANG
A Drop in the Ocean
Nicolas Pilcher, Steve Oakley and Ghazally Ismail

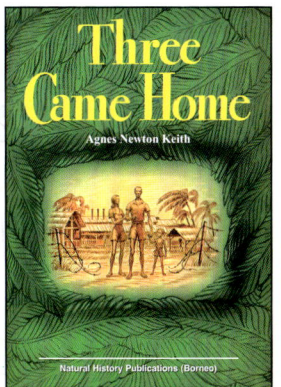

Three Came Home
Agnes Newton Keith

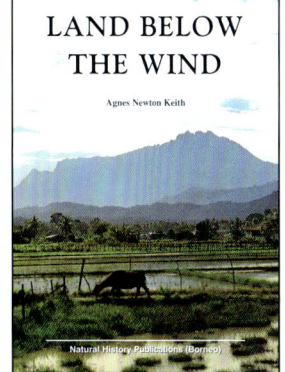

LAND BELOW THE WIND
Agnes Newton Keith

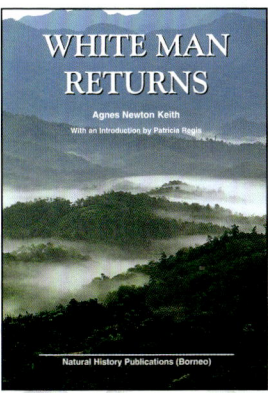

WHITE MAN RETURNS
Agnes Newton Keith
With an Introduction by Patricia Regis

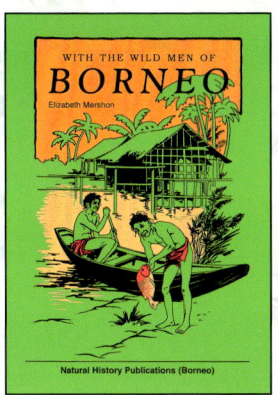

WITH THE WILD MEN OF BORNEO
Elizabeth Mershon

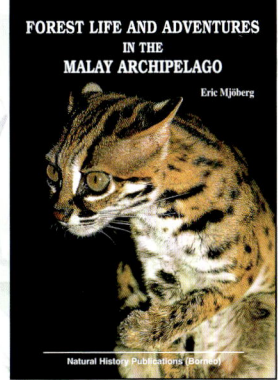

FOREST LIFE AND ADVENTURES IN THE MALAY ARCHIPELAGO
Eric Mjöberg

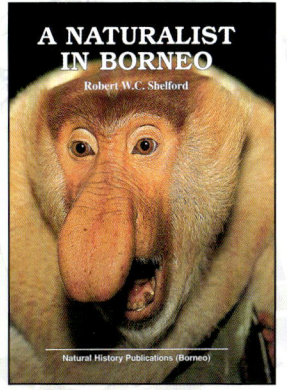

A NATURALIST IN BORNEO
Robert W.C. Shelford

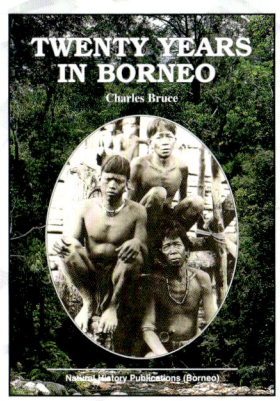

TWENTY YEARS IN BORNEO
Charles Bruce

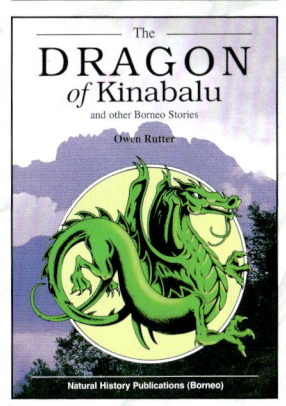

The DRAGON of Kinabalu
and other Borneo Stories
Owen Rutter

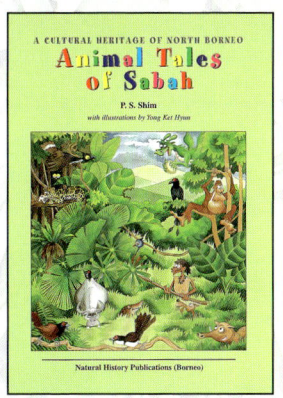

A CULTURAL HERITAGE OF NORTH BORNEO
Animal Tales of Sabah
P. S. Shim
with illustrations by Tong Kei Hyun

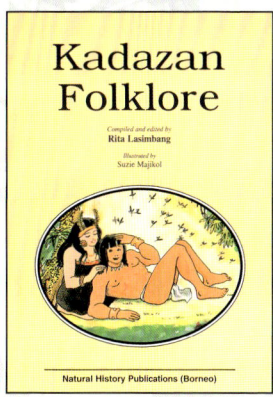

Kadazan Folklore
Compiled and edited by Rita Lasimbang
Illustrated by Suzie Majikol

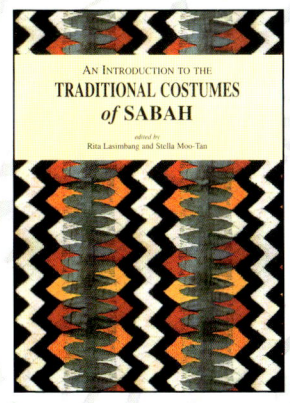

AN INTRODUCTION TO THE TRADITIONAL COSTUMES of SABAH
edited by Rita Lasimbang and Stella Moo-Tan

Water Land Cities
Sabah, Malaysian Borneo

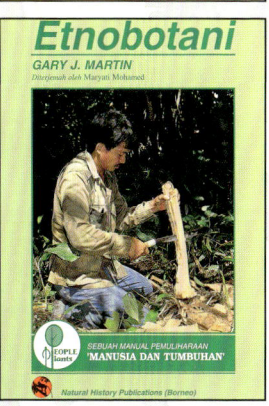

Etnobotani
GARY J. MARTIN
Diterjemah oleh Maryati Mohamed
SEBUAH MANUAL PEMULIHARAAN 'MANUSIA DAN TUMBUHAN'